APPLIED RESEARCH IN HYDRAULICS AND HEAT FLOW

APPLIED RESEARCH IN HYDRAULICS AND HEAT FLOW

Kaveh Hariri Asli, PhD, and Soltan Ali Ogli Aliyev, PhD

Apple Academic Press

TORONTO NEW JERSEY

Apple Academic Press Inc. 3333 Mistwell Crescent Oakville, ON L6L 0A2 Canada	Apple Academic Press Inc. 9 Spinnaker Way Waretown, NJ 08758 USA

International Standard Book Number-13: 978-1-926895-82-6 (Hardcover)

Library of Congress Control Number: 2014903519

Library and Archives Canada Cataloguing in Publication

Asli, Kaveh Hariri, author
Applied research in hydraulics and heat flow/Kaveh Hariri Asli, PhD and Soltan Ali Ogli Aliyev, PhD

Includes bibliographical references and index.
ISBN 978-1-926895-82-6 (bound)

1. Fluid mechanics--Mathematical models--Handbooks, manuals, etc.
2. Heat--Transmission--Mathematical models--Handbooks, manuals, etc.
3. Hydrodynamics--Mathematical models--Handbooks, manuals, etc.
4. Mechanical engineering--Mathematical models--Handbooks, manuals, etc.
5. Mechanical engineering--Research--Handbooks, manuals, etc. ,
I. Aliyev, Soltan Ali Ogli, author II. Title.

QA901.A84 2014 532 C2014-901263-2

ABOUT THE AUTHORS

Kaveh Hariri Asli, PhD

Kaveh Hariri Asli, PhD, is affiliated with the Department of Mathematics and Mechanics, National Academy of Science of Azerbaijan (AMEA) in Baku, Azerbaijan. He is a professional mechanical engineer with over 30 years of experience in practicing mechanical engineering design and teaching. He is author of over 50 articles and reports in the fields of fluid mechanics, hydraulics, automation and control systems. Dr. Hariri has consulted for a number of major corporations.

Soltan Ali Ogli Aliyev, PhD

Soltan Ali Ogli Aliyev, PhD, is Deputy Director of the Department of Mathematics and Mechanics at the National Academy of Science of Azerbaijan (AMEA) in Baku, Azerbaijan. He served as a professor at several universities. He is the author and editor of several books as well as of a number of papers published in various journals and conference proceedings.

CONTENTS

LIST OF ABBREVIATIONS

ACQ	ammonium copper quaternary
CCA	chromated copper arsenate
CFD	computational fluid dynamics
CM	condition base maintenance
DVCM	discrete vapor cavity model
EDL	electric double layer
FD	finite differences
FE	finite elements
FSI	fluid-structure interaction
FSP	fiber saturation point
FV	finite volume
FVM	finite volume method
GIS	geography information systems
MOC	method of characteristics
PLC	program logic control
Re. No	Reynolds number
RMS	root-mean-square
RTC	real-time control
UFW	unaccounted for water
VCF	velocity correction factor
WCM	wave characteristic method

LIST OF SYMBOLS

k = Permeability [cm^3 (liquid)/(cm atm sec)]

V = Volume of liquid flowing through the specimen (cm^3)

t = Time (sec)

A = Cross-sectional area of the specimen perpendicular to the direction of flow (cm^3)

ΔP = Pressure difference between ends of the specimen (atm)

L = Length of specimen parallel to the direction of flow (cm)

K_g = Superficial gas permeability [cm^3 (gas)/(cm atm sec)]

V = Volume of gas flowing through the specimen (cm^3 (gas))

P = Pressure at which V is measured (atm)

$\overline{P}$ = Average pressure across the specimen (atm)

J_f = Liquid free water flow flux, kg/$m^2 \cdot$s

K_l = Specific permeability of liquid water, $m^3(liquid)/m$

ρ_l = Density of liquid water, kg/m^3

μ_l = Viscosity of liquid water, $p_a \cdot s$

p_c = Capillary pressure, p_a

χ = Water transfer distance, m

$\partial p_c / \partial \chi$ = Capillary pressure gradient, p_a / m

J_{vf} = Water vapor flow flux, kg/m^2 ·s

K_V = Specific permeability of water vapor, $m^3(vapor)/m$

ρ_v, μ_v = Density and viscosity of water vapor respectively, kg/m^3 and $p_a \cdot s$

$\partial p_V / \partial \chi$ = Vapor partial pressure gradient, p_a / m

ρ_S = Basic density of wood, kg/m^3

MC = Moisture content of wood, %

t = Time, s

$\partial(MC)/\partial t$ = The rate of moisture content change, %/s

x = Water transfer distance, m

ρ_S = Basic density of wood, kg/ m^3

D_V = Water vapor diffusion coefficient, m^2/s

V = Water flow or discharge $\left(m^3/s\right)$, $\left(lit/s\right)$

C = The wave velocity $\left(m/s\right)$

$E_{жс}$ = Modulus of elasticity of the liquid (water), $E_{жс} = 2.10^9\,Pa\ \left(kg/m^2\right)$

E = Modulus of elasticity for pipeline material Steel, $E = 2.11^9\,Pa,\ \left(kg/m^2\right)$

d = Outer diameter of the pipe (m)

δ = Wall thickness (mm)

V_0 = Liquid with an average speed $\left(m/s\right)$

T = Time (S)

h_0 = Ordinate denotes the free surface of the liquid (m)

u = Fluid velocity $\left(m/s\right)$

λ = Wavelength

$(hu)_x$ = Amplitude a

$\frac{\partial h}{\partial t}dx$ = Changing the volume of fluid between planes in a unit time

h_0 = Phase velocity $\left(m/s\right)$

v_Φ = Expressed in terms of frequency

f = Angular frequency

ω = Wave number

$\mathbf{\Phi}$ = A function of frequency and wave vector

$v_\Phi(k)$ = Phase velocity or the velocity of phase fluctuations $\left(m/s\right)$

$\lambda(k)$ = Wavelength

k = Waves with a uniform length, but a time-varying amplitude

$k_{**}(\omega)$ = Damping vibrations in length

ω = Waves with stationary in time but varying in length amplitudes

p_{si0}= Saturated vapor pressure of the components of the mixture at an initial temperature of the mixture T_0: (Pa)

μ_2, μ_1 = Molecular weight of the liquid components of the mixture

B = Universal gas constant

p_i = The vapor pressure inside the bubble (Pa)

T_{ki}= Temperature evaporating the liquid components (^oC)

l_i = Specific heat of vaporization

D = Diffusion coefficient volatility of the components

N_{k_0}, N_{c_0} = Molar concentration of 1-th component in the liquid and steam

c_l and c_{pv}= The specific heats of liquid and vapor at constant pressure, respectively

a_l = Thermal diffusivity

ρ_v = Vapor density (kg/m^2)

$R = r = R(t)$ = Radius of the bubble (mm)

λ_l = Coefficient of thermal conductivity

ΔT = Overheating of the liquid (^oC)

β = Positive and has a pronounced maximum at $k_0 = 0{,}02$

p_1 and p_2 = The pressure component vapor in the bubble (Pa)

p_∞ = The pressure of the liquid away from the bubble (Pa)

σ = Surface tension coefficient of the liquid

ν_1 = Kinematic viscosity of the liquid

k_R = The concentration of the first component at the interface

n_i = The number of moles

V = Volume (m^3)

B = Gas constant

T_v = The temperature of steam (^oC)

$\rho_i^/$ = The density of the mixture components in the vapor bubble (kg/m^2)

μ_i = Molecular weight

P_{si} = Saturation pressure (Pa)

l_i = Specific heat of vaporization

k = The concentration of dissolved gas in liquid

v_{Φ} = Speed of long waves

h = Liquid level is above the bottom of the channel

ξ = Difference of free surface of the liquid and the liquid level is above the bottom of the channel (a deviation from the level of the liquid free surface)

u = Fluid velocity $\left(m/s\right)$

τ = Time period

a = Distance of the order of the amplitude

k = Wave number

$v_{\Phi}(k)$ = Phase velocity or the velocity of phase fluctuations

$\lambda(k)$ = Wavelength

$\omega_{**}(k)$ = Damping the oscillations in time

λ = Coefficient of combination

q = Flow rate $\left(m^3/s\right)$

μ = Fluid dynamic viscosity $\left(kg/m.s\right)$

γ = Specific weight $\left(N/m^3\right)$

j = Junction point (m)

y = Surge tank and reservoir elevation difference (m)

k = Volumetric coefficient $\left(GN/m^2\right)$

T = Period of motion

A = Pipe cross-sectional area $\left(m^2\right)$

dp = Static pressure rise (m)

h_p = Head gain from a pump (m)

h_L = Combined head loss (m)

E_v = Bulk modulus of elasticity (Pa), $\left(kg/m^2\right)$

α = Kinetic energy correction factor

P = Surge pressure (Pa)

g = Acceleration of gravity $\left(m/s^2\right)$

K = Wave number

T_P = Pipe thickness (m)

E_P = Pipe module of elasticity (Pa) $\left(kg/m^2\right)$

E_W = Module of elasticity of water (Pa), $\left(kg/m^2\right)$

C_1 = Pipe support coefficient

$Y\max$ = $Max.$ Fluctuation

R_0 = Radiuses of a bubble (mm)

D = Diffusion factor

β = Cardinal influence of componential structure of a mixture

N_{k_0}, N_{c_0} = Mole concentration of 1-th component in a liquid and steam

γ = Adiabatic curve indicator

c_l, c_{pv} = Specific thermal capacities of a liquid at constant pressure

a_l = Thermal conductivity factor

ρ_v = Steam density $\left(kg/m^3\right)$

R = Vial radius (mm)

λ_l = Heat conductivity factor

k_0 = Values of concentration, therefore

w_l = Velocity of a liquid on a bubble surface $\left(m/s\right)$

p_1 and p_2 = Pressure steam component in a bubble (Pa)

p_∞ = Pressure of a liquid far from a bubble (Pa)

σ and v_1 = Factor of a superficial tension of kinematics viscosity of a liquid

B = Gas constant

T_v = Temperature of a mixture $\left({}^oC\right)$

$\rho_i^/$ = Density a component of a mix of steam in a bubble $\left(kg/m^3\right)$

μ_i = Molecular weight

j_i = The stream weight

i = Components from an $(i=1,2)$ inter-phase surface in $r=R(t)$

w_i = Diffusion speeds of a component on a bubble surface $\left(m/s\right)$

l_i = Specific warmth of steam formation

k_R = Concentration 1$^{\text{-th}}$ components on an interface of phases

T_0, T_{ki} = Liquid components boiling temperatures of a binary mixture at initial pressure P_0, (^oC)

D = Diffusion factor

λ_l = Heat conductivity factor

N_{ul} = Parameter of Nusselt

a_l = Thermal conductivity of liquids

c_l = Factor of a specific thermal capacity

Pe_l = Number of Pekle

Sh = Parameter of Shervud

Pe_D= Diffusion number the Pekle

ρ = Density of the binary mix $\left(kg/m^3\right)$

t = Time (S)

λ_0 = Unit of length

V = Velocity $\left(m/s\right)$

S = Length (m)

D = Diameter of each pipe (mm)

R = Pipe radius (mm)

ν = Fluid dynamic viscosity $\left(kg/m.s\right)$

h_p = Head gain from a pump (m)

h_L = Combined head loss (m)

C = Velocity of Surge wave $\left(m/s\right)$

P/γ = Pressure head (m)

Z = Elevation head (m)

$V^2/2g$ = Velocity head (m)

γ = Specific weight $\left(N/m^3\right)$

Z = Elevation (m)

H_P = Surge wave head at intersection points of characteristic lines (m)

V_P = Surge wave velocity at pipeline points- intersection points of characteristic lines $\left(m/s\right)$

V_{ri} = Surge wave velocity at right hand side of intersection points of characteristic lines $\left(m/s\right)$

H_{ri} = Surge wave head at right hand side of intersection points of characteristic lines (m)

V_{le} = Surge wave velocity at left hand side of intersection points of characteristic lines $\left(m/s\right)$

H_{le} = Surge wave head at left hand side of intersection points of characteristic lines (m)

p = Pressure (bar), $\left(N/m^2\right)$

dv = Incremental change in liquid volume with respect to initial volume

$\left(d\rho/\rho\right)$ = incremental change in liquid density with respect to initial density

SUPERSCRIPTS

C^- = Characteristic lines with negative slope

C^+ = Characteristic lines with positive slope

SUBSCRIPTS

$Min.$ = Minimum

$Max.$ = Maximum

$Lab.$ = Laboratory

PREFACE

Applied Research in Hydraulics and Heat Flow covers modern subjects of mechanical engineering such as fluid mechanics, heat transfer, and flow control in complex systems as well as new aspects related to mechanical engineering education. The chapters will help to enhance the understanding of both the fundamentals of mechanical engineering and their application to the solution of problems in modern industry.

The book includes the most popular applications-oriented approach to engineering fluid mechanics and heat transfer. It offers a clear and practical presentation of all basic principles of fluid mechanics and heat transfer, tying theory directly to real devices and systems used in mechanical and chemical engineering. It presents new procedures for problem-solving and design, including measurement devices and computational fluid mechanics and heat transfer.

This book is suitable for students, both in final undergraduate mechanical engineering courses and at the graduate level. *Applied Research in Hydraulics and Heat Flow* also serves as a useful reference for academics, hydraulic engineers, and professionals in related with mechanical engineering who want to review basic principles and their applications in hydraulic engineering systems.

This fundamental treatment of engineering hydraulics balances theory with practical design solutions to common engineering problems. The authors examine the most common topics in hydraulics, including hydrostatics, pipe flow, pipelines, pipe networks, pumps, hydraulic structures, water measurement devices, and hydraulic similitude and model studies.

INTRODUCTION

This book uses many computational methods for mechanical engineering problems. Proposed methods allowed for any arbitrary combination of devices in system in fluid mechanics system and heat and mass transfer rates. Methods are used by scale models and prototype system. This is not only a platform for solving of problems, but there is a wealth of information available to help address various technical aspects of troubleshooting of mechanical system failure. The user will find key websites cited throughout the book, which are useful for equipment selection, as well as for troubleshooting mechanical operational problems.

Most chapters include a section of recommended resources that the authors have relied upon in their own consulting practice over the years and believe you will also. Although the authors' intent was not to create a college textbook, there is value in using this volume with engineering students, either as a supplemental text or a primary text on mechanical system technologies. If used as such, instructors will need to gauge the level of understanding of students before specifying the book for a course, as well as integrate the sequence and degree of coverage provided in this volume, for, admittedly, such a broad and complex subject, it is impossible to provide uniform coverage of all areas in a single volume.

— Kaveh Hariri Asli, PhD, and Soltan Ali Ogli Aliyev, PhD

CHAPTER 1

MODELING FOR HEAT FLOW PROCESS

CONTENTS

1.1 INTRODUCTION

When faced with a drying problem on an industrial scale, many factors have to be taken into account in selecting the most suitable type of dryer to install and the problem requires to be analyzed from several standpoints. Even an initial analysis of the possibilities must be backed up by pilot-scale tests unless previous experience has indicated the type most likely to be suitable. The accent today, due to high labor costs, is on continuously operating unit equipment, to what extent possible automatically controlled. In any event, the selection of a suitable dryer should be made in two stages, a preliminary selection based on the general nature of the problem and the textile material to be handled, followed by a final selection based on pilot-scale tests or previous experience combined with economic considerations.

A leather industry involves a crucial energy-intensive drying stage at the end of the process to remove moisture left from dye setting. Determining drying characteristics for leather, such as temperature levels, transition times, total drying times, and evaporation rates, is vitally important so as to optimize the drying stage. Meanwhile, a textile material undergoes some physical and chemical changes that can affect the final leather quality [1–11].

In considering a drying problem, it is important to establish at the earliest stage, the final or residual moisture content of the textile material, which can be accepted. This is important in many hygroscopic materials and if dried below certain moisture content they will absorb or "regain" moisture from the surrounding atmosphere depending upon its moisture and humidity. The material will establish a condition in equilibrium with this atmosphere and the moisture content of the material under this condition is termed the equilibrium moisture content. Equilibrium moisture content is not greatly affected at the lower end of the atmospheric scale but as this temperature increases the equilibrium moisture content figure decreases, which explains why materials can in fact be dried in the presence of superheated moisture vapor. Meanwhile, drying medium temperatures and humidities assume considerable importance in the operation of direct dryers [12–21].

It should be noted that two processes occur simultaneously during the thermal process of drying a wet leather material, namely, heat transfer in order to raise temperature of the wet leather and to evaporate its moisture content together with mass transfer of moisture to the surface of the textile material and its evaporation from the surface to the surrounding atmosphere which, in convection dryers, is the drying medium. The quantity of air required to remove the moisture as liberated, as distinct from the quantity of air which will release the required amount of heat through a drop in its temperature in the course of drying, however, has to be determined from the known capacity of air to pick up moisture at a given temperature in relation to its initial content of moisture. For most practical purposes, moisture is in the form of water vapor but the same principles apply, with different values and humidity charts, for other volatile components [22–31].

Thermal drying consumes from 9–25% of national industrial energy consumption in the developed countries. In order to reduce net energy consumption in the drying operation there are attractive alternatives for drying of heat sensitive materials. Leather industry involves a crucial energy-intensive drying stage to remove the moisture left. Synthetic leather drying is the removal of the organic solvent and water. Determining drying characteristics for leathers is vitally important so as to optimize the drying stage. This chapter describes a way to determine the drying characteristics of leather with analytical method developed for this purpose. The model presented, is based on fundamental heat and mass transfer equations. Altering air velocity varies drying conditions. The work indicates closest agreement with the theoretical model. The results from the parametric study provide a better understanding of the drying mechanisms and may lead to a series of recommendations for drying optimization. Among the many processes that are performed in the leather industry, drying has an essential role: by this means, leathers can acquire their final texture, consistency and flexibility. However, some of the unit operations involved in leather industry, especially the drying process is still based on empiricism and tradition, with very little use of scientific principles. Widespread methods of leather drying all over the world are mostly convective methods requiring a lot of energy. Specific heat energy consumption increases, especially in the last period of the drying process, when the moisture content of the leather approaches the value at which the product is storable. However, optimizing the drying process using mathematical analysis of temperature and moisture distribution in the material can reduce energy consumption in a convective dryer. Thus, development of a suitable mathematical model to predict the accurate performance of the dryer is important for energy conservation in the drying process [32–40].

The manufacturing of new-generation synthetic leathers involves the extraction of the filling polymer from the polymer-matrix system with an organic solvent and the removal of the solvent from the highly porous material. In this chapter, a mathematical model of synthetic leather drying for removing the organic solvent is proposed. The model proposed adequately describes the real processes. To improve the accuracy of calculated moisture distributions a velocity correction factor (VCF) introduced into the calculations. The VCF reflects the fact that some of the air flowing through the bed does not participate very effectively in drying, since it is channelled into low-density areas of the inhomogeneous bed. This chapter discusses the results of experiments to test the deductions that increased rates of drying and better agreement between predicted and experimental moisture distributions in the drying bed can be obtained by using higher air velocities.

This chapter focuses on reviewing convective heat and mass transfer equations in the industrial leather drying process with particular reference to VCF [41–50].

1.2 MATERIALS AND METHODS

The theoretical model proposed in this model is based on fundamental equations to describe the simultaneous heat and mass transfer in porous media. It is possible to assume the existence of a thermodynamic quasi equilibrium state, where the temperatures of gaseous, liquid and solid phases are equal,

$$\text{That is, } T_S = T_L = T_G = T \,. \tag{1}$$

Liquid Mass Balance:

$$\frac{\partial(\varepsilon_L \rho_L)}{\partial t} + \nabla(\rho_L \vec{u}_L) + \dot{m} = 0 \tag{2}$$

Water Vapor Mass Balance:

$$\frac{\partial\left[(\varepsilon - \varepsilon_L) X_V \rho_G\right]}{\partial t} + \nabla\left(X_V \rho_G \vec{u}_G + \vec{J}_V\right) - \dot{m} = 0 \tag{3}$$

$$\vec{J}_V = -\rho_G (\varepsilon - \varepsilon_L) D_{EFF} \nabla X_V \tag{4}$$

Air Mass Balance:

$$\frac{\partial\left((\varepsilon - \varepsilon_L) X_A \rho_G\right)}{\partial t} + \nabla\left(X_A \rho_G \vec{u}_G - \vec{J}_V\right) = 0 \tag{5}$$

Liquid Momentum Equation (Darcy's Law):

$$\vec{u}_L = -\left(\frac{\alpha_G}{\mu_G}\right)\nabla(P_G) \tag{6}$$

Thermal Balance:

The thermal balance is governed by Eq.(7).

$$\frac{\partial\left\{\left[\rho_S C_{p_S} + (\varepsilon - \varepsilon_L)\rho_G\left(X_V C_{p_V} + X_A C_{p_A}\right) + \varepsilon_L \rho_L C_{p_L}\right]T\right\}}{\partial t} - \nabla(k_E \nabla T) + \nabla\left[\left(\rho_L \vec{u}_L C_{p_L} + \rho_G \vec{u}_G\left(X_V C_{p_V} + X_A C_{p_A}\right)\right)T\right] + (\varepsilon - \varepsilon_L)\frac{\partial P_G}{\partial t} + \dot{m}\Delta H_V = 0 \tag{7}$$

THERMODYNAMIC EQUILIERIUM-VAPOR MASS FRACTION

In order to attain thermal equilibrium between the liquid and vapor phase, the vapor mass fraction should be such that the partial pressure of the vapor (P'_V) should be equal to its saturation pressure (P_{VS}) at temperature of the mixture. Therefore, thermodynamic relations can obtain the concentration of vapor in the air/vapor mixture inside the pores. According to Dalton's Law of Additive Pressure applied to the air/vapor mixture, one can show that:

$$\rho_G = \rho_V + \rho_A \tag{8}$$

$$X_V = \frac{\rho_V}{\rho_G} \tag{9}$$

$$\rho_V = \frac{P'}{R_V T} \tag{10}$$

$$\rho_A = \frac{\left(P_G - P'_V\right)}{R_A T} \tag{11}$$

Combining Eqs. (8)–(11), one can obtain:

$$X_V = \frac{1}{1+\left(\frac{P_G R_V}{P'_V R_A}\right)-\left(\frac{R_V}{R_A}\right)} \tag{12}$$

MASS RATE OF EVAPORATION

The mass rate of evaporation was obtained in two different ways, as follows:

First of all, the mass rate of evaporation *m* was expressed explicitly by taking it from the water vapor mass balance (Eq. (2)), since vapor concentration is given by Eq. (12).

$$\dot{m} = \frac{\partial\left[(\varepsilon-\varepsilon_L)X_V\rho_G\right]}{\partial t}+\nabla\left(X_V\rho_G\vec{u}_G+\vec{J}_V\right) \tag{13}$$

Secondly, an equation to compute the mass rate of evaporation can be derived with a combination of the liquid mass balance (Eq. (1)) with a first-order-Arrhenius type equation. From the general kinetic equation:

$$\frac{\partial\alpha}{\partial t} = -kf(\alpha) \tag{14}$$

$$k = A\exp\left(-\frac{E}{RT_{SUR}}\right) \tag{15}$$

$$\alpha = 1-\frac{\varepsilon_L(t)}{\varepsilon_0} \tag{16}$$

DRYING KINETICS MECHANISM COUPLING

Using thermodynamic relations, according to Amagat's law of additive volumes, under the same absolute pressure,

$$m_V = \frac{V_V P_G}{R_V T} \tag{17}$$

$$m_A = \frac{V_A P_G}{R_A T} \tag{18}$$

$$m_V = X_V m_T \tag{19}$$

$$m_T = m_V + m_A \tag{20}$$

$$V_G = V_V + V_A \tag{21}$$

$$V_G = (\varepsilon - \varepsilon_L) V_S \tag{22}$$

Solving the set of algebraic Eqs. (17)–(22), one can obtain the vapor-air mixture density:

$$\rho_G = \frac{(m_V + m_A)}{V_G} \tag{23}$$

$$\rho_V = \frac{m_V}{V_G} \tag{24}$$

$$\rho_A = \frac{m_A}{V_G} \tag{25}$$

EQUIVALENT THERMAL CONDUCTIVITY

It is necessary to determine the equivalent value of the thermal conductivity of the material as a whole, since no phase separation was considered in the overall energy equation. The equation we can propose now which may be used to achieve the equivalent thermal conductivity of materials K_E, composed of a continued medium with a uniform disperse phase. It is expressed as follows in Eq. (26).

$$K_E = \left[\frac{k_S + \varepsilon_L k_L \left(\frac{3k_S}{2k_S + k_L} \right) + k_G (\varepsilon - \varepsilon_L) \left(\frac{3k_S}{2k_S + k_G} \right)}{1 + \varepsilon_L \left(\frac{3k_S}{2k_S + k_L} \right) + (\varepsilon - \varepsilon_L) \left(\frac{3k_S}{2k_S + k_G} \right)} \right] \tag{26}$$

$$k_G = X_V k_V + X_A k_A \tag{27}$$

EFFECTIVE DIFFUSION COEFFICIENT EQUATION

The binary bulk diffusivity D_{AV} of air-water vapor mixture is given by:

$$D_{AV} = (2.20)(10^{-5})\left(\frac{P_{ATM}}{P_G}\right)\left(\frac{T_{REF}}{273.15}\right)^{1.75} \tag{28}$$

Factor α_F can be used to account for closed pores resulting from different nature of the solid, which would increase gas outflow resistance, so the equation of effective diffusion coefficient D_{EFF} for fiber drying is:

$$D_{EFF} = \alpha_F D_{AV} \tag{29}$$

The convective heat transfer coefficient can be expressed as:

$$h = Nu_\delta\left(\frac{k}{\delta}\right) \tag{30}$$

The convective mass transfer coefficient is:

$$h_M = \left(\frac{h}{C_{PG}}\right)\left(\frac{\Pr}{Sc}\right)^{2/3} \tag{31}$$

$$\Pr = \frac{C_{PG}\mu_G}{k_G} \tag{32}$$

$$Sc = \frac{\mu_G}{\rho_G D_{AV}} \tag{33}$$

The deriving force determining the rate of mass transfer inside the fiber is the difference between the relative humidities of the air in the pores and the fiber. The rate of moisture exchange is assumed to be proportional to the relative humidity difference in this study.

The heat transfer coefficient between external air and fibers surface can be obtained by: $h = Nu_\delta\left(\frac{k}{\delta}\right)$.

The mass transfer coefficient was calculated using the analogy between heat transfer and mass transfer as $h_M = \left(\frac{h}{C_{PG}}\right)\left(\frac{\Pr}{Sc}\right)^{2/3}$. The convective heat and mass transfer coefficients at the surface are important parameters in drying processes; they are functions of velocity and physical properties of the drying medium.

Describing kinetic model of the moisture transfer during drying as follows:

$$-\frac{dX}{dt} = k(X - X_e) \tag{34}$$

where, X is the material moisture content (dry basis) during drying (kg water/kg dry solids), X_e is the equilibrium moisture content of dehydrated material (kg water/kg dry solids), k is the drying rate (min^{-1}), and t is the time of drying (min). The drying rate is determined as the slope of the falling rate-drying curve. At zero time, the moisture content (dry basis) of the dry material X (kg water/kg dry solids) is equal to X_i, and Eq. (34) is integrated to give the following expression:

$$X = X_e - (X_e - X_i)e^{-kt} \tag{35}$$

Using above equation Moisture Ratio can be defined as follows:

$$\frac{X - X_e}{X_i - X_e} = e^{-kt} \tag{36}$$

This is the Lewis's formula introduced in 1921. But using experimental data of leather dry in git seemed that there was an error in curve fitting of e^{-at}.

The experimental moisture content data were nondimensionlized using the equation:

$$MR = \frac{X - X_e}{X_i - X_e} \tag{37}$$

Where MR is the moisture ratio. For the analysis it was assumed that the equilibrium moisture content, X_e, was equal to zero.

Selected drying models, detailed in Table 1, were fitted to the drying curves (MR versus time), and the equation parameters determined using nonlinear least squares regression analysis, as shown in Table 2.

TABLE 1 Drying models fitted to experimental data.

Model	Mathematical Expression
Lewis (1921)	$MR = \exp(-at)$
Page (1949)	$MR = \exp(-at^b)$
Henderson and Pabis (1961)	$MR = a + bt + ct^2$
Quadratic function	$MR = a + bt + ct^2$
Logarithmic (Yaldiz and Eterkin, 2001)	$MR = a\exp(-bt) + c$
3rd. Degree Polynomial	$MR = a + bt + ct^2 + dt^3$
Rational function	$MR = \frac{a + bt}{1 + ct + dt^2}$
Gaussian model	$MR = a\exp(\frac{-(t-b)^2}{2c^2})$
Presentmodel	$MR = a\exp(-bt^c) + dt^2 + et + f$

TABLE 2 Estimated values of coefficients and statistical analysis for the drying models.

Model	Constants	T= 50	T= 65	T= 80
Lewis	a	0.08498756	0.1842355	0.29379817
	S	0.0551863	0.0739872	0.0874382
	r	0.9828561	0.9717071	0.9587434
Page	a	0.033576322	0.076535988	0.14847589
	b	1.3586728	1.4803604	1.5155253
	S	0.0145866	0.0242914	0.0548030
	r	0.9988528	0.9972042	0.9856112
Henderson	a	1.1581161	1.2871764	1.4922171
	b	0.098218605	0.23327801	0.42348755
	S	0.0336756	0.0305064	0.0186881
	r	0.9938704	0.9955870	0.9983375
Logarithmic	a	1.246574	1.3051319	1.5060514
	b	0.069812589	0.1847816	0.43995186

TABLE 2 *(Continued)*

Model	Constants	T= 50	T= 65	T= 80
	c	–0.15769402	–0.09391818	0.011449769
	S	0.0091395	0.0117237	0.0188223
	r	0.9995659	0.9993995	0.9985010
Quadratic function	a	1.0441166	1.1058544	1.1259588
	b	–0.068310663	–0.16107942	–0.25732004
	c	0.0011451149	0.0059365371	0.014678241
	S	0.0093261	0.0208566	0.0673518
	r	0.9995480	0.9980984	0.9806334
3rd. Degree Polynomial	a	1.065983	1.1670135	1.3629748
	b	–0.076140508	–0.2007021	–0.45309695
	c	0.0017663191	0.011932525	0.053746805
	d	–1.335923e–005	–0.0002498328	–0.0021704758
	S	0.0061268	0.0122273	0.0320439
	r	0.9998122	0.9994013	0.9961941
Rational function	a	1.0578859	1.192437	1.9302135
	b	–0.034944627	–0.083776453	–0.16891461
	c	0.03197939	0.11153663	0.72602847
	d	0.0020339684	0.01062793	0.040207428
	S	0.0074582	0.0128250	0.0105552
	r	0.9997216	0.9993413	0.9995877
Gaussian model	a	1.6081505	2.3960741	268.28939
	b	–14.535231	–9.3358707	–27.36335
	c	15.612089	7.7188252	8.4574493
	S	0.0104355	0.0158495	0.0251066
	r	0.9994340	0.9989023	0.9973314
Present model	a	0.77015136	2.2899114	4.2572457
	b	0.073835826	0.58912095	1.4688178
	c	0.85093985	0.21252159	0.39672164
	d	0.00068710356	0.0035759092	0.0019698297
	e	–0.037543605	–0.094581302	–0.03351435

TABLE 2 *(Continued)*

Model	Constants	T= 50	T= 65	T= 80
	f	0.3191907	–0.18402789	0.04912732
	S	0.0061386	0.0066831	0.0092957
	r	0.9998259	0.9998537	0.9997716

The experimental results for the drying of leather are given in Fig. 7. Fitting curves for two sample models (Lewis model and present model) and temperature of 80°C are given in Figs. 8 and 9. Two criteria were adopted to evaluate the goodness of fit of each model, the Correlation Coefficient (r) and the Standard Error (S). The standard error of the estimate is defined as follows:

$$S = \sqrt{\frac{\sum_{i=i}^{n_{po\,int\,s}} (MR_{\exp,i} - MR_{\mathrm{Pred},i})^2}{n_{po\,int\,s} - n_{param}}} \tag{38}$$

Where $MR_{\exp,i}$ is the measured value at point *i*, and $MR_{\mathrm{Pred},i}$ is the predicted value at that point, and n_{param} is the number of parameters in the particular model (so that the denominator is the number of degrees of freedom).

To explain the meaning of correlation coefficient, we must define some terms used as follow:

$$S_t = \sum_{i=1}^{n_{po\,int\,s}} (\bar{y} - MR_{\exp,i})^2 \tag{39}$$

Where, the average of the data points ($\bar{y}$) is simply given by

$$\bar{y} = \frac{1}{n_{po\,int\,s}} \sum_{i=1}^{n_{po\,int\,s}} MR_{\exp,i} \tag{40}$$

The quantity S_t considers the spread around a constant line (the mean) as opposed to the spread around the regression model. This is the uncertainty of the dependent variable prior to regression. We also define the deviation from the fitting curve as:

$$S_r = \sum_{i=1}^{n_{po\,int\,s}} (MR_{\exp,i} - MR_{pred,i})^2 \tag{41}$$

Note the similarity of this expression to the standard error of the estimate given above; this quantity likewise measures the spread of the points around the fitting function. In view of the above, the improvement (or error reduction) due to describing the data in terms of a regression model can be quantified by subtracting the two quantities [51–108]. Because the magnitude of the quantity is dependent on the scale of the data, this difference is normalized to yield

$$r = \sqrt{\frac{S_t - S_r}{S_t}} \tag{42}$$

where, r is defined as the correlation coefficient. As the regression model better describes the data, the correlation coefficient will approach unity. For a perfect fit, the standard error of the estimate will approach $S=0$ and the correlation coefficient will approach $r=1$.

The standard error and correlation coefficient values of all models are given in Figs. 10 and 11.

1.3 RESULTS AND DISCUSSION

Synthetic leathers are materials with much varied physical properties. As a consequence, even though a lot of research of simulation of drying of porous media has been carried out, the complete validation of these models are very difficult. The drying mechanisms might strongly influence by parameters such as permeability and effective diffusion coefficients. The unknown effective diffusion coefficient of vapor for fibers under different temperatures may be determined by adjustment of the model's theoretical alpha correction factor and experimental data. The mathematical model can be used to predict the effects of many parameters on the temperature variation of the fibers. These parameters include the operation conditions of the dryer, such as the initial moisture content of the fibers, heat, and mass transfer coefficients, drying air moisture content, and dryer air temperature. From Figs.1–6, it can be observed that the shapes of the experimental and calculated curves are somewhat different. It can be seen that as the actual air velocity used in this experiment increases, the value of VCF necessary to achieve reasonable correspondence between calculation and experiment becomes closer to unity; that is, a smaller correction to air velocity is required in the calculations as the actual air velocity increases. This appears to confirm the fact that the loss in drying efficiency caused by bed in homogeneity tends to be reduced as air velocity increases. Fig. 7 shows a typical heat distribution during convective drying. Table 3 relates the VCF to the values of air velocity actually used in the experiments It is evident from the table that the results show a steady improvement in agreement between experiment and calculation (as indicated by increase in VCF) for air velocities up to 1.59 m/s, above which to be no further improvement with increased flow.

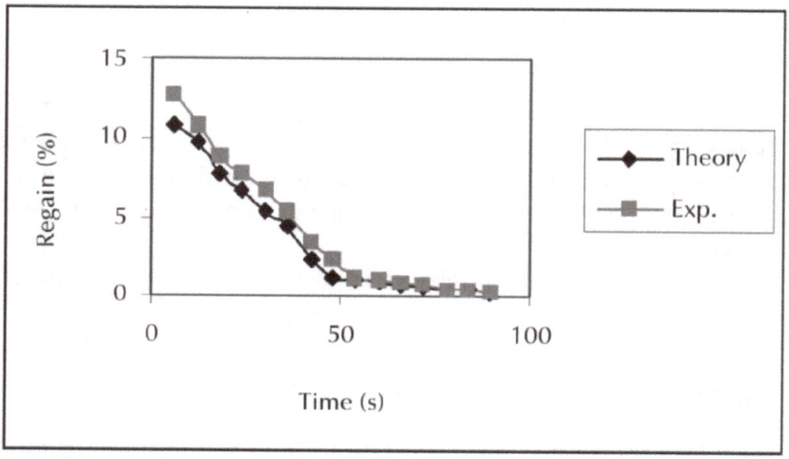

FIGURE 1 Comparison of the theoretical and experimental distribution at air velocity of 0.75 m/s and VCF= 0.39.

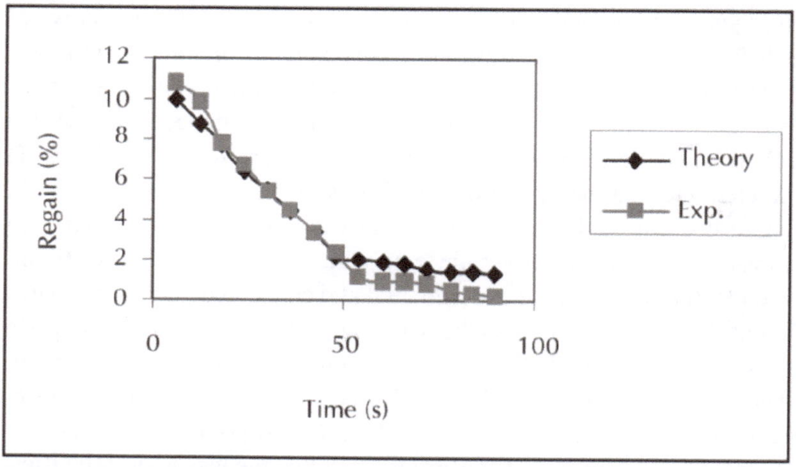

FIGURE 2 Comparison of the theoretical and experimental distribution at air velocity of 0.89 m/s and VCF= 0.44.

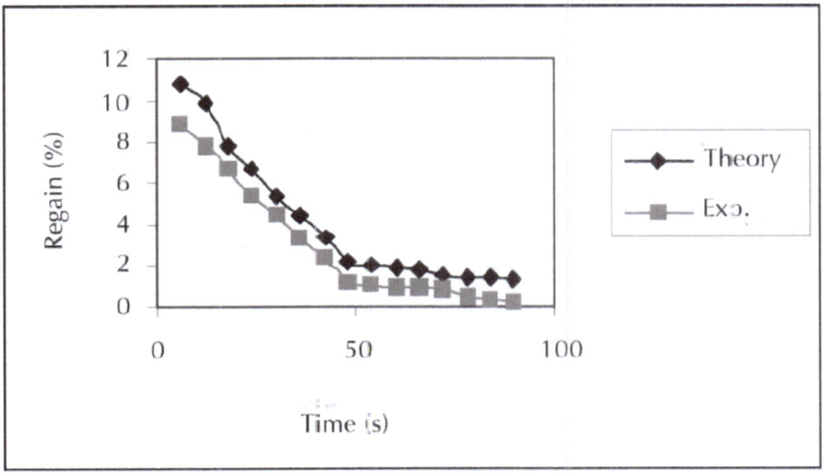

FIGURE 3 Comparison of the theoretical and experimental distribution at air velocity of0.95 m/s and VCF= 0.47.

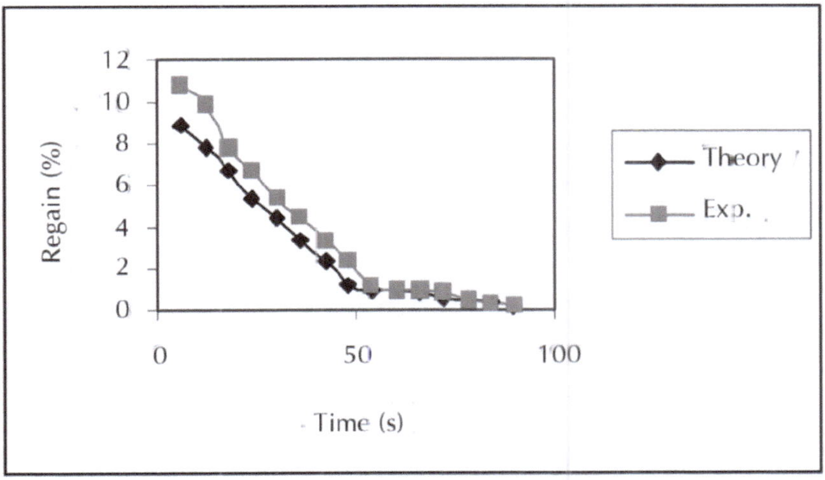

FIGURE 4 Comparison of the theoretical and experimental distribution at air velocity of 1.59 m/s and VCF= 0.62.

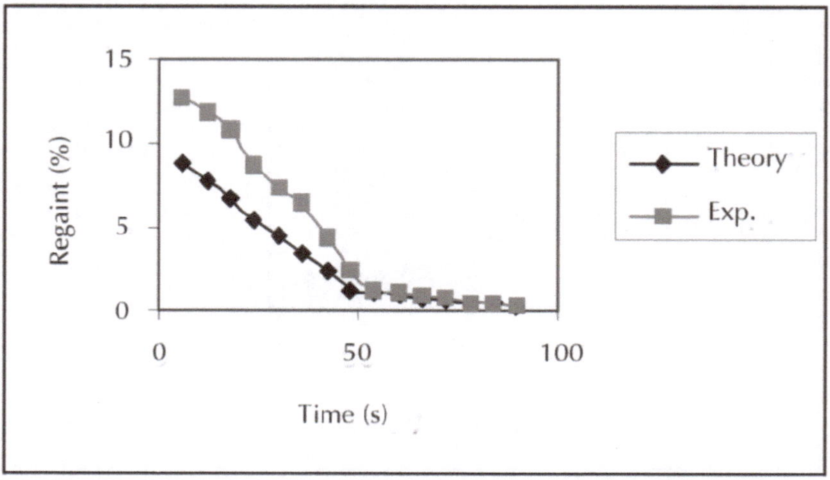

FIGURE 5 Comparison of the theoretical and experimental distribution at air velocity of 2.10 m/s and VCF= 0.62.

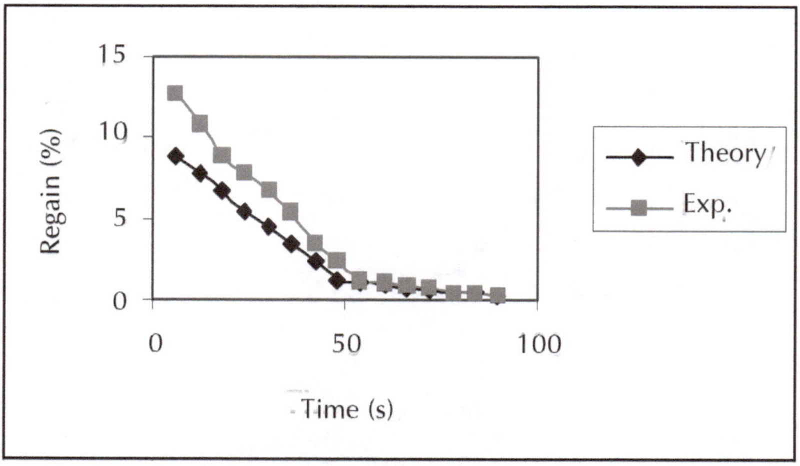

FIGURE 6 Comparison of the theoretical and experimental distribution at air velocity of 2.59 m/s and VCF= 0.61.

TABLE 3 Variation of VCF with air velocity.

Air velocity, m/s	0.75	0.89	0.95	1.59	2.10	2.59
VCF used	0.39	0.44	0.47	0.62	0.62	0.61

In this work, the analytical model has been applied to several drying experiments. The results of the experiments and corresponding calculated distributions are shown in Figs. 7 to 11. It is apparent from the curves that the calculated distribution is in reasonable agreement with the corresponding experimental one. In view of the above, it can be clearly observed that the shapes of experimental and calculated curves are somewhat similar.

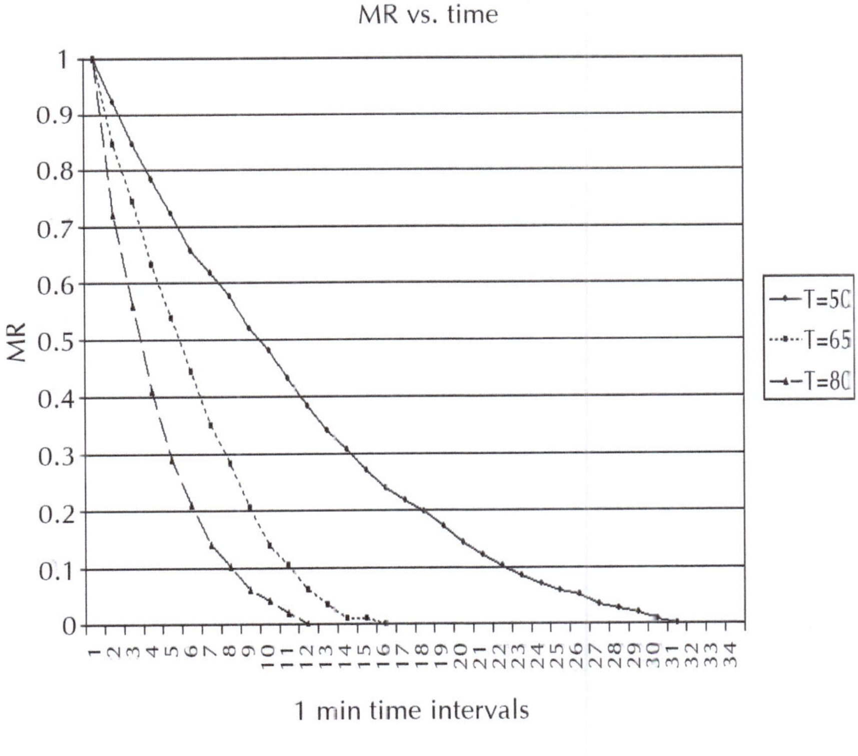

FIGURE 7 Moisture ratio vs. time.

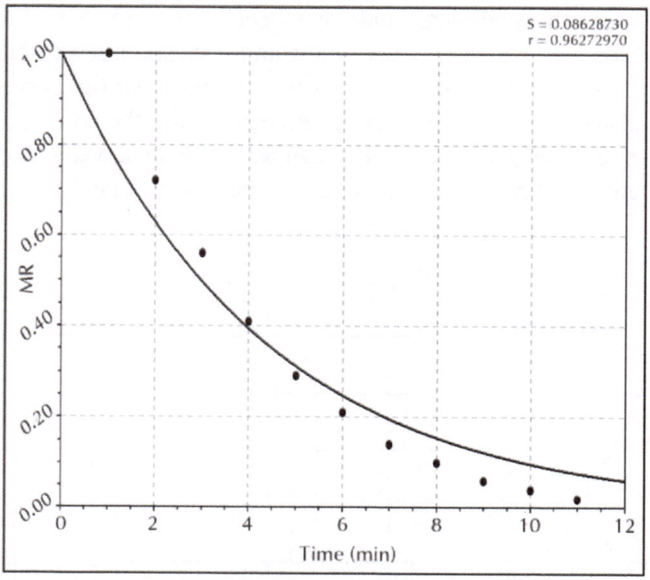

FIGURE 8 Lewis model.

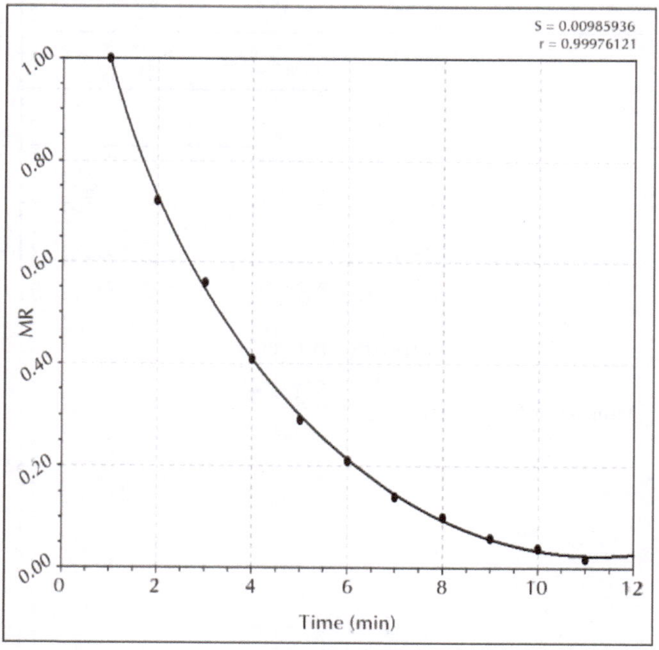

FIGURE 9 Present model.

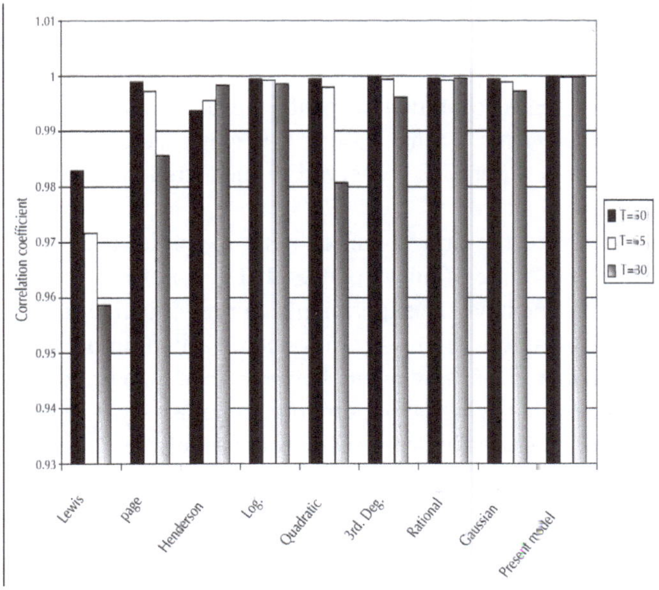

FIGURE 10 Correlation coefficient of all models.

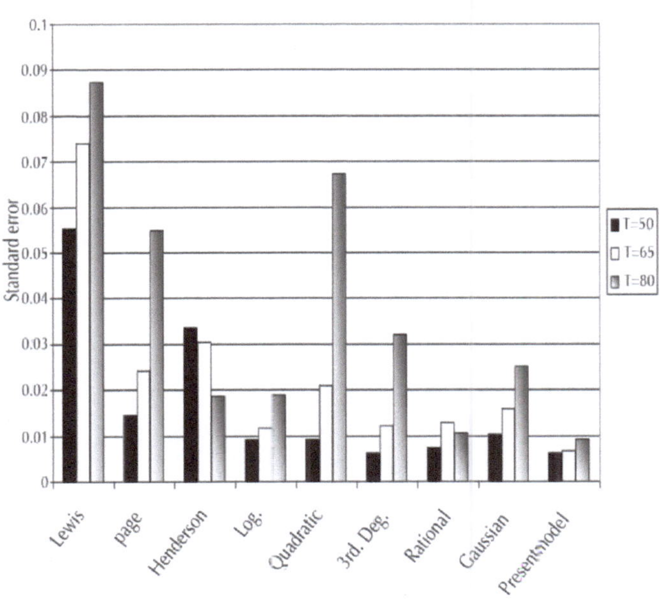

FIGURE 11 Standard error of all models.

1.4 CONCLUSIONS

In the model presented in this chapter, a simple method of predicting moisture distributions leads to prediction of drying times more rapid than those measured in experiments. From this point of view, the drying reveals many aspects which are not normally observed or measured and which may be of value in some application.

The derivation of the drying curves is an example. It is clear from the experiments over a range of air velocities that it is not possible to make accurate predictions and have the experimental curves coincide at all points with the predicted distributions simply by introducing a VCF into the calculations. This suggest that a close agreement between calculated and experimental curves over the entire drying period could be obtained by using a large value of VCF in the initial stages of drying and progressively decreasing it as drying proceeds.

KEYWORDS

- **method of characteristics**
- **regression equations**
- **safety valves**
- **water hammer**
- **water pipeline**

CHAPTER 2

FLUID AND FLUID MECHANICS

CONTENTS

2.1 INTRODUCTION

In this book, miscible liquids condition for example: velocity – pressure – temperature and the other properties is as similar and the main approach is the changes study on behavior of the fluids flow state. According to Reynolds number magnitude (RE. NO.), separation of fluid direction happened. For fluid motion modeling, 2D-component disperses fluid motion used. Modeling of two-phase liquid–liquid flows through a Kinetics static mixer by means of computational fluid dynamics (CFD) has been presented. The two-modeled phases were assumed viscous and Newtonian with the physical properties mimicking an aqueous solution in the continuous and oil in the dispersed (secondary) phase. Differential equations included in the proposed model describe the unsteady motion of a real fluid through the channels and pipes.These differential equations are derived from the following assumptions. It was assumed that the pipe is cylindrical with a constant cross-sectional area with the initial pressure. The fluid flow through the pipe is the one-dimensional. It is assumed that the characteristics of resistors, fixed for steady flows and unsteady flows are equivalent.

One of the problems in the study of fluid flow in plumbing systems is the behavior of stratified fluid in the channels. Mostly steady flows initially are ideal, then the viscous and turbulent fluid in the pipes [1–9].

2.2 MATERIALS AND METHODS

A fluid flow is compressible if its density ρ changes appreciably within the domain of interest. Typically, this will occur when the fluid velocity exceeds Mach 0.3. Hence, low velocity flows (both gas and liquids) behave incompressibly. An incompressible fluid is one whose density is constant everywhere. All fluids behave incompressibly (to within 5%) when their maximum velocities are below Mach 0.3. Mach number is the relative velocity of a fluid compared to its sonic velocity. Mach numbers less than 1 correspond to subsonic velocities, and Mach numbers > 1 corresponds to super-sonic velocities. A Newtonian fluid (1–34) is a viscous fluid whose shear stresses is a linear function of the fluid strain rate. Mathematically, this can be expressed as: $\tau_{ij} = K_{ijqp} * D_{pq}$, where τ_{ij} is the shear stress component, and D_{pq} are fluid strain rate components[10–12].

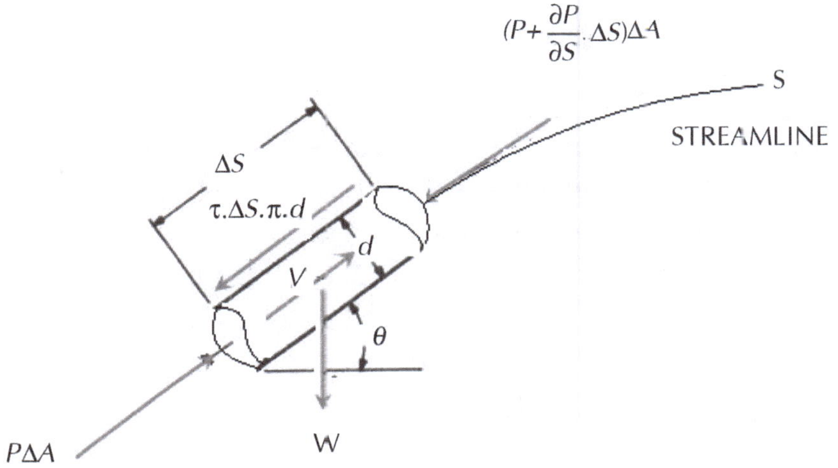

FIGURE 1 Newton second law(conservation of momentum equation)for fluid element

It is defined as the combination of momentum equation (Fig.1) and continuity equation (Fig.2) for determining the velocity and pressure in a one-dimensional flow system. The solving of these equations produces a theoretical result that usually corresponds quite closely to actual system measurements.

$$P\Delta A-(P+\frac{\partial P}{\partial S}.\Delta S)\Delta A-W.\sin\theta-\tau.\Delta S.\pi.d=\frac{W}{g}.\frac{dV}{dt}, \tag{1}$$

Both sides are divided by m and with assumption:

$$\frac{\partial Z}{\partial S}=+\sin\theta, \tag{2}$$

$$-\frac{1}{\partial}.\frac{\partial P}{\partial S}-\frac{\partial Z}{\partial S}-\frac{4\tau}{\gamma D}=\frac{1}{g}.\frac{dV}{dt}, \tag{3}$$

$$\Delta A=\frac{\Pi.D^2}{4}, \tag{4}$$

If, fluid diameter assumed equal to pipe diameter then:

$$\frac{-1}{\gamma}.\frac{\partial P}{\partial S}-\frac{\partial Z}{\partial S}-\frac{4\tau_{\circ}}{\gamma.D}, \tag{5}$$

$$\tau_{\circ}=\frac{1}{8}\rho.f.V^{2}, \tag{6}$$

$$-\frac{1}{\gamma}.\frac{\partial P}{\partial S}-\frac{\partial Z}{\partial S}-\frac{f}{D}.\frac{V^{2}}{2g}=\frac{1}{g}.\frac{dV}{dt}, \tag{7}$$

$$V^{2}=V\,|V|,\ \frac{dV}{dt}+\frac{1}{\rho}.\frac{\partial P}{\partial S}+g\frac{dZ}{dS}+\frac{f}{2D}V|V|=0, \tag{8}$$

(Euler equation)

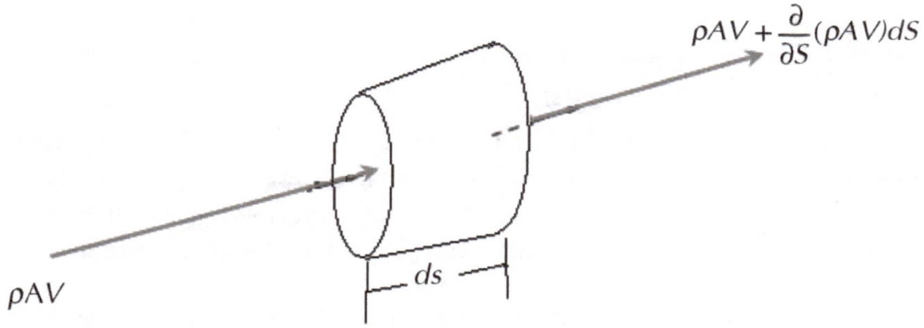

FIGURE 2 Continuity equation (conservation of mass) for fluid element.

For finding (V) and (P) we need to "conservation of mass law"(Fig.2):

$$\rho AV-\left[\rho AV-\frac{\partial}{\partial S}(\rho AV)dS\right]=\frac{\partial}{\partial t}(\rho AdS)-\frac{\partial}{\partial S}(\rho AV)dS=\frac{\partial}{\partial t}(\rho AdS), \tag{9}$$

$$-\left(\rho A\frac{\partial V}{\partial S}dS+\rho V\frac{\partial A}{\partial S}dS+AV\frac{\partial \rho}{\partial S}dS\right)=\rho A\frac{\partial}{\partial t}(dS)+\rho dS\frac{\partial A}{\partial t}+AdS\frac{\partial p}{\partial t},$$
$$\frac{1}{\rho}\left(\frac{\partial \rho}{\partial t}+V\frac{\partial \rho}{\partial S}\right)+\frac{1}{A}\left(\frac{\partial A}{\partial t}+V\frac{\partial A}{\partial S}\right)+\frac{1}{dS}.\frac{\partial}{\partial t}(dS)+\frac{\partial V}{\partial S}=\circ \tag{10}$$

With $$\frac{\partial \rho}{\partial t}+V\frac{\partial \rho}{\partial S}=\frac{d\rho}{dt}\ \text{and}\ \frac{\partial A}{\partial t}+V\frac{\partial A}{\partial S}=\frac{dA}{dt}, \tag{11}$$

$$\frac{1}{\rho}\cdot\frac{d\rho}{dt}+\frac{1}{A}\cdot\frac{dA}{dt}+\frac{\partial V}{\partial S}+\frac{1}{dS}\cdot\frac{1}{dt}(dS)=\circ, \tag{12}$$

$$K=\left(\frac{d\rho}{\left(\frac{d\rho}{\rho}\right)}\right)$$

(Fluid module of elasticity) then:

$$\frac{1}{\rho}\cdot\frac{d\rho}{dt}=\frac{1}{k}\cdot\frac{d\rho}{dt}, \tag{13}$$

Put Eq. (7) into Eq. (8) Then:

$$\frac{\partial V}{\partial S}+\frac{1}{k}\cdot\frac{d\rho}{dt}+\frac{1}{A}\cdot\frac{dA}{dt}+\frac{1}{dS}\cdot\frac{d}{dt}(dS)=\circ \tag{14}$$

$$\rho\frac{\partial V}{\partial S}+\frac{d\rho}{dt}\rho\left[\frac{1}{k}+\frac{1}{A}\cdot\frac{dA}{d\rho}+\frac{1}{dS}\cdot\frac{d}{d\rho}(dS)\right]=\circ, \tag{15}$$

$$\rho\left[\frac{1}{k}+\frac{1}{A}\cdot\frac{dA}{dt}+\frac{1}{dS}\cdot\frac{d}{d\rho}(dS)\right]=\frac{1}{C^2}, \tag{16}$$

Then

$$C^2\frac{\partial V}{\partial S}+\frac{1}{\rho}\cdot\frac{d\rho}{dt}=\circ, \tag{17}$$

(Continuity equation)

Partial differential Eqs.(4)and(10) are solved by method of characteristics "MOC":

$$\frac{dp}{dt}=\frac{\partial p}{\partial t}+\frac{\partial p}{\partial S}\cdot\frac{dS}{dt}, \tag{18}$$

$$\frac{dV}{dt}=\frac{\partial V}{\partial t}+\frac{\partial V}{\partial S}\cdot\frac{dS}{dt}, \tag{19}$$

Then:

$$\left| \begin{array}{l} \frac{\partial V}{\partial t}+\frac{1}{\rho}\frac{\partial p}{\partial S}+g\frac{dz}{dS}+\frac{f}{2D}V|V|=\circ, \\ C^{2}\frac{\partial V}{\partial S}+\frac{1}{\rho}\frac{\partial P}{\partial t}=\circ, \end{array} \right. , \tag{20}$$

By Linear combination of Eqs.(13)and(14)

$$\lambda\left(\frac{\partial V}{\partial t}+\frac{1}{\rho}\frac{\partial p}{\partial S}+g.\frac{dz}{dS}+\frac{f}{2D}V|V|\right)+c^{2}\frac{\partial V}{\partial S}+\frac{1}{\rho}\frac{\partial p}{\partial t}=\circ, \tag{21}$$

$$(\lambda\frac{\partial V}{\partial t}+C^{2}\frac{\partial V}{\partial S})+(\frac{1}{\rho}.\frac{\partial \rho}{\partial t}+\frac{\lambda}{\rho}.\frac{\partial P}{\partial S})+\lambda.g.\frac{dz}{dS}+\frac{\lambda.f}{2D}V|V|=\circ, \tag{22}$$

$$\lambda\frac{\partial V}{\partial t}+C^{2}\frac{\partial V}{\partial S}=\lambda\frac{dV}{dt}\Rightarrow\lambda\frac{dS}{dt}=C^{2}, \tag{23}$$

$$\begin{array}{l} \frac{1}{\rho}.\frac{\partial p}{\partial t}+\frac{\lambda}{\rho}.\frac{\partial \rho}{\partial S}=\frac{1}{\rho}.\frac{d\rho}{dt}\Rightarrow \\ \frac{\lambda}{\rho}=\frac{1}{\rho}.\frac{dS}{dt} \end{array} , \tag{24}$$

$\left|\frac{C^{2}}{\lambda}\right.=\lambda$(By removing $\frac{dS}{dt}$), $\lambda=\pm C$

For $\lambda=\pm C$ from Eq. (18) we have:

$$\mathrm{C}\frac{dV}{dt}+\frac{1}{\rho}.\frac{dp}{dt}+C.g.\frac{dz}{dS}+C.\frac{f}{2D}V|V|=\circ, \tag{25}$$

With dividing both sides by "C":

$$\frac{dV}{dt}+\frac{1}{c.\rho}\frac{dP}{dt}+g.\frac{dz}{dS}+\frac{f}{2D}V|V|=\circ, \tag{26}$$

For $\lambda=-C$ by Eq.(16):

$$\frac{dV}{dt}-\frac{1}{c.\rho}\frac{dp}{dt}+g\frac{dZ}{dS}+\frac{f}{2D}V|V|=\circ, \tag{27}$$

If $$p = \rho . g(H - Z), \tag{28}$$

From Eqs. (9) and (10):

$$\left| \begin{aligned} & \frac{dV}{dt} + \frac{g}{c} \cdot \frac{dH}{dt} + \frac{f}{2D} V|V| = \circ \\ & if : \frac{dS}{dt} = C, \end{aligned} \right. , \tag{29}$$

$$\left| \begin{aligned} & \frac{dV}{dt} + \frac{g}{c} \cdot \frac{dH}{dt} + \frac{f}{2D} V|V| = \circ, \\ & if : \frac{dS}{dt} = -C, \end{aligned} \right. \tag{30}$$

The method of characteristics is a finite difference technique which pressures (Figs.3 and 4) were computed along the pipe for each time step (1)–(35).

Calculation automatically subdivided the pipe into sections (intervals) and selected a time interval for computations Eqs. (22) and (24) are the characteristic equation of Eq. (21) and (23).

If: $f = 0$ Then Eq. (23) will be (Figs.3 and 4):

$\frac{dV}{dt} - \frac{g}{c} \cdot \frac{dH}{dt} = \circ$ or

$$dH = \left(\frac{C}{g} \right) dV, (Zhukousky), \tag{31}$$

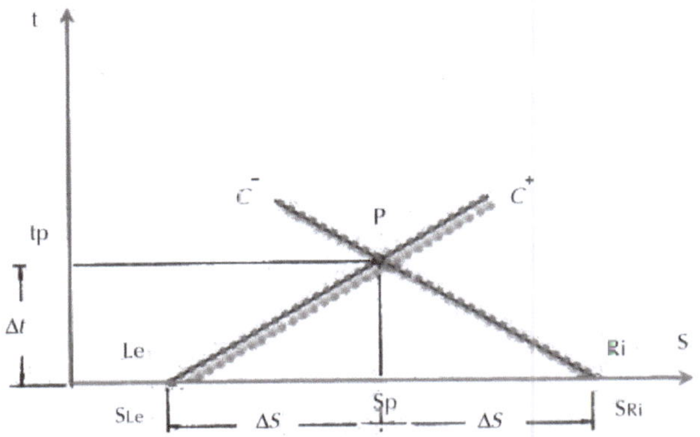

FIGURE 3 Intersection of characteristic lines with positive and negative slope.

If the pressure at the inlet of the pipe and along its length is equal to p_0, then slugging pressure undergoes a sharp increase:

$$\Delta p : p = p_0 + \Delta p\,, \tag{32}$$

The Zhukousky formula is as flowing:

$$\Delta p = \left(\frac{C.\Delta V}{g} \right), \tag{33}$$

The speed of the shock wave is calculated by the formula:

$$C = \sqrt{\frac{g.\dfrac{E_W}{\rho}}{1 + \dfrac{d}{t_W} \cdot \dfrac{E_W}{E}}}\,, \tag{34}$$

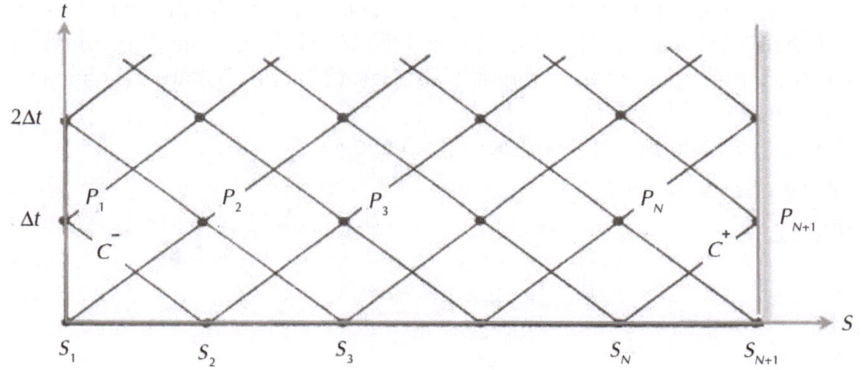

FIGURE 4 Set of characteristic lines intersection for assumed pipe.

By finite difference method of water hammer:

$$T_p - 0 = \Delta t$$

$$c^{+} : (V_p - V_{Le}) / (T_P - 0) + (\frac{g}{c})(H_p - H_{Le}) / (T_P - 0) + fV_{Le}\left|V_{Le}\right| / 2D) = 0, \tag{35}$$

$$c^{-} : (V_p - V_{Ri}) / (T_P - 0) + (\frac{g}{c})(H_p - H_{Ri}) / (T_P - 0) + fV_{Ri}\left|V_{Ri}\right| / 2D) = 0, \tag{36}$$

$$c^{+} : (V_p - V_{Le}) + (\frac{g}{c})(H_p - H_{Le}) + (f.\Delta t)(f.V_{Le}\left|V_{Le}\right| / 2D) = 0, \tag{37}$$

$$c^{-}:(V_{p}-V_{Ri})+(\frac{g}{c})(H_{p}-H_{Ri})+(f.\Delta t)(fV_{Ri}|V_{Ri}|/2D)=\circ, \quad (38)$$

$$V_{p}=\frac{1}{2}\left[(V_{Le}+V_{Ri})+\frac{g}{c}\left(H_{Le}-H_{Ri}\right)-(f.\Delta t/2D)(V_{Le}|V_{Le}|-V_{Ri}|V_{Ri}|)\right], \quad (39)$$

$$H_{p}=\frac{1}{2}\left[\frac{c}{g}(V_{Le}+V_{Ri})+(H_{Le}-H_{Ri})-\frac{c}{g}(f.\Delta t/2D)(V_{Le}|V_{Le}|-V_{Ri}|V_{Ri}|)\right], \quad (40)$$

$V_{Le}, V_{Ri}, H_{Le}, H_{Ri}, f, D$ are initial conditions parameters.

They are applied for solution at steady state condition. Water hammer equations calculation starts with pipe length "L" divided by "N" parts:

$$\Delta S=\frac{L}{N} \;\&\; \Delta t=\frac{\Delta s}{C}, \quad (41)$$

Equations (28) and (29) are solved for the range P_2 through P_N, therefore H and V are found for internal points. Hence:

At P_1 there is only one characteristic Line (c^-)

At P_{N+1} there is only one characteristic Line (c^+)

For finding H and V at P_1 and P_{N+1} the boundary conditions are used.

The Lagrangian approach was used to track the trajectory of dispersed fluid elements (drops) in the simulated static mixer. The particle history was analyzed in terms of the residence time in the mixer. Although two relaxing miscible fluids (35–50) are mixed together, their appearances in terms of colors and shapes will change due to their mixing interpenetration(Fig. 5).

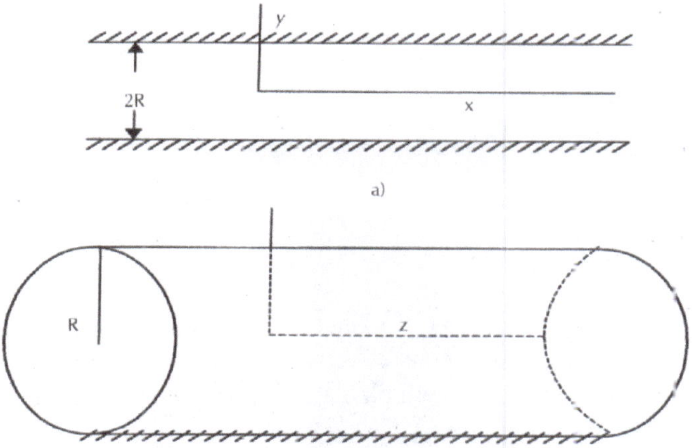

FIGURE 5 Two Dimensional fluids flow.

Use equations of motion of two relaxing fluids in pipe are as flowing:

$$u_1 = u_1(y,t)\ ,\ \ u_2 = u_2(y,t)$$

$$\left.\begin{aligned} &\rho_1 \frac{\partial u_1}{\partial t} = f_1\mu_1 \frac{\partial^2 u_1}{\partial y^2} + k(u_2 - u_1) - f_1 \frac{\partial p}{\partial x}\ , \\ &\rho_2 \frac{\partial u_2}{\partial t} = f_2\mu_2 \frac{\partial^2 u_2}{\partial y^2} + k(u_1 - u_2) - f_2 \frac{\partial p}{\partial x}\ , \\ &\frac{\partial p}{\partial y} = 0\ ,\ \ \frac{\partial p}{\partial z} = 0\ ,\ \ f_1 + f_2 = 0 \end{aligned}\right\}, \tag{42}$$

$\bar{u}$, – velocity (m/s), p – pressure, k – module of elasticity of water (kg/m²), f – Darcy-Weisbach friction factor (obtained from Moody diagram) for each pipe, μ – fluid dynamic, viscosity (kg/m.s), ρ – density (kg/ m³).

Calculation for equation of motion for relaxing fluids:

$$\left.\begin{aligned} &\theta_1 \frac{\partial \tau_1}{\partial t} + \tau_1 = \mu_1 \frac{\partial u_1}{\partial y}\ , \\ &\theta_2 \frac{\partial \tau_2}{\partial t} + \tau_2 = \mu_2 \frac{\partial u_2}{\partial y} \end{aligned}\right\}, \tag{43}$$

θ_1, θ_2 – relaxing time of fluids, define equation of motion for Interpenetration of two *2D* pressurized relaxing fluids at parallel between plates and turbulent moving in pipe as flowing:

$$\left.\begin{aligned} &\rho_1 \frac{\partial u_1}{\partial t} + \rho_1\theta_1 \frac{\partial^2 u_1}{\partial t^2} = f_1\mu_1 \frac{\partial^2 u_1}{\partial y^2} + \theta_1 k \frac{\partial (u_1 - u_2)}{\partial t} + k(u_2 - u_1) - f_1\left[\theta_1 \frac{\partial^2 p}{\partial t \partial x} + \frac{\partial p}{\partial x}\right] \\ &\rho_2 \frac{\partial u_2}{\partial t} + \rho_2\theta_2 \frac{\partial^2 u_2}{\partial t^2} = f_2\mu_2 \frac{\partial^2 u_2}{\partial y^2} + \theta_2 k \frac{\partial (u_1 - u_2)}{\partial t} + k(u_1 - u_2) - f_2\left[\theta_2 \frac{\partial^2 p}{\partial t \partial x} + \frac{\partial p}{\partial x}\right] \\ &\frac{\partial p}{\partial y} = 0\ ,\ \ \frac{\partial p}{\partial z} = 0\ , \\ &f_1 + f_2 = 0 \end{aligned}\right\} \tag{1.3}$$

From Eq.(1.3) (i.e., Eq. (3) of Chapter 1) concluded that pressure drop $\partial p / \partial x$ it is not effective but time is effective.

Assumed that at first time both plan are stopped and pressure at coordination for this time is low.

$$t=0\begin{cases} u_1=0\ ,\ u_2=0 \\ \partial u_1/\partial t=0\ ,\ \partial u_2/\partial t=0 \end{cases}$$
$$\left.\begin{array}{l} y=h \quad (t>0) \quad u_1=0 \quad u_2=0 \\ y=-h \quad (t>0) \quad u_1=0 \quad u_2=0 \end{array}\right\}, \tag{44}$$

At time t condition with Laplace rule, with Eqs.(1.3)and(1.4), we have:

$$\left.\begin{array}{l} \dfrac{d^2\bar{u}_1}{dy^2}-\alpha_1\bar{u}_1+\beta_1\bar{u}_2=\dfrac{1}{\mu_1}\dfrac{\partial P}{\partial x} \\ \dfrac{d^2\bar{u}_2}{dy^2}-\alpha_2\bar{u}_2+\beta_2\bar{u}_1=\dfrac{1}{\mu_2}\dfrac{\partial P}{\partial x} \end{array}\right\}, \tag{45}$$

With:

$$\left.\begin{array}{l} y=h \quad \bar{u}_1=0\ ,\ \bar{u}_2=0 \\ y=-h \quad \bar{u}_1=0\ ,\ \bar{u}_2=0 \end{array}\right\}, \tag{46}$$

Where:

$$\left.\begin{array}{l} \alpha_1=\dfrac{\rho_1(\theta_1 s^2+s)+k(\theta_1 s+1)}{f_1\mu_1}, \\ \beta_1=\dfrac{k(\theta_1 s+1)}{f_1\mu_1}, \\ \alpha_2=\dfrac{\rho_2(\theta_2 s^2+s)+k(\theta_2 s+1)}{f_2\mu_2}, \\ \beta_2=\dfrac{k(\theta_2 s+1)}{f_2\mu_2}, \end{array}\right\}, \tag{47}$$

Calculation $\partial p/\partial x=A=const$ and with product of Eq.(1.5) into N flowing differential equation received:

$$\frac{d^2}{dy^2}(N\bar{u}_1+\bar{u}_2)-(\alpha_2-N\beta_1)(N\bar{u}_1+\bar{u}_2)=\left[\frac{N(1+\theta_1 s)}{\mu_1}+\frac{1+\theta_2 s}{\mu_2}\right]A, \tag{48}$$

$$N_{1,2} = \frac{-(\alpha_1 - \alpha_2) \pm \sqrt{(\alpha_1 - \alpha_2)^2 + 4\beta_1\beta_2}}{2\beta_1} \text{.,} \tag{49}$$

Equation (1.6) calculated with Eq.(1.8):

$$A\bar{u}_1 + \bar{u}_2 = -A\left[\frac{N}{\mu_1} + \frac{1}{\mu_2}\right]\left[1 - \frac{ch\sqrt{\alpha_2 - N\beta_1 y}}{ch\sqrt{\alpha_2 - N\beta_1 h}}\right], \tag{50}$$

N calculation with two meaning:

$$N_1\bar{u}_1 + \bar{u}_2 = -A\left[\frac{N_1}{\mu_1} + \frac{1}{\mu_2}\right]\left[1 - \frac{ch\sqrt{\alpha_2 - N_1\beta_1 y}}{ch\sqrt{\alpha_2 - N_1\beta_1 h}}\right], \tag{51}$$

$$N_2\bar{u}_1 + \bar{u}_2 = -A\left[\frac{N_2}{\mu_1} + \frac{1}{\mu_2}\right]\left[1 - \frac{ch\sqrt{\alpha_2 - N_2\beta_1 y}}{ch\sqrt{\alpha_2 - N_2\beta_1 h}}\right], \tag{52}$$

Where for equation velocity find:

$$\bar{u}_1 = \frac{A}{N_2 - N_1}\left\{\frac{\frac{N_1}{\mu_1} + \frac{1}{\mu_2}}{\alpha_2 - N_1\beta_1}\left[1 - \frac{ch\sqrt{\alpha_2 - N_1\beta_1 y}}{ch\sqrt{\alpha_2 - N_1\beta_1 h}}\right] - \frac{\frac{N_2}{\mu_1} + \frac{1}{\mu_2}}{\alpha_2 - N_2\beta_1}\left[1 - \frac{ch\sqrt{\alpha_2 - N_2\beta_1 y}}{ch\sqrt{\alpha_2 - N_2\beta_1 h}}\right]\right\},$$

$$\bar{u}_2 = \frac{A}{N_1 - N_2}\left\{\frac{\frac{N_2}{\mu_1} - \frac{\beta_2}{\beta_1}\frac{1}{\mu_2}}{\alpha_2 - N_1\beta_1}\left[1 - \frac{ch\sqrt{\alpha_2 - N_1\beta_1 y}}{ch\sqrt{\alpha_2 - N_1\beta_1 h}}\right] - \frac{\frac{N_1}{\mu_2} - \frac{\beta_2}{\beta_1}\frac{1}{\mu_1}}{\alpha_2 - N_2\beta_1}\left[1 - \frac{ch\sqrt{\alpha_2 - N_2\beta_1 y}}{ch\sqrt{\alpha_2 - N_2\beta_1 h}}\right]\right\}.$$

$$\bar{u}_i = \frac{1}{2\pi}\int_{\sigma - i\infty}^{\sigma + i\infty}\frac{A}{N_2 - N_1}\left\{\frac{\frac{N_1}{\mu_1} - \frac{1}{\mu_2}}{\mu_2 - N_1\beta_1}\left[1 - \frac{ch\sqrt{\alpha_2 - N_1\beta_1 y}}{ch\sqrt{\alpha_2 - N_1\beta_1 h}}\right] - \right.$$

$$\left. - \frac{\frac{N_2}{\mu_1} - \frac{1}{\mu_2}}{\alpha_2 - N_2\beta_1}\left[1 - \frac{ch\sqrt{\alpha_2 - N_2\beta_1 y}}{ch\sqrt{\alpha_2 - N_2\beta_1 h}}\right]\right\} . e^{st}\frac{ds}{s}, \tag{53}$$

In this calculation we have:

$$s = s_1 \quad , \quad s = s_2 \quad , \quad s = s_3 \quad , \quad s = s_4 \;\therefore s_1, s_2, s_3, s_4,$$

$$s_{1n} = \gamma_{1n} \quad , \quad s_{2n} = \gamma_{2n} \quad , \quad s_{3n} = \gamma_{3n} \quad , \quad s_{4n} = \gamma_{4n} \quad , \quad \gamma_{in}$$

Proportional to forth procedure:

$$\alpha_2 - N_1\beta_1 = -\frac{\pi^2}{h^2}\left(n+\frac{1}{2}\right)^2 , \tag{54}$$

$$\alpha_2 - N_2\beta_1 = -\frac{\pi^2}{h^2}\left(n+\frac{1}{2}\right)^2 , \tag{55}$$

In this state for velocity we have:

$$u_1 = -\frac{\frac{1}{f_1\mu_1 f_2\mu_2}}{\frac{1}{f_1\mu_1}+\frac{1}{f_2\mu_2}} A\left\{-\frac{1}{2}(y^2-h^2)+\frac{\left(\frac{1}{\mu_1}-\frac{1}{\mu_2}\right)f_2\mu_2}{\left(\frac{1}{f_1\mu_1}+\frac{1}{f_2\mu_2}\right)k}\times\right.$$

$$\times\left[1-\frac{ch\sqrt{\left(\frac{1}{f_1\mu_1}+\frac{1}{f_2\mu_2}\right)}ky}{ch\sqrt{\left(\frac{1}{f_1\mu_1}+\frac{1}{f_2\mu_2}\right)}kh}\right]\Bigg\}+\frac{4A}{\delta}\sum_{i=1}^{4}\sum_{n=1}^{\infty}\frac{(-1)^{n+1}}{\left(n+\frac{1}{2}\right)}\cos\pi\left(n+\frac{1}{2}\right)\frac{y}{h}\times$$

$$\times\left\{\frac{\dfrac{\left[\frac{\pi^2}{h^2}\left(n+\frac{1}{2}\right)+\frac{\left(\theta^2\gamma_{in}+1\right)\left(\rho^2\gamma_{in}+k\right)}{f_2\mu_2}\right]\frac{1}{\mu_1}+\frac{k\left(\theta_1\gamma_{in}+1\right)}{f_1\mu_1}\frac{1}{\mu_2}}{\frac{\left(\theta_1\gamma_{in}+1\right)\left(\rho_1\gamma_{in}+k\right)}{f_1\mu_1}+\frac{\left(\theta_2\gamma_{in}+1\right)\left(\rho_2\gamma_{in}+k\right)}{f_2\mu_2}-2\frac{\pi^2}{h^2}\left(n+\frac{1}{2}\right)^2}\cdot\dfrac{1}{\gamma_{in}}}{2\gamma_{in}\left(\frac{\theta_1\rho_1}{f_1\mu_1}+\frac{\theta_2\rho_2}{f_2\mu_2}\right)+\frac{\theta_1 k+\rho_1}{f_1\mu_1}+\frac{\theta_2 k+\rho_2}{f_2\mu_2}+\dfrac{\frac{\left(\theta_1\gamma_{in}+1\right)\left(\rho_1\gamma_{in}+k\right)}{f_1\mu_1}}{2\frac{\pi^2}{h^2}\left(n+\frac{1}{2}\right)^2+\frac{\left(\theta_1\gamma_{in}+1\right)\left(\rho_1\gamma_{in}+k\right)}{f_1\mu_1}}}\right.$$

$$
\left.\frac{e^{-\gamma_{in} t}}{\dfrac{-\dfrac{(\theta_2\gamma_{in}+1)(\rho_2\gamma_{in}+k)}{f_2\mu_2}\left(2\gamma_{in}\left(\dfrac{\theta_1\rho_1}{f_1\mu_1}+\dfrac{\theta_2\rho_2}{f_2\mu_2}\right)-\dfrac{\theta_2 k+\rho_2}{f_2\mu_2}-\dfrac{\theta_1 k+\rho_1}{f_1\mu_1}\right)}{\left(\dfrac{(\theta_2\gamma_{in}+1)(\rho_2\gamma_{in}+k)}{f_2\mu_2}\right)}}\right], \qquad (56)
$$

$$
u_2 = -\frac{\dfrac{1}{f_1\mu_1 f_2\mu_2}}{\dfrac{1}{f_1\mu_1}+\dfrac{1}{f_2\mu_2}} A\left\{-\frac{1}{2}\left(y^2+h^2\right)+\frac{\left(\dfrac{1}{\mu_1}-\dfrac{1}{\mu_2}\right)f_1\mu_1}{\left(\dfrac{1}{f_1\mu_1}-\dfrac{1}{f_2\mu_2}\right)k}\times\right.
$$

$$
\left.\times\left[1-\frac{ch\sqrt{\left(\dfrac{1}{f_1\mu_1}+\dfrac{1}{f_2\mu_2}\right)ky}}{ch\sqrt{\left(\dfrac{1}{f_1\mu_1}+\dfrac{1}{f_2\mu_2}\right)kh}}\right]\right\}+\frac{4A}{\pi}\sum_{i=1}^{4}\sum_{n=1}^{\infty}\frac{(-1)^{n+1}}{\left(n+\dfrac{1}{2}\right)}\cos\left[\pi\left(n+\frac{1}{2}\right)\frac{y}{h}\right]\times
$$

$$
\times\left\{\frac{\dfrac{\left[\dfrac{\pi^2}{h^2}\left(n+\dfrac{1}{2}\right)+\dfrac{(\theta^2\gamma_{in}+1)(\rho_1\gamma_{in}+k)}{f_1\mu_1}\right]\dfrac{1}{\mu_2}+\dfrac{k(\theta_1\gamma_{in}+1)}{f_2\mu_2}\dfrac{1}{\mu_1}}{\dfrac{(\theta_1\gamma_{in}+1)(\rho_1\gamma_{in}+k)}{f_1\mu_1}+\dfrac{(\theta_2\gamma_{in}+1)(\rho_2\gamma_{in}+k)}{f_2\mu_2}-2\dfrac{\pi^2}{h^2}\left(n+\dfrac{1}{2}\right)^2}\cdot\dfrac{1}{\gamma_{in}}}{2\gamma_{in}\left(\dfrac{\theta_1\rho_1}{f_1\mu_1}+\dfrac{\theta_2\rho_2}{f_2\mu_2}\right)+\dfrac{\theta_1 k+\rho_1}{f_1\mu_1}+\dfrac{\theta_2 k+\rho_2}{f_2\mu_2}+\dfrac{\left(\dfrac{(\theta_1\gamma_{in}+1)(\rho_1\gamma_{in}+k)}{f_1\mu_1}\right)}{2\dfrac{\pi^2}{h^2}\left(n+\dfrac{1}{2}\right)^2+\dfrac{(\theta_1\gamma_{in}+1)(\rho_1\gamma_{in}+k)}{f_1\mu_1}}}\right.
$$

$$\left.\frac{e^{-\gamma_{in} t}}{-\frac{(\theta_2\gamma_{in}+1)(\rho_2\gamma_{in}+k)}{f_2\mu_2}\Bigg]\Bigg[2\gamma_{in}\left(\frac{\theta_1\rho_1}{f_1\mu_1}+\frac{\theta_2\rho_2}{f_2\mu_2}\right)-\frac{\theta_2 k+\rho_2}{f_2\mu_2}-\frac{\theta_1 k+\rho_1}{f_1\mu_1}+\frac{4k^2(\theta_1\theta_2\gamma_{in}+\theta_1+\theta_z)}{f_1\mu_1 f_2\mu_2}}{+\frac{(\theta_1\gamma_{in}+1)(\rho_1\gamma_{in}+k)}{f_1\mu_1}-\frac{(\theta_2\gamma_{in}+1)(\rho_2\gamma_{in}+k)}{f_2\mu_2}}\right\}, \quad (57)$$

When $\theta_1=\theta_2=0$ from Eqs.(9)and(10) we have:

$$\theta_1=\theta_2=0$$

$$\mu_1=\mu_2 \quad , \quad \rho_{1i}=\rho_{2i} \, ,$$

$$u=u_1=u_2=\frac{A}{2\mu}\left(h^2-y^2\right)-\frac{16h^2 A}{\pi\mu}\sum_{n=1}^{\infty}\frac{(-1)^n}{(2n+1)^3}\cos\frac{(2n+1)}{2h}y.e^{-\frac{\pi^2}{h^2}\left(n+\frac{1}{2}\right)^2\frac{\mu}{\rho}t}$$

At condition $t\rightarrow\infty$ for unsteady motion of fluid, it is easy for calculation table procedure.

$$\left.\begin{aligned}\rho_1\frac{\partial u_1}{\partial t}&=f_1\mu_1\left(\frac{\partial^2 u_1}{\partial r^2}+\frac{1}{r}\frac{\partial u_1}{\partial r}\right)+k(u_2-u_1)-f_1\frac{\partial P}{\partial z}\\ \rho_2\frac{\partial u_2}{\partial t}&=f_2\mu_2\left(\frac{\partial^2 u_2}{\partial r^2}+\frac{1}{r}\frac{\partial u_2}{\partial r}\right)+k(u_1-u_2)-f_2\frac{\partial P}{\partial z}\end{aligned}\right\}, \quad (58)$$

For every relaxing phase we have:

$$\left.\begin{aligned}\theta_1\frac{\partial \tau_1}{\partial t}+\tau_1&=\mu_1\frac{\partial u_1}{\partial r},\\ \theta_2\frac{\partial \tau_2}{\partial t}+\tau_2&=\mu_2\frac{\partial u_2}{\partial r},\end{aligned}\right\}, \quad (59)$$

Start and limiting conditions:

$$\begin{aligned}&t=0 && u_1=0\,,\,u_2=0\,,\\ &r=R(t>0) && u_1=0\,,\,u_2=0\,.\end{aligned} \quad (60)$$

In condition of differential Eq.(1.11) by $\partial\tau_1/\partial t$ from Eq.(1.12) and with τ_1 concluded:

$$\left.\begin{array}{l} \rho_1\left(\frac{\partial u_1}{\partial t}+\theta_1\frac{\partial^2 u_1}{\partial t^2}\right)=f_1\mu_1\left(\frac{\partial^2 u_1}{\partial t^2}+\frac{1}{r}\frac{\partial u_1}{\partial r}\right)+k\left[\theta_1\frac{\partial}{\partial t}(u_2-u_1)+(u_2-u_1)\right]- \\ -f_1\left[\theta_1\frac{\partial^2 p}{\partial t\,\partial z}+\frac{\partial p}{\partial z}\right], \\ \rho_2\left(\frac{\partial u_2}{\partial t}+\theta_2\frac{\partial^2 u_2}{\partial t^2}\right)=f_2\mu_2\left(\frac{\partial^2 u_2}{\partial r^2}+\frac{1}{r}\frac{\partial u_2}{\partial r}\right)+k\left[\theta_2\frac{\partial}{\partial t}(u_1-u_2)+(u_1-u_2)\right]- \\ -f_2\left[\theta_2\frac{\partial^2 p}{\partial t\,\partial z}+\frac{\partial p}{\partial z}\right]. \end{array}\right\} \tag{61}$$

Data condition Eq. (1.13) and integration. In this condition Laplace is toward Eq. (1.14). Then solution find in the form of velocity equation, *1D* fluid viscosity in round pipe is:

$$u_1=-\frac{A}{f_1\mu_1\left(\frac{1}{f_1\mu_1}+\frac{1}{f_2\mu_2}\right)}\left\{\frac{1}{4}\left(r^2-R^2\right)\frac{1}{f_2\mu_2}+\frac{\left(\frac{1}{\mu_2}-\frac{1}{\mu_1}f_2\mu_2\right)}{k\left(\frac{1}{f_1\mu_1}+\frac{1}{f_2\mu_2}\right)}\times\right.$$

$$\left.\times\left[1-\frac{I_0\sqrt{\left(\frac{1}{f_1\mu_1}+\frac{1}{f_2\mu_2}\right)kr}}{I_0\sqrt{\left(\frac{1}{f_1\mu_1}+\frac{1}{f_2\mu_2}\right)kR}}\right]\right\}+\sum_{i=1}^{4}\sum_{n=1}^{\infty}\frac{4A}{\alpha_n}\frac{J_0\left(\alpha_n\frac{r}{R}\right)}{J_1(\alpha_n)}\times$$

$$\times\left\{\frac{\frac{\alpha_n^2}{R^2}+\frac{(\theta_2\gamma_{in}+1)(-\rho_2\gamma_{in}+k)}{f_1\mu_1}\frac{1}{\mu_1}+\frac{k(\theta_1\gamma_{in}+1)}{f_1\mu_1}\frac{1}{\mu_2}}{\frac{(\theta_1\gamma_{in}+1)(-\rho_1\gamma_{in}+k)}{f_1\mu_1}+\frac{(\theta_2\gamma_{in}+1)(-\rho_2\gamma_{in}+k)}{f_2\mu_2}\pm 2\frac{\alpha_n^2}{R^2}}\right.\times$$

$$\times \frac{e^{-\gamma_{in} t}}{\gamma_{in}} \Bigg/ \left\{ 2\gamma_{in}\left(\frac{\theta_1\rho_1}{f_1\mu_1}+\frac{\theta_2\rho_2}{f_2\mu_2}\right) \pm \frac{\theta_1 k+\rho_1}{f_1\mu_1}+\frac{\theta_2 k+\rho_2}{f_2\mu_2}+\right.$$

$$+\left[\left(\frac{(\theta_1\gamma_{in}+1)(\rho_1\gamma_{in}+k)}{f_1\mu_1}-\frac{(\theta_2\gamma_{in}+1)(\rho_2\gamma_{in}+k)}{f_2\mu_2}\right)\left(2\gamma_{in}\left(\frac{\theta_1\rho_1}{f_1\mu_1}+\frac{\theta_2\rho_2}{f_2\mu_2}\right)-\right.\right.$$

$$\left.\left.-\frac{\theta_2 k+\rho_2}{f_2\mu_2}-\frac{\theta_1 k+\rho_1}{f_1\mu_1}+4k^2\frac{\theta_1\theta_2\gamma_{in}+\theta_1+\theta_2}{f_1\mu_1 f_2\mu_2}\right)\right]\Bigg/\left[2\frac{\alpha_n^2}{R}+\right.$$

$$\left.\left.+\frac{(\theta_1\gamma_{in}+1)(-\rho_1\gamma_{in}-k)}{f_1\mu_1}+\frac{(\theta_2\gamma_{in}+1)(-\rho_2\gamma_{in}+k)}{f_2\mu_2}\right]\right\}, \tag{62}$$

$$u_2=-\frac{A}{f_1\mu_1\left(\frac{1}{f_1\mu_1}+\frac{1}{f_2\mu_2}\right)}\left\{\frac{1}{4}\left(r^2-R^2\right)\frac{1}{f_2\mu_2}+\frac{\left(\frac{1}{\mu_2}-\frac{1}{\mu_1}\right)f_2\mu_2}{k\left(\frac{1}{f_1\mu_1}+\frac{1}{f_2\mu_2}\right)}\times\right.$$

$$\left.\times\left[1-\frac{I_0\sqrt{\left(\frac{1}{f_1\mu_1}+\frac{1}{f_2\mu_2}\right)kr}}{I_0\sqrt{\left(\frac{1}{f_1\mu_1}+\frac{1}{f_2\mu_2}\right)kR}}\right]\right\}+\sum_{i=1}^{4}\sum_{n=1}^{\infty}\frac{4A}{\alpha_n}\frac{J_0\left(\alpha_n\frac{r}{R}\right)}{J_1(\alpha_n)}\times$$

$$\times\left\{\frac{\frac{\alpha_n^2}{R^2}+\frac{(\theta_2\gamma_{in}+1)(\rho_1\gamma_{in}+k)}{f_1\mu_1}\frac{1}{\mu_2}+\frac{k(\theta_2\gamma_{in}+1)}{f_2\mu_2}\frac{1}{\mu_1}}{\frac{(\theta_1\gamma_{in}+1)(-\rho_1\gamma_{in}-k)}{f_1\mu_1}+\frac{(\theta_2\gamma_{in}+1)(-\rho_2\gamma_{in}-k)}{f_2\mu_2}\pm 2\frac{\alpha_n^2}{R^2}}\times\right.$$

$$\times\frac{e^{-\gamma_{in}t}}{\gamma_{in}}\Bigg/\left\{2\gamma_{in}\left(\frac{\theta_1\rho_1}{f_1\mu_1}+\frac{\theta_2\rho_2}{f_2\mu_2}\right)+\frac{\theta_1k+\rho_1}{f_1\mu_1}+\frac{\theta_2k-\rho_2}{f_2\mu_2}\pm\right.$$

$$\pm\left[\left(\frac{(\theta_1\gamma_{in}+1)(\rho_1\gamma_{in}+k)}{f_1\mu_1}-\frac{(\theta_2\gamma_{in}+1)}{f_2\mu_2}\right)\left(2\gamma_{in}\left(\frac{\theta_1\rho_1}{f_1\mu_1}+\frac{\theta_2\rho_2}{f_2\mu_2}\right)+\right.\right.$$

$$\left.\left.+\frac{\theta_2k+\rho_2}{f_2\mu_2}-\frac{\theta_1k+\rho_1}{f_1\mu_1}+4k^2\frac{\theta_1\theta_2\gamma_{in}+\theta_1+\theta_2}{f_1\mu_1f_2\mu_2}\right)\right]\Bigg/\left[-2\frac{\alpha_n^2}{R}+\right.$$

$$\left.\left.+\frac{(\theta_1\gamma_{in}+1)(\rho_1\gamma_{in}+k)}{f_1\mu_1}-\frac{(\theta_2\gamma_{in}+1)(\rho_2\gamma_{in}+k)}{f_2\mu_2}\right]\right\}, \qquad (63)$$

When $\theta_1=\theta_2=0$ from Eqs.(1.15)and(1.16) we have Eq. (1.4) in condition:

$$\begin{cases}\theta_1=\theta_2=0\\ \mu_1=\mu_2=\mu\\ \rho_{1i}=\rho_{2i}=\rho\end{cases}$$

One of the problems in the study of fluid flow in plumbing systems is the behavior of stratified fluid in the channels. Mostly steady flows initially are ideal, then the viscous and turbulent fluid in the pipes.

At the deep pool filled with water, and on its surface to create a disturbance, then the surface of the water will begin to propagate. Their origin is explained by the fact that the fluid particles are located near the cavity.

The fluid particles create disturbance, which will seek to fill the cavity under the influence of gravity. The development of this phenomenon is led to the spread of waves on the water. The fluid particles in such a wave do not move up and down around in circles. The waves of water are neither longitudinal nor transverse. They seem to be a mixture of both. The radius of the circles varies with depth of moving fluid particles. They reduce to as long as they do not become equal to zero.

If we analyze the propagation velocity of waves on water, it will be reveal that the velocity of waves depends on length of waves. The speed of long waves is proportional to the square root of the acceleration of gravity multiplied by the wavelength:

$$v_{\Phi} = \sqrt{g\lambda}$$

The cause of these waves is the force of gravity.

For short waves the restoring force due to surface tension force, and therefore, the speed of these waves is proportional to the square root of the private. The numerator of which is the surface tension, and in the denominator – the product of the wavelength to the density of water:

$$v_{\Phi} = \sqrt{\sigma / \lambda\rho}\,, \tag{64}$$

Suppose there is a channel with a constant slope bottom, extending to infinity along the axis *Ox*.

And let the feed in a field of gravity flows, incompressible fluid. It is assumed that the fluid is devoid of internal friction. Friction neglects on the sides and bottom of the channel. The liquid level is above the bottom of the channel *h*. A small quantity compared with the characteristic dimensions of the flow, the size of the bottom roughness, etc.

Free liquid surface h_0 (Fig. 5), which is in equilibrium in the gravity field – is flat. As a result of any external influence, liquid surface in a location removed from its equilibrium position. There is a movement spreading across the entire surface of the liquid in the form of waves, called gravity.

They are caused by the action of gravity field. This type of waves occurs mainly on the liquid surface. They capture the inner layers, the deeper for the smaller liquid surface.

Let
$$h = \xi + h_0, \tag{65}$$

Where h_0 – ordinate denotes the free surface of the liquid.

ξ – a deviation from the level of the liquid free surface,

h – depth of the fluid and

z – vertical coordination of any point in the water column. We assume that the fluid flow is characterized by a spatial variable x and time dependent t.

Thus, it is believed that the fluid velocity u has a nonzero component u_x, which will be denoted by u (other components can be neglected). In addition, the level of h depends only on x and t.

Let us consider such gravitational waves, in which the speed of moving particles are so small that for the Euler equation, it can be neglected the $(u\nabla)u$ compared with $\partial u/\partial t$.

During the time period τ, committed by the fluid particles in the wave, these particles pass the distance of the order of the amplitude a.

Therefore, the speed of their movement will be $u \sim a/\tau$.

Rate u varies considerably over time intervals of the order τ and for distances of the order λ along the direction of wave propagation, λ – Wavelength.

Therefore, the derivative of the velocity time – order u/τ and the coordinates – order u/λ.

Thus, the condition:

$$(u\nabla)u < \partial u/\partial t$$

Equivalent to the requirement

$$\frac{1}{\lambda}\left(\frac{a}{\lambda}\right)^2 << \frac{a}{\tau}\frac{1}{\tau} \quad a << \lambda \text{, or,} \tag{66}$$

That is, amplitude of the wave must be small compared with the wavelength. Consider the propagation of waves in the channel Ox directed along the axis for fluid flow along the channel.

Channel cross section can be of any shape and change along its length with changes in liquid level, cross-sectional area of the liquid in the channel denoted by: $h = h(x,t)$.

The depth of the channel and basin are assumed to be small compared with the wavelength. We write the Euler equation in the form of:

$$\frac{\partial u}{\partial t} = -\frac{1}{\rho}\frac{\partial p}{\partial x}, \tag{67}$$

$$\frac{1}{\rho}\frac{\partial p}{\partial z} = -g, \tag{68}$$

where ρ – Density,

p – Pressure,

g – Acceleration of free fall.

Quadratic in velocity members omitted, since the amplitude of the waves is still considered low. From the second equation we have that at the free surface:

$$z = h(x,t)$$

Where: $p = p_0$ should be satisfied:

$$p = p_0 + \rho g(h - z), \tag{69}$$

Substituting this expression in Eq. (67), we obtain:

$$\frac{\partial u}{\partial t} = -g\frac{\partial h}{\partial x}, \tag{70}$$

to determine u and h we use the continuity equation for the case under consideration.

Consider the volume of fluid contained between two planes of the cross-section of the canal at a distance dx from each other per unit time through a cross-section x enter the amount of fluid, equal to $(hu)_x$.

At the same time through the section:

$$x + dx$$

There is forth coming $(hu)_{x+dx}$.

Therefore, the volume of fluid between the planes is changed to

$$(hu)_{x+dx} - (hu)_x = \frac{\partial(hu)}{\partial x}dx, \tag{71}$$

By virtue of incompressibility of the liquid is a change could occur only due to changes in its level. Changing the volume of fluid between these planes in a unit time is equal

$$\frac{\partial h}{\partial t}dx$$

Consequently, we can write:

$$\frac{\partial(hu)}{\partial x}dx = -\frac{\partial h}{\partial t}dx \text{ and } \frac{\partial(hu)}{\partial x} + \frac{\partial h}{\partial t} = 0,\ t > 0,\ -\infty < x < \infty \text{ or,} \tag{72}$$

Since $h = h_0 + \xi$ wherea h_0 denotes the ordinate of the free liquid surface, in a state of relative equilibrium and evolving the influence of gravity is:

$$\frac{\partial \xi}{\partial t} + h_0 \frac{\partial u}{\partial x} = 0 \qquad (73)$$

Thus, we obtain the following system of equations describing the fluid flow in the channel:

$$\frac{\partial \xi}{\partial t} + h_0 \frac{\partial u}{\partial x} = 0, \quad \frac{\partial u}{\partial t} + g \frac{\partial \xi}{\partial x} = 0, \ t > 0, \ -\infty < x < \infty, \qquad (74)$$

2.2.1 VELOCITY PHASE OF THE HARMONIC WAVE

The phase velocity h_0 expressed in terms of frequency v_Φ and wavelength f (or the angular frequency) λ and wave number $\omega = 2\pi f$ formula $k = 2\pi / \lambda$.

The concept of phase velocity can be used if the harmonic wave propagates without changing shape.

This condition is always performed in linear environments. When the phase velocity depends on the frequency, it is equivalent to talk about the velocity dispersion. In the absence of any dispersion the waves assumed with a rate equal to the phase velocity.

Experimentally, the phase velocity at a given frequency can be obtained by determining the wavelength of the interference experiments. The ratio of phase velocities in the two media can be found on the refraction of a plane wave at the plane boundary of these environments. This is because the refractive index is the ratio of phase velocities.

It is known that the wave number k satisfies the wave equation is not any values ω but only if their relationship. To establish this connection is sufficient to substitute the solution of the form:

$$\exp\left[i(\omega t - kx)\right], \qquad (75)$$

in the wave equation.

The complex form is the most convenient and compact. We can show that any other representation of harmonic solutions, including in the form of a standing wave leads to the same connection between ω and k.

Substituting the wave solution into the equation for a string, we can see that the equation becomes an identity for:

$$\omega^2 = k^2 v_\Phi^2, \qquad (76)$$

Exactly the same relation follows from the equations for waves in the gas, the equations for elastic waves in solids and the equation for electromagnetic waves in vacuum.

The presence of energy dissipation [Loytsyanskiy L.G., Fluid, Moscow: Nauka, 1970, p.904] leads to the appearance of the first derivatives (forces of friction) in the wave equation. The relationship between frequency and wave number becomes the domain of complex numbers.

For example, the telegraph equation (for electric waves in a conductive line) yields:

$$\omega^2 = k^2 v_\Phi^2 + i \times \omega R / L \tag{77}$$

The relation connecting between a frequency and wave number (wave vector), in which the wave equation has a wave solution is called a dispersion relation, the dispersion equation or dispersion.

This type of dispersion relation determines the nature of the wave. Since the wave equations are partial differential equations of second order in time and coordinates, the dispersion is usually a quadratic equation in the frequency or wave number.

The simplest dispersion equations presented above for the canonical wave equation are also two very simple solutions:

$$\omega = +kv_\Phi \text{ and } \omega = -kv_\Phi, \tag{78}$$

We know that these two solutions represent two waves traveling in opposite directions. By its physical meaning the frequency is a positive value so that the two solutions must define two values of the wave number, which differ in sign. The Act permits the dispersion, generally speaking, the existence of waves with all wave numbers that is of any length, and, consequently, any frequencies. The phase velocity of these waves:

$$v_\Phi = \omega / k, \tag{79}$$

Coincides with the most velocity, which appears in the wave equation and is a constant, which depends only on the properties of the medium.

The phase velocity depends on the wave number, and, consequently, on the frequency. The dispersion equation for the telegraph equation is an algebraic quadratic equation has complex roots. By analogy with the theory of oscillations, the presence of imaginary part of the frequency means the damping or growth of waves. It can be noted that the form of the dispersion law determines the presence of damping or growth.

In general terms, the dispersion can be represented by the equation: $\Phi(\omega, \bar{k}) = 0$ where

Φ – A function of frequency and wave vector.

By solving this equation for ω you can obtain an expression for the phase velocity):

$$v_\Phi = \omega / k = f\left(\omega, \vec{k}\right), \tag{80}$$

By definition, the phase velocity is a vector directed normal to phase surface.

Then, more correctly write the last expression in the following form:

$$\vec{v_\Phi} = \frac{\lambda}{T} = \frac{\omega}{k^2} \cdot \vec{k} = f\left(\omega, \vec{k}\right), \tag{81}$$

2.2.2 DISPERSIVE PROPERTIES OF MEDIA

The most important subject of research in wave physics, which has the primary practical significance.

If we refer to dimensionless parameters and variables:

$$\tau = t\sqrt{\frac{g}{h_0}},\ X = \frac{x}{h_0},\ U = u\frac{1}{\sqrt{gh_0}},\ \delta = \frac{\xi}{h_0}, \tag{82}$$

The system of Eq.(2.4.8)becomes

$$\frac{\partial \delta}{\partial \tau} + \frac{\partial U}{\partial X} = 0,\ \frac{\partial U}{\partial \tau} + \frac{\partial \delta}{\partial X} = 0,\ t > 0, -\infty < X < \infty, \tag{83}$$

Consider plane harmonic longitudinal waves, that is, we seek the solution of Eq. (83) as the real part of the following complex expressions:

$$\Psi = \Psi^0 \exp\left[i(k_* X + \omega_* \tau)\right],\ \Psi^0 = \Psi^0_* + i\Psi^0_{**},$$

$$k_* = k + ik_{**},\ \omega_* = \omega + i\omega_{**}, \tag{84}$$

where:

$$\Psi = \delta, U \text{, a } \Psi^0 = \delta^0, U^0$$

determines the amplitude of the perturbations of displacement and velocity.

There are two types of solutions.

Type I. Solution or wave of the first type, when:
$k_* = k$ – A real positive number $(k > 0, k_{**} = 0)$.
In this case we have:

$$\Psi = \left(\Psi_*^0 + i\Psi_{**}^0\right)\exp\left[i\left(kX + \omega\tau + i\omega_{**}\tau\right)\right] = \left(\Psi_*^0 + i\Psi_{**}^0\right)\exp\left(-\omega_{**}\tau\right)\times$$

$$\left[\cos\left(kX + \omega\tau\right) + i\sin\left(kX + \omega\tau\right)\right], \quad (85)$$

$$\mathrm{Re}\{\Psi\} = \exp\left(-\omega_{**}\tau\right)\left|\Psi^0\right|\sin\left[\phi + \left(kX + \omega\tau\right)\right], \quad (86)$$

$$\left|\Psi^0\right| = \sqrt{\Psi_*^{0\,2} + \Psi_{**}^{0\,2}}, \quad \varphi = arctg\left(-\Psi_*^0 / \Psi_{**}^0\right)$$

Thus, the decision of the first type is a sinusoidal coordinate and $\omega_{**} > 0$ decaying exponentially in time perturbation, which is called k wave:

$$\Psi(k) = \left|\Psi^0\right|\exp\left[-\omega_{**}(k)\tau\right]\sin\left\{f + \frac{2\pi\left[X + v_\Phi(k)\tau\right]}{\lambda(k)}\right\} \quad (87)$$

where:

$$v_\Phi(k) = \omega(k)/k, \ \lambda(k) = 2\pi / k, \quad (88)$$

φ – Initial phase

Here,

$v_\Phi(k)$ – phase velocity or the velocity of phase fluctuations,

$\lambda(k)$ – Wavelength,

$\omega_{**}(k)$ – damping the oscillations in time.

In other words, k is the waves that have uniform length, but time-varying amplitude. These waves are analog of free oscillations.

Type II. Decisions, or wave, the second type, when:

$\omega_* = \omega$ – a

Real positive number $(\omega > 0, \omega_{**} = 0)$.
In this case we have:

$$\psi = \left(\psi_*^0 + i\psi_{**}^0\right)\exp\left[i\left(kX + \omega\tau + ik_{**}z\right)\right] = \left(\Psi_*^0 + i\Psi_{**}^0\right)\exp\left(-k_{**}X\right)\times$$

$$\left[\cos\left(kX + \omega\tau\right) + i\sin\left(kX + \omega\tau\right)\right], \tag{89}$$

$$\mathrm{Re}\{\Psi\} = \exp\left(-k_{**}X\right)\left|\Psi^0\right|\sin\left[\phi + \left(kX + \omega\tau\right)\right], \tag{90}$$

Thus, the solution of the second type is a sinusoidal oscillation in time (excited, for example, any stationary source of external monochromatic vibrations at) $X = 0$, decaying exponentially along the length of the amplitude.

Such disturbances, which are analogous to a wave of forced oscillations, called ω – waves:

$$\Psi(\omega) = \left|\Psi^0(\omega)\right|\exp\left(-k_{**}(\omega)X\right)\sin\left\{\phi + \frac{2\pi\left[X + v_\delta(\omega)\tau\right]}{\lambda(\omega)}\right\}, \tag{91}$$

$$v_\Phi(\omega) = \omega / k(\omega), \tag{92}$$

$$\lambda(\omega) = 2\pi / k(\omega)$$

Here, $k_{**}(\omega)$ – damping vibrations in length.

In other words, ω – waves with stationary in time but varying in length amplitudes.

Cases $k < 0, k_{**} > 0$ and $k > 0, k_{**} < 0$ consistent with attenuation of amplitude of the disturbance regime in the direction of phase fluctuations or phase velocity.

Let us obtain the characteristic equation, linking k_* and ω_*. After substituting (92) in the system of Eq. (91) we obtain:

$$\delta^0 \frac{\omega_*}{k_*} + U^0 = 0,\ U^0 \frac{\omega_*}{k_*} + \delta^0 = 0, \tag{93}$$

From the condition of the existence of a system of linear homogeneous algebraic Eq. (93) with respect to perturbations of a nontrivial solution implies the desired characteristic, or dispersion, which has one solution:

$$v_\Phi = \sqrt{gh_0}, \tag{94}$$

Thus, we obtain a solution representing a sinusoidal in time and coordinate free undammed oscillations [13–106].

Such behaviors of the waves are due to the absence of any dissipation in the fluid. The fluid is incompressible and ideal. There is no heat-mass transfer. Equation (94) with respect to perturbations take the form of wave equations:

$$\frac{\partial^2 \xi}{\partial t^2} = gh_0 \frac{\partial^2 \xi}{\partial x^2} \text{ and } \frac{\partial^2 u}{\partial t^2} = gh_0 \frac{\partial^2 u}{\partial x^2}, \quad (95)$$

Note that in gas dynamics $v_\Phi = \sqrt{gh_0}$ equivalent to the speed of sound.

2.3 CONCLUSION

Thus, we obtain a solution representing a sinusoidal in time and coordinate free undammed oscillations. Such behaviors of the waves are due to the absence of any dissipation in the fluid. The fluid is incompressible and ideal. There is no heat mass transfer.

KEYWORDS

- **dispersed fluid**
- **hydraulic system**
- **ideal fluid**
- **incompressible fluid**
- **pipe wall**

CHAPTER 3

TWO PHASES FLOW AND VAPOR BUBBLE

CONTENTS

3.1 INTRODUCTION

In the two phases flow is extremely important to the concept of volume concentration. This is the relative volume fraction of one phase in the volume of the pipe[1]. Such an environment typical fluid is a high density and little compressibility. This property contributes to the creation of various forms of transient conditions [2–3]. Fig.1 shows an experimental setup, which investigated the formation of different modes of fluid flow with gas bubbles and steam.

3.2 MATERIALS AND METHODS

The experimental results show that the bubble flow usually occurs at low concentrations of vapor. It includes three main types of flow regimes in microgravity bubble, slug and an annular.

TABLE 1 Simulators, models and problems.

Cases	Range of problems
Steady or gradually varying turbulent flow	Rapidly varying or transient flow
Incompressible, Newtonian, single-phase fluids	Slightly compressible, two-phase fluids (vapor and liquid) and two-fluid systems (air and liquid)
Full pipes	Closed-conduit pressurized systems with air intake and release at discrete points

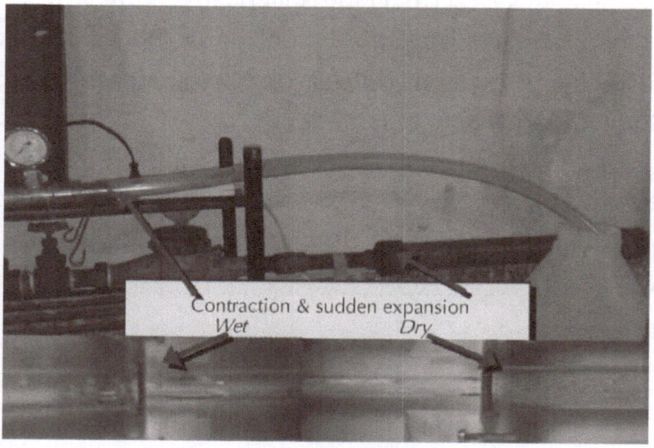

FIGURE 1 Snapshot laboratory setup for studying the structure of flow in different configurations tube.

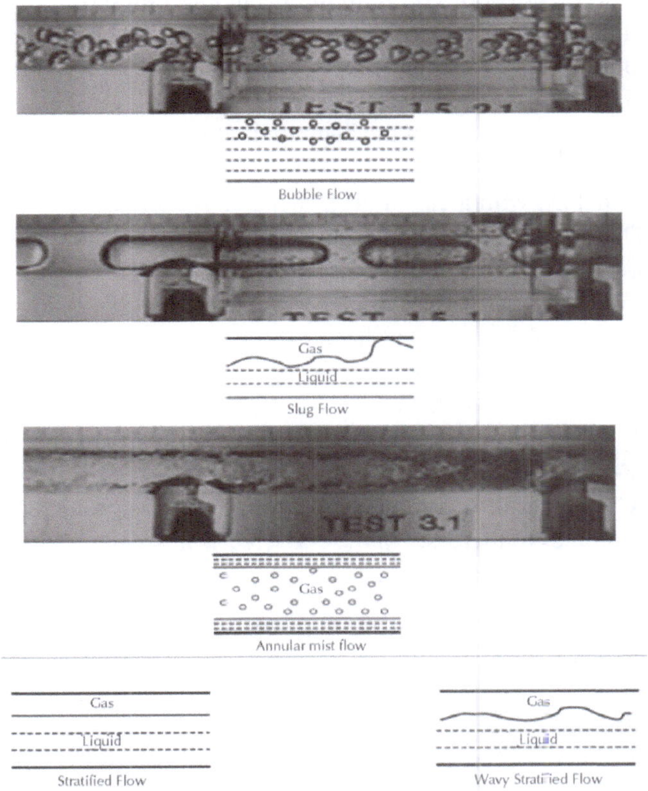

FIGURE 2 Types of flows.

Fig.2 shows snapshots of the flow pattern for various configurations of the tube[4–5]. The flow enters the tee at the bottom of the picture, and then is divided at the entrance to the tube. The inner diameter of the tee is 1.27 cm (Fig. 1.).The narrowing of the flow is achieved by reducing the diameter of the hose.

Within the reduction is a liquid recirculation zone called the "vena contraction" Wet picture that, when the liquid is recirculated to the "vena contraction."However, there are conditions, whereby the gas phase of the contract is caught vena. Fig.3 shows the different flow regimes observed in the experimental setup[6].

Increased flow is achieved by increasing the size of the pipe. Again, there is an area of the liquid recirculation near the "corner" a sudden expansion.

Depending on the level of consumption of bubbly liquid or gas, it falls into the trap in this area.

In the inlet fluid moves out of the pipe diameter of 12.7 (mm) in the pipe diameter of 25(mm).

Normal extension occurs at the beginning of the flow. Soon comes the expansion section, and the flow rate continues to increase. Two-phase jet stream created, ultimately, with areas of air flowing above and below the bubble region [6]. The behavior of gas-liquid mixture in the expansion is proportional to increasing the diameter of the pipe[7–8].

It is shown (Figs. 1–3) that in place of the sudden expansion of a transition flow regime of the turbulent flow. Depending on the flow or gas bubble mixture it falls into the trap in this area.

Experiments were conducted on the pipe, whose diameter is suddenly doubled [9]. In this case the region of turbulence of fluid are observed around the "corner" a sudden expansion. The expansion is observed at the beginning of the flow. As a result of turbulence flow gap expansion increases and the flow rate continues to increase. In the end, creates a stream of two phase flow, with air fields, the current above and below the bubble area [10–11].

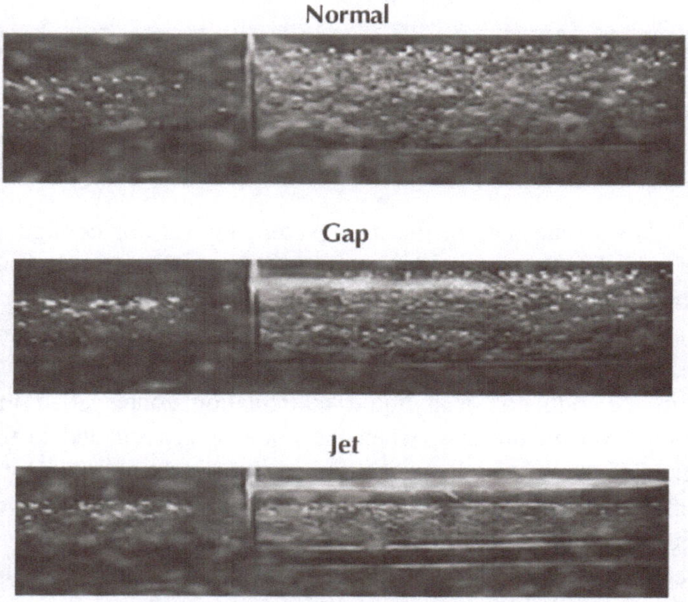

FIGURE 3 Narrowing and sudden expansion flow level.

With this experimental setup is shown that the formation of different modes of two phase flow depends on the relative concentration of these phases and the flow rate. Figs.1 and 2 shows a diagram of core flow of vapor-liquid flow regimes, in particular, the bubble, stratified, slug, stratified, and the wave dispersion circular flow.

In these experiments the mode of vapor-liquid flow in a pipe, when the bubbles are connected in long steam field, whose dimensions are commensurate with the diameter of the pipe[12–13].This flow is called the flow of air from the tube. In the transition from moderate to high-speed flow, when the concentration of vapor and liquid are approximately equal, the flow regime is often irregular and even chaotic[14–15]. By the simulated conditions, It is assumed that the electricity suddenly power off without warning (i.e., no time to turn the diesel generators or pumps)[15–16].

3.3 RESULTS

Such situations are the strong reason of the installation of pressure sensors, equipped with high-speed data loggers. Therefore, the following items are consequences, which may result in these situations.

3.3.1 EFFECTS OF TRANSIENTS

Hydraulic transients can lead to the following physical phenomena. High or low transient pressures that may arise in the piping and connections in the share of second. They often alternate from highest to lowest levels and vice versa. High pressures are a consequence of the collapse of steam bubbles or cavities is similar to steam pump cavitations. It can yield the tensile strength of the pipes. It can also penetrate the groundwater into the pipeline[17–18].

3.3.2 HIGH TRANSIENT FLOWS

High-speed flows are also very fast pulse pressure. It leads to temporary but very significant transient forces in the bends and other devices that can make a connection to deform. Even strain buried pipes under the influence of cyclical pressures may lead to deterioration of joints and lead to leakage. In the low-pressure pumping stations at downstream a very rapid closing of the valve, known as shut off valve, may lead to high pressure transient flows.

3.3.3 WATER COLUMN SEPARATION

Water column, usually are separated with sharp changes in the profile or the local high points. It is because of the excess of atmospheric pressure. The spaces between the columns are filled with water or the formation of steam (e.g., steam at ambient temperature) or air, if allowed admission into the pipe through the valve. Collapse of cavitation bubbles or steam can cause the dramatic impact of rising pressure on the transition process.

If the water column is divided very quickly, it could in turn lead to rupture of the pipeline. Vapors cavitation may also lead to curvature of the pipe. High pressure wave can also be caused by the rapid removal of air from the pipeline.

Steam bubbles or cavities are generated during the hydraulic transition. The level of hydraulic pressure (EGD) or pressure in some areas could fall low enough to reach the top of the pipe. It leads to subatmospheric pressure or even full-vacuum pressures. Part of the water may undergo a phase transition, changing from liquid to steam, while maintaining the vacuum pressure.

This leads to a temporary separation of the water column. When the system pressure increases, the columns of water rapidly approach to each other. The pair reverts to the liquid until vapor cavity completely dissolved. This is the most powerful and destructive power of water hammer phenomenon.

3.3.4 GLOBAL REGULATION OF STEAM PRESSURE

If system pressure drops to vapor pressure of the liquid, the fluid passes into the vapor, leading to the separation of liquid columns. Consequently, the vapor pressure is a fundamental parameter for hydraulic transient modeling. The vapor pressure varies considerably at high temperature or altitude.

Fortunately, for typical water pipelines and networks, the pressure does not reach such values. If the system is at high altitude or if it is the industrial system, operating at high temperatures or pressures, it should be guided by a table or a state of vapor pressure curve vaporliquid.

3.3.5 VIBRATION

Pressure fluctuations associated with the peculiarities of the system, as well as the peculiarities of its design. The pump must be assumed just as one of the promoters of the system. Effect of pressure relief valve on the fluctuations of the liquid often turned out three times more damaging than the effect of the pump. Such monitoring fluid flow controlling means usually has more negative impact than the influence of the pump.

Rapid changes in the transition pressures can lead to fluctuations or resonance. It can damage the pipe, resulting in leakage or rupture. Experiments show that the flow in the pipe will very small, say, located at $24\left(KM/_h\right)$.

This corresponds to about 0.45% of the velocity of pressure. In this case, the flow fluctuations can be easily accumulated and redeemed until the next perturbation. Fluctuations of the flow are no fluctuations of pressure.

In the case when the source of the pressure fluctuations is the so-called, "Acceleration factor", one can say that in order to accelerate the mass of fluid in the system until the new rate of additional efforts.

3.4 BASIC EQUATIONS DESCRIBING THE SPHERICALLY SYMMETRIC MOTION OF A BUBBLE BINARY SOLUTION

The dynamics and heat and mass transfer of vapor bubble in a binary solution of liquids, in [8], was studied for significant thermal, diffusion and inertial effect.

It was assumed that binary mixture with a density ρ_l, consisting of components 1 and 2, respectively, the density ρ_1 and ρ_2.

Moreover:

$$\rho_1 + \rho_2 = \rho_l,$$

Where the mass concentration of component 1 of the mixture.

Also Ref. [8]consider a two-temperature model of interphase heat exchange for the bubble liquid. This model assumes homogeneity of the temperature in phases.

The intensity of heat transfer for one of the dispersed particles with an endless stream of carrier phase will be set by the dimensionless parameter of Nusselt Nu_l.

Bubble dynamics described by the Rayleigh equation:

$$R\dot{w}_l + \frac{3}{2}w_l^2 = \frac{p_1 + p_2 - p_\infty - 2\sigma / R}{\rho_l} - 4\nu_1 \frac{w_l}{R} \tag{1}$$

wherep_1 and p_2 – the pressure component of vapor in the bubble,

p_∞ – the pressure of the liquid away from the bubble,

σ and ν_1 – surface tension coefficient of kinematic viscosity for the liquid.

Consider the condition of mass conservation at the interface.

Mass flow j_i^{TH} component $(i=1,2)$ of the interface $r = R(t)$ in j_i^{TH} phase per unit area and per unit of time and characterizes the intensity of the phase transition is given by:

$$j_i = \rho_i\left(\dot{R} - w_l - w_i\right),\ (i=1,2), \tag{2}$$

where:

w_i – The diffusion velocity component on the surface of the bubble. The relative motion of the components of the solution near the interface is determined by Fick's law:

$$\rho_1 w_1 = -\rho_2 w_2 = -\rho_l D \left.\frac{\partial k}{\partial r}\right|_R \tag{3}$$

If we add Eq.(2), while considering that:
$\rho_1 + \rho_2 = \rho_l$ and draw the Eq.(3), we obtain

$$\dot{R} = w_l + \frac{j_1 + j_2}{\rho_l}, \tag{4}$$

Multiplying the first Eq.(2) on ρ_2, the second in ρ_1 and subtract the second equation from the first. In view of Eq. (3) we obtain

$$k_R j_2 - (1 - k_R) j_1 = -\rho_l D \left.\frac{\partial k}{\partial r}\right|_R$$

where – the concentration of the first component at the interface. With the assumption of homogeneity of parameters inside the bubble changes in the mass of each component due to phase transformations can be written as,

$$\frac{d}{dt}\left(\frac{4}{3}\pi R^3 \rho_i'\right) = 4\pi R^2 j_i \text{ or}$$

$$\frac{R}{3}\dot{\rho}_i' + \dot{R}\rho_i' = j_i,\ (i=1,2), \tag{5}$$

Express the composition of a binary mixture in mole fractions of the component relative to the total amount of substance in liquid phase

$$N = \frac{n_1}{n_1 + n_2}, \tag{6}$$

The number of moles i^{TH} component n_i, which occupies the volume V, expressed in terms of its density:

$$n_i = \frac{\rho_i V}{\mu_i}, \tag{7}$$

Substituting Eq. (7) in Eq. (6), we obtain

$$N_1(k) = \frac{\mu_2 k}{\mu_2 k + \mu_1 (1 - k)}, \tag{8}$$

By law, Raul partial pressure of the component above the solution is proportional to its molar fraction in the liquid phase, that is,

$$p_1 = p_{S1}(T_v) N_1(k_R),\ p_2 = p_{S2}(T_v)\left[1 - N_1(k_R)\right], \tag{9}$$

Equations of state phases have the form:

$$p_i = BT_v \rho_i^{/} / \mu_i,\ (i = 1,2), \tag{10}$$

where: B – gas constant, T_v – the temperature of steam, ρ_i – the density of the mixture components in the vapor bubble, μ_i – molecular weight, p_{si} – saturation pressure.

The boundary conditions $r = \infty$ and on a moving boundary can be written as

$$k\big|_{r=\infty} = k_0,\ k\big|_{r=R} = k_R,\ T_l\big|_{r=\infty} = T_0,\ T_l\big|_{r=R} = T_v, \tag{11}$$

$$j_1 l_1 + j_2 l_2 = \lambda_l D \frac{\partial T_l}{\partial r}\bigg|_{r=R}, \tag{12}$$

where: l_i – specific heat of vaporization.

By the definition of Nusselt parameter – the dimensionless parameter characterising the ratio of particle size and the thickness of thermal boundary layer in the phase around the phase boundary and determined from additional considerations or from experience.

The heat of the bubble's intensity with the flow of the carrier phase will be further specified as:

$$\left(\lambda_l \frac{\partial T_l}{\partial r}\right)_{r=R} = Nu_l \cdot \frac{\lambda_l (T_0 - T_v)}{2R}, \tag{13}$$

In [16] obtained an analytical expression for the Nusselt parameter:

$$Nu_l = 2\sqrt{\frac{\omega R_0^2}{a_l}} = 2\sqrt{\frac{R_0}{a_l}\sqrt{\frac{3\gamma p_0}{\rho_l}}} = 2\sqrt{\sqrt{3\gamma} \cdot Pe_l}\,, \tag{14}$$

where: $a_l = \dfrac{\lambda_l}{\rho_l c_l}$ – thermal diffusivity of fluid,

$Pe_l = \dfrac{R_0}{a_l}\sqrt{\dfrac{p_0}{\rho_l}}$ – Peclet number.

The intensity of mass transfer of the bubble with the flow of the carrier phase will continue to ask by using the dimensionless parameter Sherwood *Sh*:

$$\left(D \frac{\partial k}{\partial r}\right)_{r=R} = Sh \cdot \frac{D(k_0 - k_R)}{2R}$$

where: D – diffusion coefficient, k – the concentration of dissolved gas in liquid.The subscripts 0 and R refer to the parameters in an undisturbed state and at the interface.

We define a parameter in the form of Sherwood [16]

$$Sh = 2\sqrt{\frac{\omega R_0^2}{D}} = 2\sqrt{\frac{R_0}{D}\sqrt{\frac{3\gamma p_0}{\rho_l}}} = 2\sqrt{\sqrt{3\gamma} \cdot Pe_D}\,, \tag{15}$$

where $Pe_D = \dfrac{R_0}{D}\sqrt{\dfrac{p_0}{\rho_l}}$ – diffusion Peclet number.

The system of Eqs.(1)–(15) is a closed system of equations describing the dynamics and heat transfer of insoluble gas bubbles with liquid[19–58].

3.5 THE BRAKING EFFECT OF THE INTENSITY OF PHASE TRANSFORMATIONS IN BOILING BINARY SOLUTIONS

If we use Eqs. (7)–(9), we obtain relations for the initial concentration of component 1:

$$k_0 = \frac{1-\chi_2^0}{1-\chi_2^0+\mu\left(\chi_1^0-1\right)}, \mu = \mu_2 / \mu_1, \chi_i^0 = p_{si0} / p_0, i = 1,2, \tag{16}$$

where: μ_2, μ_1 – Molecular weight of the liquid components of the mixture, p_{si0} – saturated vapor pressure of the components of the mixture at an initial temperature of the mixture T_0, which were determined by integrating the Clausius–Clapeyron relation. The parameter χ_i^0 is equal to

$$\chi_i^0 = \exp\left[\frac{l_i\mu_i}{B}\left(\frac{1}{T_{ki}}-\frac{1}{T_0}\right)\right], \tag{17}$$

Gas-phase liquid components in the derivation of Eq. (2) seemed perfect gas equations of state:

$$p_i = \rho_i BT_i / \mu_i.$$

where B – universal gas constant, p_i – the vapor pressure inside the bubble T_1 to the temperature in the ratio of Eq. (2), T_{ki} – temperature evaporating the liquid components of binary solution at an initial pressure p_0, l_i – specific heat of vaporization.

The initial concentration of the vapor pressure of component p_0 is determined from the relation:

$$c_0 = \frac{k_0\chi_1^0}{k_0\chi_1^0+\left(1-k_0\right)\chi_2^0}, \tag{18}$$

In this chapter the problem of radial motions of a vapor bubble in binary solution was solved. It was investigated at various pressure drops in the liquid for different initial radii R_0 for a bubble. It is of great practical interest of aqueous solutions of ethanol and ethylene glycol.

It was revealed an interesting effect. The parameters characterized the dynamics of bubbles in aqueous ethyl alcohol. It was studied in the field of variable pressure lie between the limiting values of the parameter p_0 for pure components [59–68].

The pressure drops and consequently the role of diffusion are assumed unimportant.The pressure drop along with the heat dissipation is included diffusion dissipation. The rate of growth and collapse of the bubble is much higher than in the corresponding pure components of the solution under the same conditions. A completely different situation existed during the growth and collapse of vapor bubble in aqueous solutions of ethylene glycol.

In this case, the effect of diffusion resistance, leaded to inhibition of the rate of phase transformations. The growth rate and the collapse of the bubble is much smaller than the corresponding values, but for the pure components of the solution. Further research and calculations have to give a physical explanation for the observed effect. The influence of heat transfer and diffusion on damping of free oscillations of a vapor bubble binary solution.

It was found that the dependence of the damping rate of oscillations of a bubble of water solutions of ethanol, methanol, and toluene monotonic on k_0.

It was mentioned for the aqueous solution of ethylene glycol similar dependence with a characteristic minimum at:

$$k_0 \approx 0,02 .$$

Moreover, for:

$$0,01 \leq k_0 \leq 0,3$$

decrement, binary solution has less damping rates for pulsations of a bubble in pure (one-component) water and ethylene glycol.

This means that in the range of concentrations of water:

$$0,01 \leq k_0 \leq 0,3$$

Pulsations of the bubble (for water solution of ethylene glycol) decay much more slowly and there is inhibition of the process of phase transformations. A similar process was revealed and forced oscillations of bubbles in an acoustic field.

In this book the influence of nonstationary heat and mass transfer processes was investigated in the propagation of waves in a binary solution of liquids with bubbles. The influence of component composition and concentration of binary solution was investigated on the dispersion, dissipation and attenuation of monochromatic waves in two-phase, two-component media.

The aqueous solution of ethyl alcohol in aqueous ethylene glycol decrements showed perturbations less relevant characteristics of pure components of the solution.

Unsteady interphase heat transfer revealed in calculation, the structure of stationary shock waves in bubbly binary solutions. The problem signifies on effect a violation of monotonicity behavior of the calculated curves for concentration, indicating the presence of diffusion resistance.

In some of binary mixtures, it is seen the effect of diffusion resistance. It is led to inhibition of the intensity of phase transformations.

The physical explanation revealed the reason for an aqueous solution of ethylene glycol. Pronounced effect of diffusion resistance is related to the solution with limited ability. It diffuses through the components of $D = 10^{-9}\left(m^2 / \mathrm{sec}\right)$, where D – diffusion coefficient volatility of the components is very different, and thus greatly different concentrations of the components in the solution and vapor phase.

In the case of aqueous solution of ethanol volatility component are roughly the same $\chi_1^0 \approx \chi_2^0$.

In accordance with Eq. (3) $c_0 \approx k_0$, so the finiteness of the diffusion coefficient does not lead to significant effects in violation of the thermal and mechanical equilibrium phases.

Figs. 4 and 5 show the dependence $k_0(c_0)$ of ethyl alcohol and ethylene glycol's aqueous solutions. From Fig. 4 it is clear that almost the entire range of k_0, $k_0 \approx c_0$.

At the same time for an aqueous solution of ethylene glycol, by the calculations and Fig. 2 $0{,}01 \le k_0 \le 0.3$ $k_0 \le c_0$, and when k0>3 $k_0 \sim c_0$.

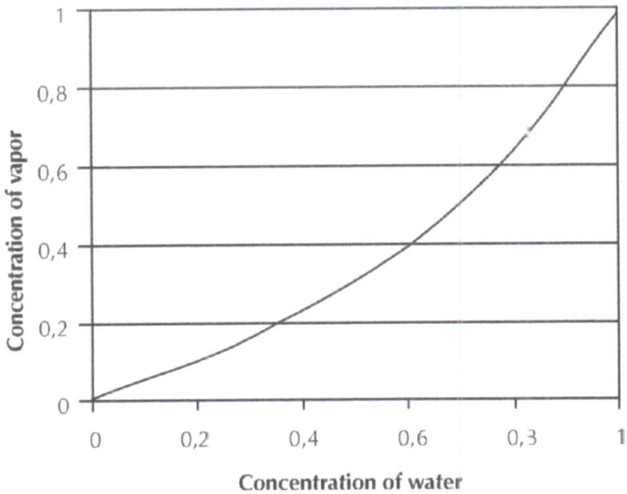

FIGURE 4 The dependence ($k_0(c_0)$) for an aqueous solution of ethanol.

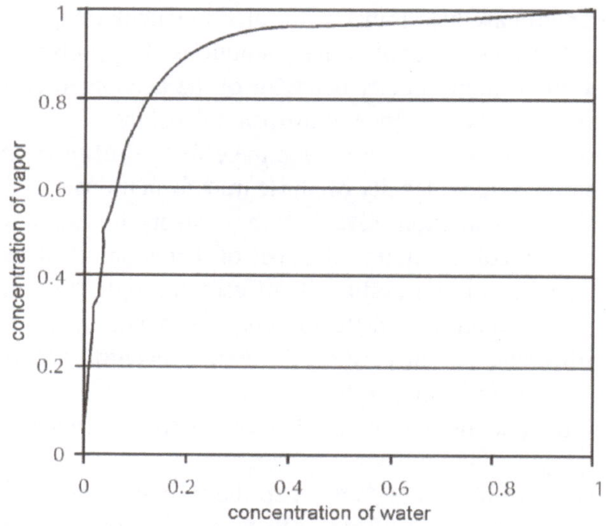

FIGURE 5 Dependence of ($k_0(c_0)$) for an aqueous solution of ethylene glycol.

Figs. 6 and 7 show the boiling point of the concentration for the solution of two systems. When $k_0 = 1$, $c_0 = 1$ and get clean water to steam bubbles. It is for boiling of a liquid at $t_0 = 373°C$.

If $k_0 = 1$, $c_0 = 1$ and have correspondingly pure bubble ethanol $t_0 = 350°C$ and ethylene glycol $t_0 = 370°C$.

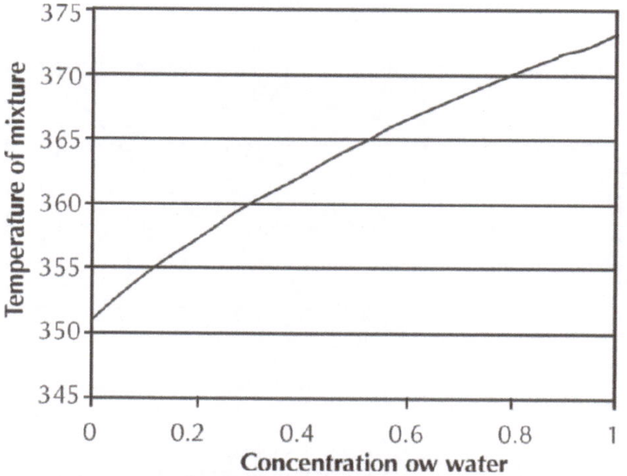

FIGURE 6 The dependence of the boiling temperature of the concentration of the solution to an aqueous solution of ethanol.

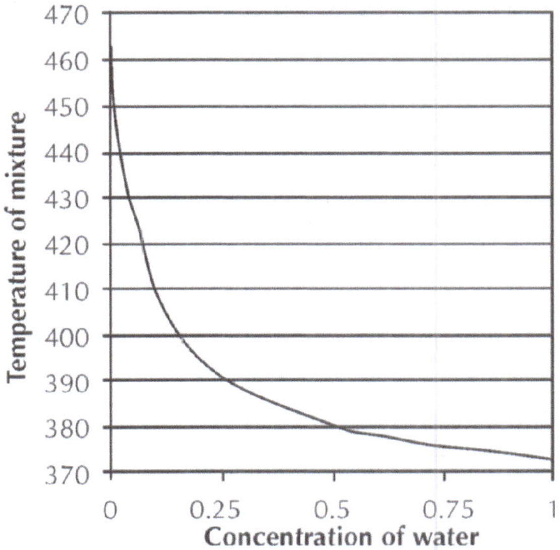

FIGURE 7 The dependence of the boiling point of the concentration of the solution to an aqueous solution of ethylene glycol.

It should be noted that in all the above works regardless of the above problems in the mathematical description of the cardinal effects of component composition of the solution shows the value of the parameter β equal

$$\beta=\left(1-\frac{1}{\gamma}\right)\frac{\left(c_0-k_0\right)\left(N_{c_0}-N_{k_0}\right)}{k_0\left(1-k_0\right)}\frac{c_l}{c_{pv}}\left(\frac{c_{pv}T_0}{L}\right)^2\sqrt{\frac{a_l}{D}}, \tag{19}$$

where: N_{k0}, N_{c0} – molar concentration of 1-th component in the liquid and steam.

$$N_{k_0}=\frac{\mu k_0}{\mu k_0+1-k_0},$$

$$N_{c_0}=\frac{\mu c_0}{\mu c_0+1-c_0}$$

where γ – Adiabatic index, c_l and c_{pv} respectively the specific heats of liquid and vapor at constant pressure, a_l – thermal diffusivity.

$$L=l_1c_0+l_2\left(1-c_0\right)$$

We also note that option Eq. (4) is a self-similar solution describing the growth of a bubble in a superheated solution. This solution has the form:

$$R = 2\sqrt{\frac{3}{\pi}}\frac{\lambda_l \Delta T \sqrt{t}}{L\rho_v \sqrt{a_l}(1+\beta)}, \tag{20}$$

where ρ – vapor density, t – time R – radius of the bubble, λ_l is the coefficient of thermal conductivity, ΔT – overheating of the liquid.

Figs. 8 and 9 show the dependence $\beta(K_0)$ for the above binary solutions. For aqueous ethanol β is negative for any value of concentration and dependence on K_0 is monotonic.

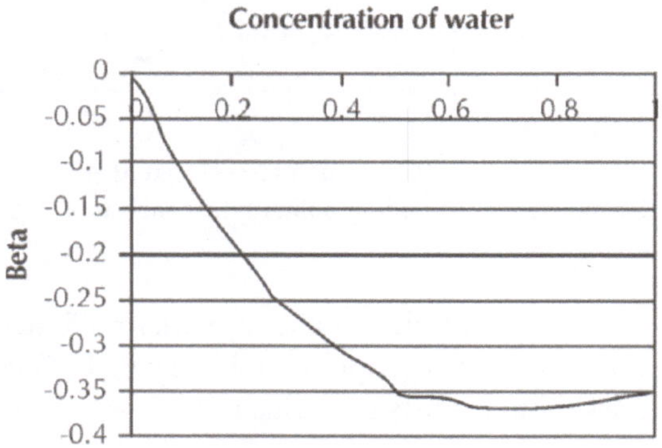

FIGURE 8 The dependence $\beta(K_0)$ for an aqueous solution of ethanol.

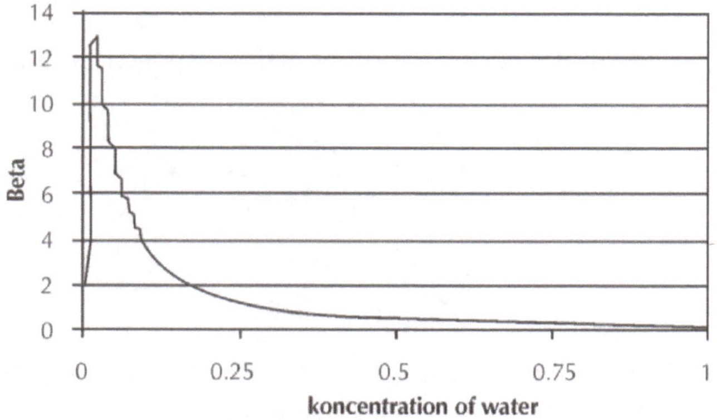

FIGURE 9 Dependence of $\beta(K_0)$ for an aqueous solution of ethylene glycol.

For an aqueous solution of ethylene glycol β – is positive and has a pronounced maximum at $k_0 = 0{,}02$.

As a result of this chapter, at low-pressure drops (superheating and super cooling of the liquid, respectively), diffusion does not occur in aqueous solutions of ethyl alcohol. By approximate equality of k_0 and c_0 all calculated dependence lie between the limiting curves for the case of one-component constituents of the solution.

They are included dependence of pressure, temperature, vapor bubble radius, the intensity of phase transformations, and so from time to time).

The pressure difference becomes important diffusion processes. Mass transfer between bubble and liquid is in a more intensive mode than in single-component constituents of the solution.

In particular, the growth rate of the bubble in a superheated solution is higher than in pure water and ethyl alcohol. It is because of the negative β according to Eq. (5).

In an aqueous solution of ethylene glycol, there is the same perturbations due to significant differences between k_0 and c_0. It is especially when $0{,}01 \le k_0 \le 0{,}3$,the effect of diffusion inhibition contributes to a significant intensity of mass transfer. In particular, during the growth of the bubble, the rate of growth in solution is much lower than in pure water and ethylene glycol. It is because of the positive β by Eq. (5).

Moreover, the maximum braking effect is achieved at the maximum value of β, when $k = 0{,}02$.

A similar pattern is observed at the pulsations and the collapse of the bubble. Dependence of the damping rate of fluctuations in an aqueous solution of ethyl alcohol from the water concentration is monotonic.Aqueous solution of ethylene glycol dependence of the damping rate has a minimum at

$$k_0 = 0{,}02,\ 0{,}01 < k_0 < 0{,}3.$$

The function decrement is small respectively large difference between k_0 and c_0 and β takes a large value. These ranges of concentrations in the solution have significant effect of diffusion inhibition.

For aqueous solutions of glycerin, methanol, toluene, etc., calculations are performed. Comparison with experimental data confirms the possibility of theoretical prediction of the braking of Heat and Mass Transfer.

It was analyzed the dependence of the parameter β, decrement of oscillations of a bubble from the equilibrium concentration of the mixture components. Therefore, in every solutions, it was determined the concentration of the components of a binary mixture.

Figs.10 and 11 are illustrated by theoretical calculations. These figures defined on the example of aqueous solutions of ethyl alcohol and ethylene glycol (antifreeze used in car radiators). It is evident that the first solution is not suitable to the task.

The aqueous solution of ethylene glycol with a certain concentration is theoretically much more slowly boils over with clean water and ethylene glycol. This confirms the reliability of the method.

Calculations show that such a solution is almost never freezes. The same method can offer concrete solutions for cooling of hot parts and components of various machines and mechanisms.

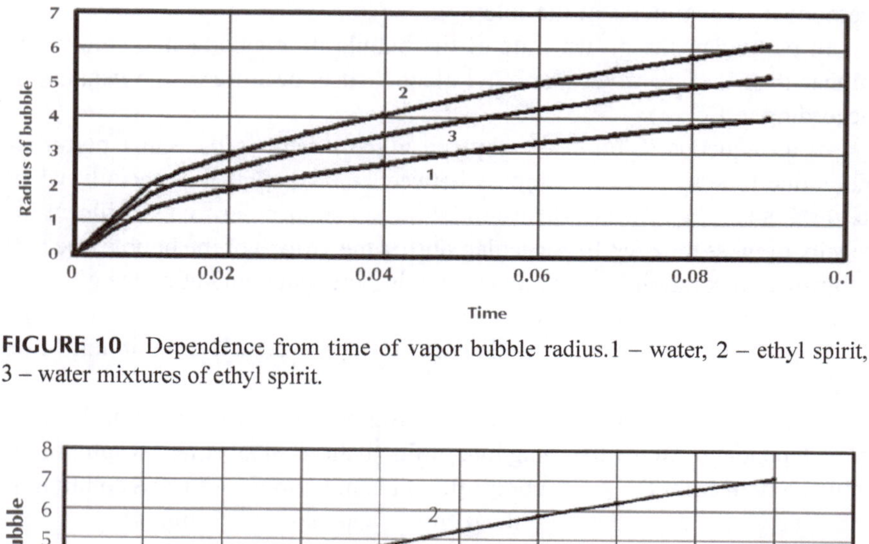

FIGURE 10 Dependence from time of vapor bubble radius. 1 – water, 2 – ethyl spirit, 3 – water mixtures of ethyl spirit.

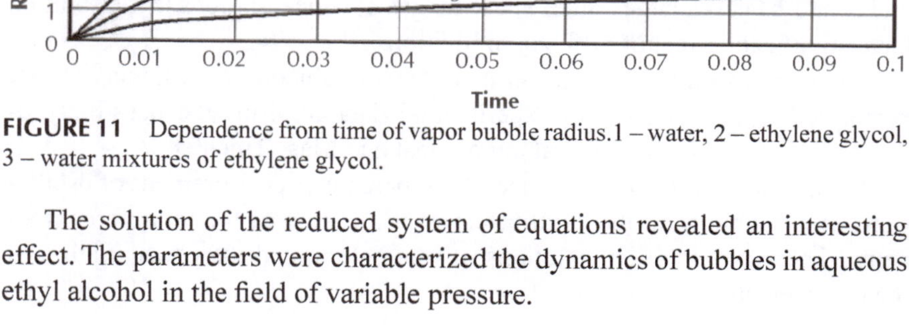

FIGURE 11 Dependence from time of vapor bubble radius. 1 – water, 2 – ethylene glycol, 3 – water mixtures of ethylene glycol.

The solution of the reduced system of equations revealed an interesting effect. The parameters were characterized the dynamics of bubbles in aqueous ethyl alcohol in the field of variable pressure.

They lied between the limiting values of relevant parameters for the pure components. It was for the case, which pressure drops and consequently the role of diffusion was unimportant.

A completely different situation is observed during the growth and collapse of vapor bubble in aqueous solutions of ethylene glycol. The effect of diffusion resistance, leads to inhibition of the rate of phase transformations. For pure components of the solution, the growth and the collapse rate of the bubble is much smaller than the corresponding values.

3.6 THE STRUCTURE OF THE PRESSURE WAVE FOR A SIMPLE PIPELINE SYSTEM

The wave dynamics of dispersed two-phase mixtures, in contrast to homogeneous media, is determined by processes of interaction between phases. The essential difference between mechanical and physical-chemical properties of the phases leads to the fact that the external disturbance has on the carrier and the dispersed phase of different actions. As a result of the wave front of finite intensity phase is no longer in equilibrium and the resulting relaxation processes can significantly influence the course.

The rates of relaxation processes determine the structure of individual relaxation zones for elementary waves, and the whole flow as a whole.

For example, the rapid mechanical fragmentation of drops for shock and detonation waves leads to the formation of a large number of secondary droplets, the surface area that is several orders of magnitude higher than those for the source of aerosol.

That leads to rapid evaporation of the liquid, mixing the vapor on gaseous oxidizer and the formation of a homogeneous air-fuel mixture. The basis for studying the dynamics of disperses mixtures is a description of the mechanical motion of the mixture "in general" under review at the scale of the whole problem. Extensive interactions of the phases (including those caused by the deformation, fragmentation, evaporation and mixing), going at the scale of individual particles, determined by this macro scale motion, providing, in turn, a significant inverse effect on him.

Currently, conventional mathematical hydrodynamics model of heterogeneous environments is a model multispeed continuum, the most complete exposition in the works.

Investigate the structure of stationary shock waves in binary mixtures, propagating with the velocity of pressure wave. This speed is a fundamental parameter for modeling hydraulic transients. Consider the effect of the dynamics of vapor bubbles in gas-liquid two component mixtures on the propagation of shock waves, the effect of unsteady forces on transient flows, division of water columns in the vapor or steam bubbles, control steam pressure in cavities to pump up the direction of the pipeline. Solving this problem is extremely important to study all the conditions under which the piping system having adverse transients, in particular in the pumps and valves. In addition, it is important to develop methods of protection and devices to be used during the design and construction of separate parts of the system, as well as to identify their practical shortcomings.

In solving problems for the management of transients caused by abrupt change in pressure, suggests two possible strategies. The first strategy is to minimize the possibility of transient conditions during the design, determine the appropriate methods of flow control to eliminate the possibility of emergency and unusual situations in the system. The second strategy is to provide for the establishment of security device to control the possible transients due to events beyond control, such as equipment failures and power supply.

Figs. 12 and 13 show the curves of pressure of time, received both experimental and theoretical methods. It is seen that the results of theoretical calculations agree well with experimental data.

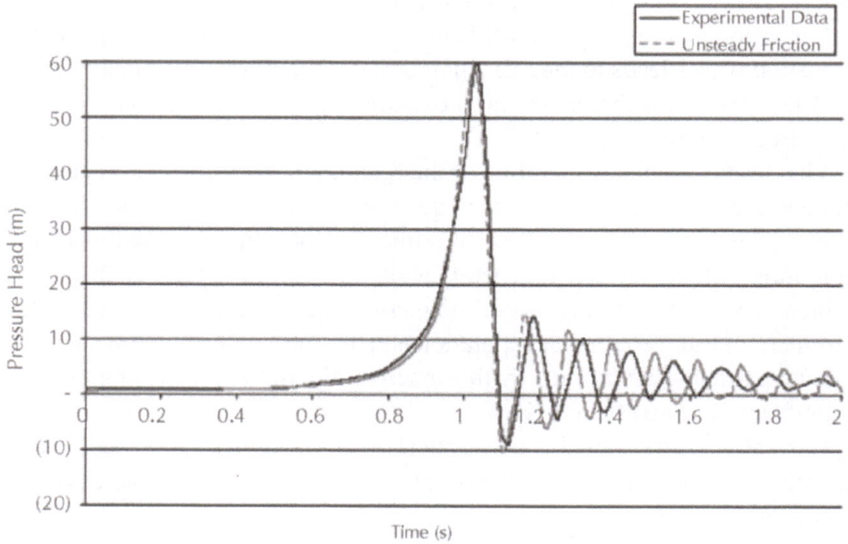

FIGURE 12 The structure of the pressure wave for a simple pipeline system using the model of equilibrium and no equilibrium friction.

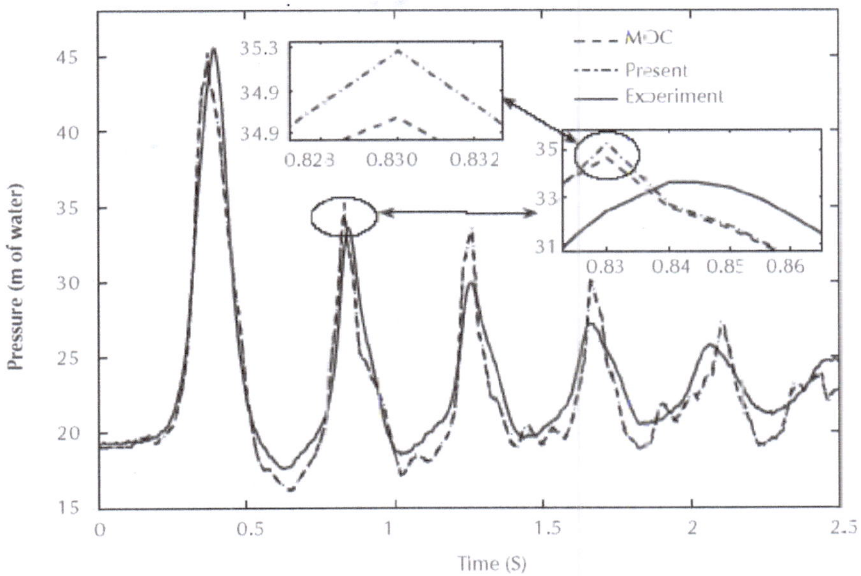

FIGURE13 Comparison the results with experimental data.

In the experiments, the pipeline was equipped with a valve at the end of the main pipes, joined by timers to record time of closing. The characteristics of water hydraulic impact measured and recorded by extensometers in the computer memory. Pumping of water to the system was carried out by the pool, which allowed the pressure to stabilize the inlet.

The experiments were performed for three cases:

1. A simple positive hydraulic shock for a straight pipe of constant diameter. The measured characteristics were the basis for assessing the impact of changes in diameter and the local diversion to the distribution of water hammer.
2. Water hammer in pipelines with diameters change: contraction and expansion.
3. Water hammer in pipelines with a local diversion in two scenarios: the outflow from the brain to super compression pool and a free outflow from the brain (to atmospheric pressure, with the ability to absorb air in the negative phase). This was the main reason for the air intake in the negative phase.

Separation columns for pump power off on pipeline were carried out for two cases:

(a) With the surge tank and the assumption of local diversion: In this case, the air was sucked into the pipeline; and
(b) Without the surge tank, provided the local diversion.

It can be recorded pressure wave and the wave velocity in fast transients up to 5 ms(in this study to 1 s). The assessment procedure was used to analyze Curve data, which were obtained on real systems. Software regression curve providing customized features and provided a regression analysis. Thus, the regression model (first model) was used in the final procedure. This model was compared with the results of the model performance (second model). The calculation results allowed us to develop a technical solution for the management of transition processes in the pipeline system. Fig.14 shows a scheme of how the developed device.

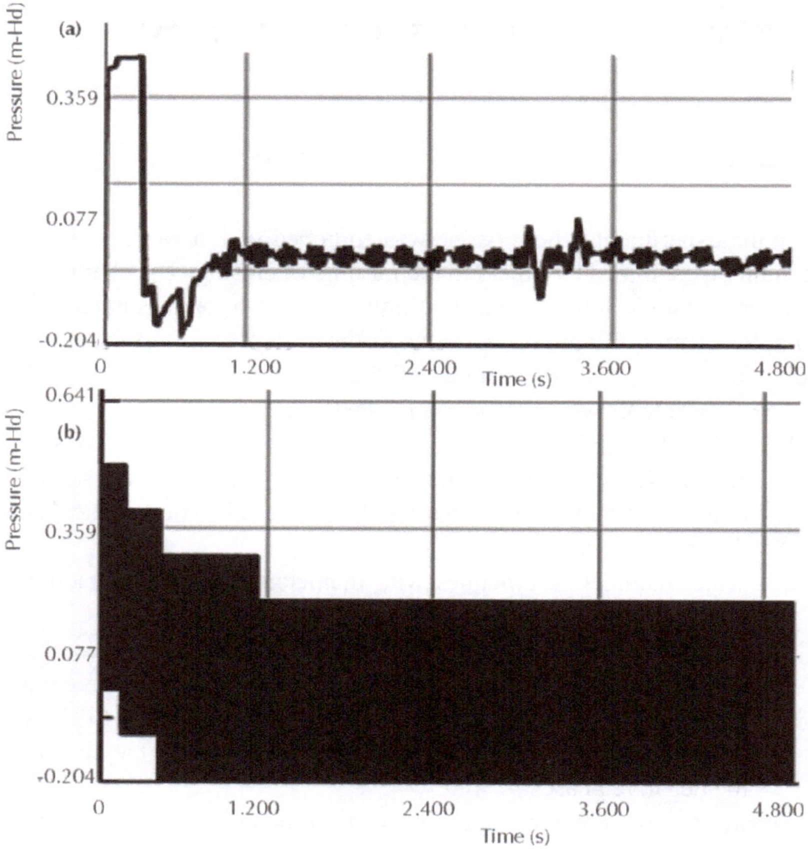

FIGURE 14 The working of the developed device for water supply.

The surge tank with double bottom manages transitions, converting accumulated potential energy of water to kinetic energy. In critical situations during periods of rapid change in the nature of the flow of water from the reservoir flows into the system piping. The tank is usually located in the pumping station or highest point in the water system profile. In the vertical jumpers reservoir can be drilled various holes to control the supply of water from the system into the reservoir. But this design results in very little loss of flux, spilling from the tank. If there is overflow and leak, the tank can also act as a protective device against unintentional pressure increase.

Thus, the device may serve as a way of protection from the overflow. There is another important problem. The problem is related to stability in the reservoir. It is necessary for avoiding rapid rising or lowering the water level in the reservoir.

Therefore, the surface area of the tank must be much greater cross section of the pipeline [70–106].

3.7 CONCLUSION

It was evaluated by a comprehensive approach the strength of pipeline systems consisting of technical diagnostics segment of the pipe and the mathematical modeling of the pipeline. Results implicated on prediction of the selected model.

It was used a mathematical model for determining the nature of pulse propagation of pressure in pipelines. It was concluded the presence of leaks, bends, and local resistance. The model provides a basis for developing a new method for determining leakage and unauthorized connections. It gives a clear idea about the nature of pulse propagation of pressure in pipelines.

The experimental setup and performed on studies confirmed the validity of the proposed mathematical model, and comparing the results of calculations and experiments showed satisfactory agreement.

Industrial testing method for detection of leaks and to combat water hammer for the water pipeline showed the effectiveness of the proposed methods.

It was confirmed the two temperature model theoretically on anomalous braking speed of phase transformations in boiling of binary mixtures.

It was investigated a comparison on accuracy of the numerical methods, regression model and the method of characteristics for the analysis of transient flows. It was shown that the method of characteristics is computationally more efficient for the analysis of large water pipelines.

KEYWORDS

- **bubble binary solution**
- **hydraulic transition**
- **Nusselt parameter**
- **vapor bubble**
- **vena contraction**
- **water hammer**

CHAPTER 4

DYNAMIC MODELING FOR HEAT AND MASS TRANSFER

CONTENTS

4.1 INTRODUCTION

In the two phases flow is extremely important to the concept of volume concentration. This is the relative volume fraction of one phase in the volume of the pipe [1]. Such an environment typical fluid is a high density and little compressibility. This property contributes to the creation of various forms of transient conditions [2–3]. Figure 1 shows an experimental setup, which investigated the formation of different modes of fluid flow with gas bubbles and steam. The experimental results show that the bubble flow usually occurs at low concentrations of vapor. It includes three main types of flow regimes in microgravity bubble, slug and an annular (Fig. 2).

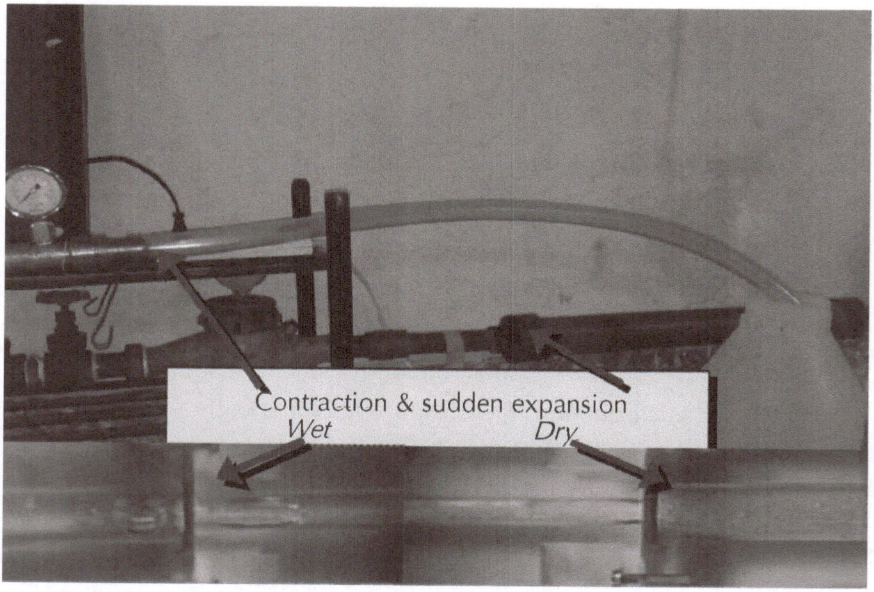

FIGURE 1 Snapshot laboratory setup for studying the structure of flow in different configurations tube.

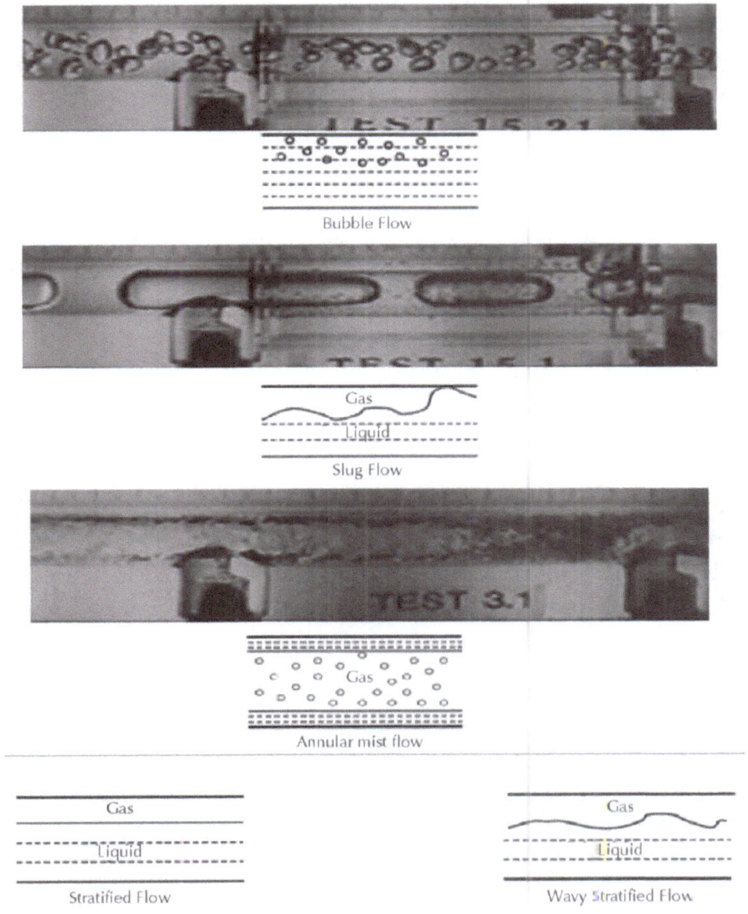

FIGURE 2 Types of flows.

Figure 2 shows snapshots of the flow pattern for various configurations of the tube [4–5]. The flow enters the tee at the bottom of the picture, and then is divided at the entrance to the tube. The inner diameter of the tee is 1.27 cm (Fig. 1). The narrowing of the flow is achieved by reducing the diameter of the hose.

Within the reduction is a liquid recirculation zone, called the "vena contraction" Wet picture that, when the liquid is recirculated to the "vena contraction." However, there are conditions, whereby the gas phase of the contract is caught vena. Figure 3 shows the different flow regimes observed in the experimental setup [6].

Increased flow is achieved by increasing the size of the pipe. Again, there is an area of the liquid recirculation near the "corner" a sudden expansion. Depending on the level of consumption of bubbly liquid or gas, it falls into the trap in this area.

In the inlet fluid moves out of the pipe diameter of 12.7 *(mm)* in the pipe diameter of 25 *(mm)*.

Normal extension occurs at the beginning of the flow. Soon comes the expansion section, and the flow rate continues to increase. Two-phase jet stream created, ultimately, with areas of air flowing above and below the bubble region [6]. The behavior of gas-liquid mixture in the expansion is proportional to increasing the diameter of the pipe [7–8].

It is shown that in place of the sudden expansion of a transition flow regime of the turbulent flow. Depending on the flow or gas bubble mixture it falls into the trap in this area.

Experiments were conducted on the pipe, whose diameter is suddenly doubled [9]. In this case the region of turbulence of fluid are observed around the "corner" a sudden expansion. The expansion is observed at the beginning of the flow. As a result of turbulence flow gap expansion increases and the flow rate continues to increase. In the end, creates a stream of two-phase flow, with air fields, the current above and below the bubble area [10–11].

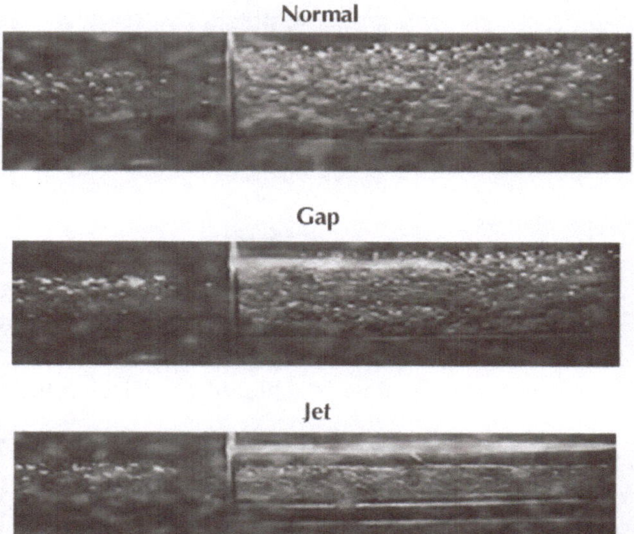

FIGURE 3 Narrowing and sudden expansion flow level.

With this experimental setup is shown that the formation of different modes of two-phase flow depends on the relative concentration of these phases and the flow rate.

Figures 1 and 2 shows a diagram of core flow of vapor-liquid flow regimes, in particular, the bubble, stratified, slug, stratified, and the wave dispersion circular flow.

In these experiments the mode of vapor-liquid flow in a pipe, when the bubbles are connected in long steam field, whose dimensions are commensurate with the diameter of the pipe [12–13].

This flow is called the flow of air from the tube. In the transition from moderate to high-speed flow, when the concentration of vapor and liquid are approximately equal, the flow regime is often irregular and even chaotic [14–15]. By the simulated conditions, It is assumed that the electricity suddenly power off without warning (i.e., no time to turn the diesel generators or pumps) [15–16].

Such situations are the strong reason of the installation of pressure sensors, equipped with high-speed data loggers. Therefore, the following items are consequences, which may result in these situations.

4.2 MATERIALS AND METHODS

Hydraulic transients can lead to the following physical phenomena—high or low transient pressures that may arise in the piping and connections in the share of second. They often alternate from highest to lowest levels and *vice versa*.

High pressures are a consequence of the collapse of steam bubbles or cavities are similar to steam pump cavitations. It can yield the tensile strength of the pipes. It can also penetrate the groundwater into the pipeline [17–18].

4.2.1 HIGH TRANSIENT FLOWS

High-speed flows are also very fast pulse pressure. It leads to temporary but very significant transient forces in the bends and other devices that can make a connection to deform. Even strain buried pipes under the influence of cyclical pressures may lead to deterioration of joints and lead to leakage. In the low-pressure pumping stations at downstream a very rapid closing of the valve, known as shut off valve, may lead to high pressure transient flows.

4.2.2 WATER COLUMN SEPARATION

Water column, usually are separated with sharp changes in the profile or the local high points. It is because of the excess of atmospheric pressure. The spaces between the columns are filled with water or the formation of steam (e.g., steam at ambient temperature) or air, if allowed admission into the pipe through the valve. Collapse of cavitation bubbles or steam can cause the dramatic impact of rising pressure on the transition process.

If the water column is divided very quickly, it could in turn lead to rupture of the pipeline. Vapors cavitation may also lead to curvature of the pipe. High-pressure wave can also be caused by the rapid removal of air from the pipeline.

Steam bubbles or cavities are generated during the hydraulic transition. The level of hydraulic pressure (EGD) or pressure in some areas could fall low enough to reach the top of the pipe. It leads to subatmospheric pressure or even full-vacuum pressures. Part of the water may undergo a phase transition, changing from liquid to steam, while maintaining the vacuum pressure.

This leads to a temporary separation of the water column. When the system pressure increases, the columns of water rapidly approach to each other. The pair reverts to the liquid until vapor cavity completely dissolved. This is the most powerful and destructive power of water hammer phenomenon.

4.2.3 GLOBAL REGULATION OF STEAM PRESSURE

If system pressure drops to vapor pressure of the liquid, the fluid passes into the vapor, leading to the separation of liquid columns. Consequently, the vapor pressure is a fundamental parameter for hydraulic transient modeling. The vapor pressure varies considerably at high temperature or altitude.

Fortunately, for typical water pipelines and networks, the pressure does not reach such values. If the system is at high altitude or if it is the industrial system, operating at high temperatures or pressures, it should be guided by a table or a state of vapor pressure curve vaporliquid.

4.2.4 VIBRATION

Pressure fluctuations associated with the peculiarities of the system, as well as the peculiarities of its design. The pump must be assumed just as one of the promoters of the system. Effect of pressure relief valve on the fluctuations of the liquid often turned out three times more damaging than the effect of the pump. Such monitoring fluid flow controlling means usually has more negative impact than the influence of the pump.

Rapid changes in the transition pressures can lead to fluctuations or resonance. It can damage the pipe, resulting in leakage or rupture. Experiments show that the flow in the pipe will be uniform if its fluctuations are very small, say, located at 24 $\left(KM/_{h}\right)$.

4.3 RESULTS

This corresponds to about 0.45% of the velocity of pressure. In this case, the flow fluctuations can be easily accumulated and redeemed until the next perturbation. Fluctuations of the flow are no fluctuations of pressure.

4.3.1 *BASIC EQUATIONS DESCRIBING THE SPHERICALLY SYMMETRIC MOTION OF A BUBBLE BINARY SOLUTION*

The dynamics and heat and mass transfer of vapor bubble in a binary solution of liquids, in Ref. [8], was studied for significant thermal, diffusion and inertial effect. It was assumed that binary mixture with a density ρ_l, consisting of components 1 and 2, respectively, the density ρ_1 and ρ_2. Moreover:

$$\rho_1 + \rho_2 = \rho_l,$$

where, the mass concentration of component one of the mixture [19–20] also consider a two temperature model of interphase heat exchange for the bubble liquid. This model assumes homogeneity of the temperature in phases [21, 22].

The intensity of heat transfer for one of the dispersed particles with an endless stream of carrier phase will be set by the dimensionless parameter of Nusselt Nu_l.

Bubble dynamics described by the Rayleigh equation:

$$R\dot{w}_l + \frac{3}{2}w_l^2 = \frac{p_1 + p_2 - p_\infty - 2\sigma / R}{\rho_l} - 4v\frac{w_l}{R} \tag{1}$$

where, p_1 and p_2 = the pressure component of vapor in the bubble, p_∞ = The pressure of the liquid away from the bubble, σ and v_1 = surface tension coefficient of kinematic viscosity for the liquid. Consider the condition of mass conservation at the interface [23].

Mass flow j_i^{TH} component $(i=1,2)$ of the interface $r=R(t)$ in j_i^{TH} phase per unit area and per unit of time and characterizes the intensity of the phase transition is given by [24]:

$$j_i = \rho_i\left(\overset{\bullet}{R} - w_l - w_i\right), (i=1,2) \quad (2)$$

where w_i = The diffusion velocity component on the surface of the bubble. The relative motion of the components of the solution near the interface is determined by Fick's law [25, 26].

$$\rho_1 w_1 = -\rho_2 w_2 = -\rho_l D \left.\frac{\partial k}{\partial r}\right|_R \quad (3)$$

If we add Eq. (2), while considering that: $\rho_1 + \rho_2 = \rho_l$ and draw the Eq. (3), we obtain [27, 28]

$$\overset{\bullet}{R} = w_l + \frac{j_1 + j_2}{\rho_l} \quad (4)$$

Multiplying the first Eq. (2) on ρ_2, the second in ρ_1 and subtract the second equation from the first. In view of (3.2.3) we obtain:

$$k_R j_2 - \left(1-k_R\right) j_1 = -\rho_l D \left.\frac{\partial k}{\partial r}\right|_R$$

Here k_R = the concentration of the first component at the interface.

With the assumption of homogeneity of parameters inside the bubble changes in the mass of each component due to phase transformations can be written as:

$$\frac{d}{dt}\left(\frac{4}{3}\pi R^3 \rho_i'\right) = 4\pi R^2 j_i \text{ or } \frac{R}{3}\overset{\bullet}{\rho_i'} + \overset{\bullet}{R}\rho_i' = j_i,\ (i=1,2) \quad (5)$$

Express the composition of a binary mixture in mole fractions of the component relative to the total amount of substance in liquid phase:

$$N = \frac{n_1}{n_1 + n_2} \quad (6)$$

The number of moles i^{TH} component n_i, which occupies the volume V, expressed in terms of its density:

$$n_i = \frac{\rho_i V}{\mu_i} \tag{7}$$

Substituting Eq. (3.2.7) in Eq. (3.2.6), we obtain:

$$N_1(k) = \frac{\mu_2 k}{\mu_2 k + \mu_1 (1-k)} \tag{8}$$

By law, Raul partial pressure of the component above the solution is proportional to its molar fraction in the liquid phase, that is:

$$p_1 = p_{S1}(T_v) N_1(k_R) \text{ and } p_2 = p_{S2}(T_v)[1 - N_1(k_R)] \tag{9}$$

Equations of state phases have the form:

$$p_i = BT_v \rho_i^{/} / \mu_i, \; (i = 1,2), \tag{10}$$

where, B = Gas constant, T_v = The temperature of steam, $\rho_i^{/}$ = The density of the mixture components in the vapor bubble, μ_i = Molecular weight, p_{Si} = Saturation pressure.

The boundary conditions $r = \infty$ and on a moving boundary can be written as:

$$k\big|_{r=\infty} = k_0, k\big|_{r=R} = k_R, \; T_l\big|_{r=\infty} = T_0, T_l\big|_{r=R} = T_v \tag{11}$$

$$j_1 l_1 - j_2 l_2 = \lambda_l D \frac{\partial T_l}{\partial r}\bigg|_{r=R} \tag{12}$$

where, l_i = specific heat of vaporization [29, 30].

By the definition of Nusselt parameter the dimensionless parameter characterising the ratio of particle size and the thickness of thermal boundary layer in the phase around the phase boundary and determined from additional considerations or experience [31, 32].

The heat of the bubble's intensity with the flow of the carrier phase will be further specified as:

$$\left(\lambda_l \frac{\partial T_l}{\partial r}\right)_{r=R} = Nu_l \cdot \frac{\lambda_l (T_0 - T_v)}{2R} \tag{13}$$

From Ref. [33] an analytical expression obtained for the Nusselt parameter:

$$Nu_l = 2\sqrt{\frac{\omega R_0^2}{a_l}} = 2\sqrt{\frac{R_0}{a_l}\sqrt{\frac{3\gamma p_0}{\rho_l}}} = 2\sqrt{\sqrt{3\gamma}\cdot Pe_l} \tag{14}$$

where, $a_l = \frac{\lambda_l}{\rho_l c_l}$ = thermal diffusivity of fluid,

$$Pe_l = \frac{R_0}{a_l}\sqrt{\frac{p_0}{\rho_l}} = \text{Peclet number.}$$

The intensity of mass transfer of the bubble with the flow of the carrier phase will continue to ask by using the dimensionless parameter Sherwood Sh:

$$\left(D\frac{\partial k}{\partial r}\right)_{r=R} = Sh\cdot\frac{D(k_0 - k_R)}{2R}$$

where, D = Diffusion coefficient, k = The concentration of dissolved gas in liquid,

The subscripts 0 and R refer to the parameters in an undisturbed state and at the interface.

We define a parameter in the form of Sherwood [33]:

$$Sh = 2\sqrt{\frac{\omega R_0^2}{D}} = 2\sqrt{\frac{R_0}{D}\sqrt{\frac{3\gamma p_0}{\rho_l}}} = 2\sqrt{\sqrt{3\gamma}\cdot Pe_D} \tag{15}$$

where, $Pe_D = \frac{R_0}{D}\sqrt{\frac{p_0}{\rho_l}}$ = diffusion Peclet number.

The system of Equations (1–15) is a closed system of equations describing the dynamics and heat transfer of insoluble gas bubbles with liquid.

4.3.2 THE BRAKING EFFECT OF THE INTENSITY OF PHASE TRANSFORMATIONS IN BOILING BINARY SOLUTIONS

If we use Eqs. (3.2.7)–(3.2.9), we obtain relations for the initial concentration of component 1:

$$k_0 = \frac{1-\chi_2^0}{1-\chi_2^0+\mu(\chi_1^0-1)},\ \mu = \mu_2/\mu_1,\ \chi_i^0 = p_{si0}/p_0,\ i = 1,2 \tag{16}$$

where, μ_2, μ_1 = Molecular weight of the liquid components of the mixture. p_{si0} = saturated vapor pressure of the components of the mixture at an initial temperature of the mixture T_0, which were determined by integrating the ClausiusClapeyron relation. The parameter χ_i^0 is equal to:

$$\chi_i^0 = \exp\left[\frac{l_i \mu_i}{B}\left(\frac{1}{T_{ki}} - \frac{1}{T_0}\right)\right], \tag{17}$$

Gas phase liquid components in the derivation of Eq. (17) seemed perfect gas equations of state:

$$p_i = \rho_i B T_i / \mu_i.$$

where, B = Universal gas constant, p_i = The vapor pressure inside the bubble T_i to the temperature in the ratio of Eq. (17), T_{ki} = Temperature evaporating the liquid components of binary solution at an initial pressure p_0, l_i = Specific heat of vaporization.

The initial concentration of the vapor pressure of component F_0 is determined from the relation:

$$c_0 = \frac{k_0 \chi_1^0}{k_0 \chi_1^0 + (1 - k_0)\chi_2^0} \tag{18}$$

The problem of radial motions of a vapor bubble in binary solution [34] was solved. It was investigated at various pressure drops in the liquid for different initial radii R_0 for a bubble. It is of great practical interest of aqueous solutions of ethanol and ethylene glycol.

It was revealed an interesting effect. The parameters characterized the dynamics of bubbles in aqueous ethyl alcohol. It was studied in the field of variable pressure lie between the limiting values of the parameter p_0 for pure components.

The pressure drops and consequently the role of diffusion are assumed unimportant. The pressure drop along with the heat dissipation is included diffusion dissipation. The rate of growth and collapse of the bubble is much higher than in the corresponding pure components of the solution under the same conditions. A completely different situation existed during the growth and collapse of vapor bubble in aqueous solutions of ethylene glycol.

In this case, the effect of diffusion resistance, leaded to inhibition of the rate of phase transformations. The growth rate and the collapse of the bubble is much smaller than the corresponding values but for the pure components of the solution. Further research and calculations have to give a physical explanation for the observed effect. In Refs. [35–38] studied the influence of heat transfer and diffusion on damping of free oscillations of a vapor bubble binary solution.

It was found that the dependence of the damping rate of oscillations of a bubble of water solutions of ethanol, methanol, and toluene monotonic on k_0.

It was mentioned for the aqueous solution of ethylene glycol similar dependence with a characteristic minimum at:

$$k_0 \approx 0.02$$

Moreover, for $0.01 \le k_0 \le 0.3$ decrement, binary solution has less damping rates for pulsations of a bubble in pure (one-component) water and ethylene glycol.

This means that in the range of concentrations of water:

$$0.01 \le k_0 \le 0.3$$

Pulsations of the bubble (for water solution of ethylene glycol) decay much more slowly and there is inhibition of the process of phase transformations. A similar process was revealed and forced oscillations of bubbles in an acoustic field [35].

The influence of nonstationary heat and mass transfer [39] processes was investigated in the propagation of waves in a binary solution of liquids with bubbles. The influence of component composition and concentration of binary solution was investigated on the dispersion, dissipation, and attenuation of monochromatic waves in two phases, two component media.

The aqueous solution of ethyl alcohol in aqueous ethylene glycol decrements showed perturbations less relevant characteristics of pure components of the solution.

The unsteady interphase heat transfer revealed in Ref. [40] calculation, the structure of stationary shock waves in bubbly binary solutions. The problem signifies on effect a violation of monotonicity behavior of the calculated curves for concentration, indicating the presence of diffusion resistance.

In some of binary mixtures, it is seen the effect of diffusion resistance. It is led to inhibition of the intensity of phase transformations.

The physical explanation revealed the reason for an aqueous solution of ethylene glycol. The pronounced effect of diffusion resistance is related to the solution with limited ability. It diffuses through the components of $D = 10^{-9}\left(m^2/\sec\right)$.

D = Diffusion coefficient volatility of the components is very different, and thus greatly different concentrations of the components in the solution and vapor phase.

In the case of aqueous solution of ethanol volatility component are roughly the same $\chi_1^0 \approx \chi_2^0$

In accordance with Eq. (3) $c_0 \approx k_0$, so the finiteness of the diffusion coefficient does not lead to significant effects in violation of the thermal and mechanical equilibrium phases.

Figures 4 and 5 show the dependence $k_0(c_0)$ of ethyl alcohol and ethylene glycol's aqueous solutions. From Fig. 4 it is clear that almost the entire range of k_0, $k_0 \approx c_0$.

At the same time for an aqueous solution of ethylene glycol, by the calculations and Fig. 4 $0.01 \le k_0 \le 0.3$, $k_0 \le c_0$, and when $k_0 > 0.3$, $k_0 \approx c_0$.

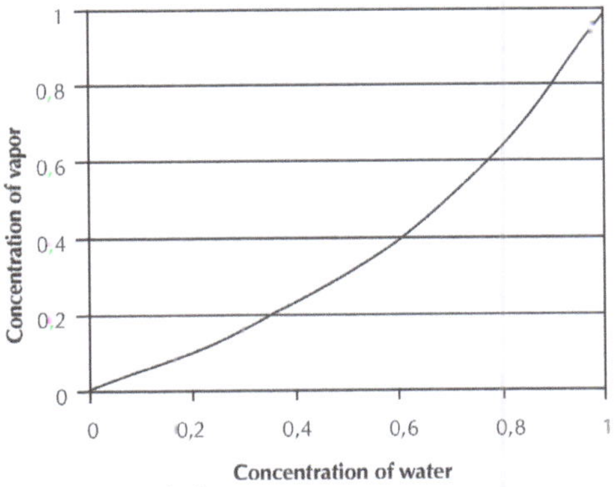

FIGURE 4 The dependence ($k_0(c_0)$) for an aqueous solution of ethanol.

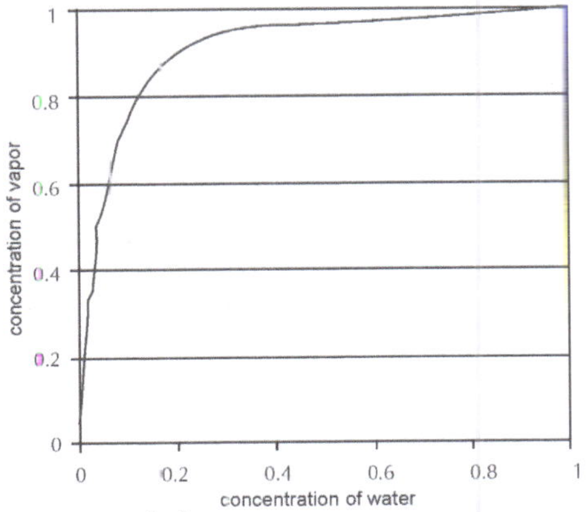

FIGURE 5 Dependence of ($k_0(c_0)$) for an aqueous solution of ethylene glycol.

Figures 6 and 7 show the boiling point of the concentration for the solution of two systems when, $k_0 = 1$, $c_0 = 1$ and get clean water to steam bubbles. It is for boiling of a liquid at

$$T_0 = 373^0 K.$$

If $k_0 = 0$, $c_0 = 0$ and have correspondingly pure bubble ethanol $(T_0 = 350^0 K)$ and ethylene glycol $(T_0 = 470^0 K)$.

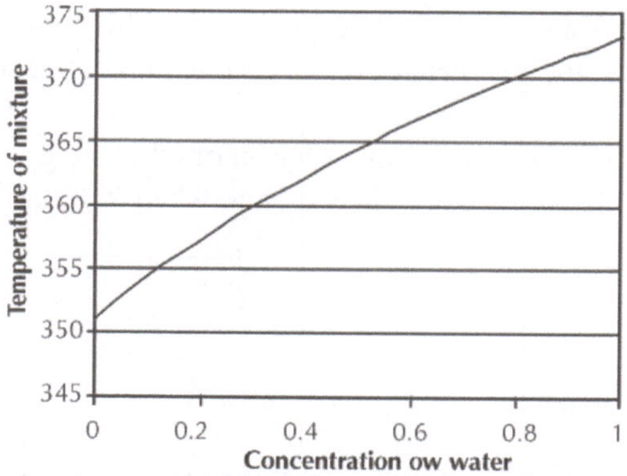

FIGURE 6 The dependence of the boiling temperature of the concentration of the solution to an aqueous solution of ethanol.

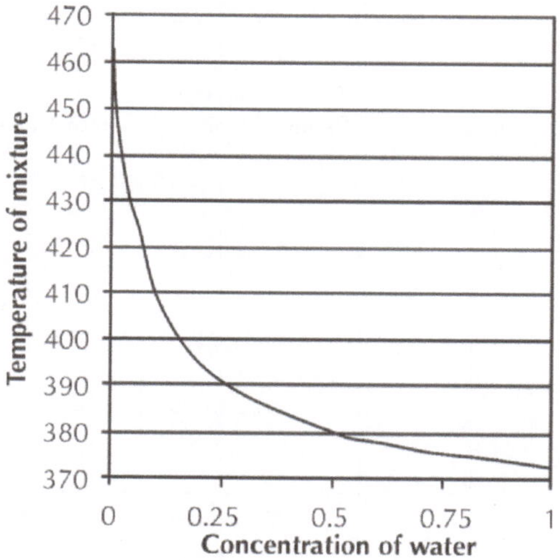

FIGURE 7 The dependence of the boiling point of the concentration of the solution to an aqueous solution of ethylene glycol.

It should be noted that, all works regardless of the problems in the mathematical description of the cardinal effects of component composition of the solution shows the value of the parameter β equal:

$$\beta=\left(1-\frac{1}{\gamma}\right)\frac{(c_0-k_0)(N_{c_0}-N_{k_0})}{k_0(1-k_0)}\frac{c_l}{c_p}\left(\frac{c_p T_0}{L}\right)^2\sqrt{\frac{a_l}{D}} \tag{19}$$

where, N_{k_0}, N_{c_0} = Molar concentration of 1th component in the liquid and steam

$$N_{k_0}=\frac{\mu k_0}{\mu k_0+1-k_0},$$

$N_{c_0}=\dfrac{\mu c_0}{\mu c_0+1-c_0}$ γ = Adiabatic index,

c_l and c_{pv}, respectively, the specific heats of liquid and vapor at constant pressure, a_l = Thermal diffusivity.

$$L=l_1c_0+l_2(1-c_0)$$

We also note that option (Eq. (4)) is a self-similar solution describing the growth of a bubble in a superheated solution. This solution has the form [41]:

$$R=2\sqrt{\frac{3}{\pi}}\frac{\lambda_l\Delta T\sqrt{t}}{L\rho_v\sqrt{a_l(1+\beta)}} \tag{20}$$

where, ρ_v = vapor density, t = time, R = radius of the bubble, λ_l = the coefficient of thermal conductivity, ΔT = Overheating of the liquid.

Figures 8 and 9 shows the dependence $\beta(k_0)$ for the binary solutions. For aqueous ethanol β is negative for any value of concentration and dependence on k_0 is monotonic.

For an aqueous solution of ethylene glycol β = is positive and has a pronounced maximum at $k_0 = 0.02$.

As a result of present work at low-pressure drops (superheating and super cooling of the liquid, respectively), diffusion does not occur in aqueous solutions of ethyl alcohol. By approximate equality of k_0 and c_0 all calculated dependence lie between the limiting curves for the case of one-component constituents of the solution.

They are included dependence of pressure, temperature, vapor bubble radius, the intensity of phase transformations, and so from time to time). The pressure difference becomes important diffusion processes. Mass transfer between bubble and liquid is in a more intensive mode than in single component constituents of the solution. In particular, the growth rate of the bubble in a superheated solution is higher than in pure water and ethyl alcohol. It is because of the negative β according to Eq. (17).

In an aqueous solution of ethylene glycol, there is the same perturbations due to significant differences between k_0 and c_0. It is especially when $0.01 \leq k_0 \leq 0.3$, the effect of diffusion inhibition contributes to a significant intensity of mass transfer. In particular, during the growth of the bubble, the rate of growth in solution is much lower than in pure water and ethylene glycol. It is because of the positive β by Eq. (17).

Moreover, the maximum braking effect is achieved at the maximum value of β, when $k_0 = 0.02$. A similar pattern is observed at the pulsations and the collapse of the bubble. The dependence of the damping rate of fluctuations in an aqueous solution of ethyl alcohol from the water concentration is monotonic is shown in [35]. Aqueous solution of ethylene glycol dependence of the damping rate has a minimum at $k_0 = 0.02$, $0.01 \leq k_0 \leq 0.3$.

The function decrement is small respectively large difference between k_0 and c_0 and β takes a large value. These ranges of concentrations in the solution have significant effect of diffusion inhibition. For aqueous solutions of glycerin, methanol, toluene, and so on, calculations are performed. The comparison with experimental data confirms the possibility of theoretical prediction of the braking of heat and mass transfer.

It was analyzed the dependence of the parameter β, decrement of oscillations of a bubble from the equilibrium concentration of the mixture components. Therefore, in every solutions, it was determined the concentration of the components of a binary mixture.

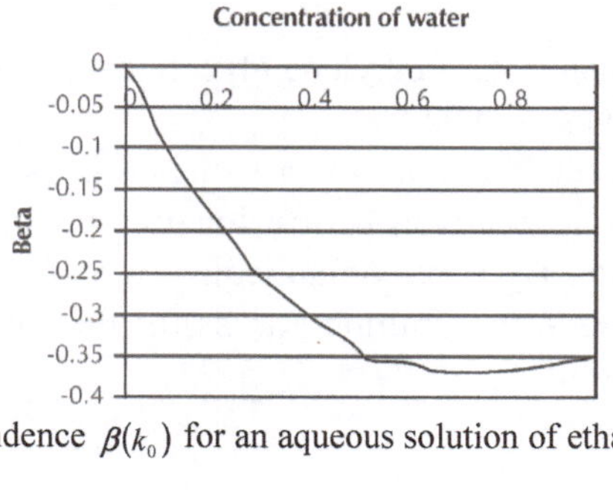

FIGURE 8 The dependence $\beta(k_0)$ for an aqueous solution of ethanol.

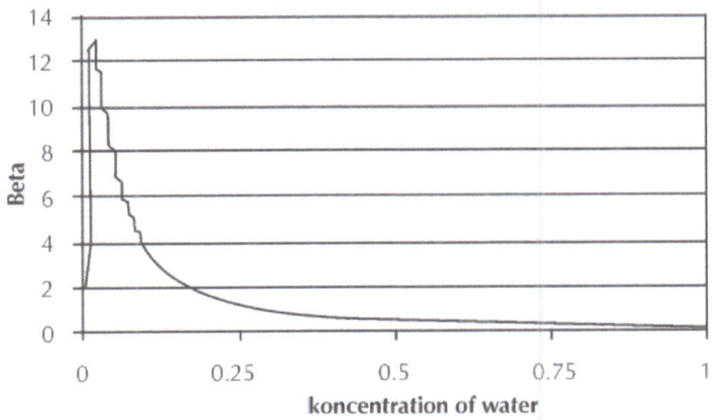

FIGURE 9 Dependence of $\beta(k_0)$ for an aqueous solution of ethylene glycol.

The Figures 10 and 11 are illustrated by theoretical calculations. These figures defined on the example of aqueous solutions of ethyl alcohol and ethylene glycol (antifreeze used in car radiators). It is evident that the first solution is not suitable to the task.

The aqueous solution of ethylene glycol with a certain concentration is theoretically much more slowly boils over with clean water and ethylene glycol. This confirms the reliability of the method.

The calculations shows that solution is never freezes. The same method can offer concrete solutions for cooling of hot parts and components of various machines and mechanisms.

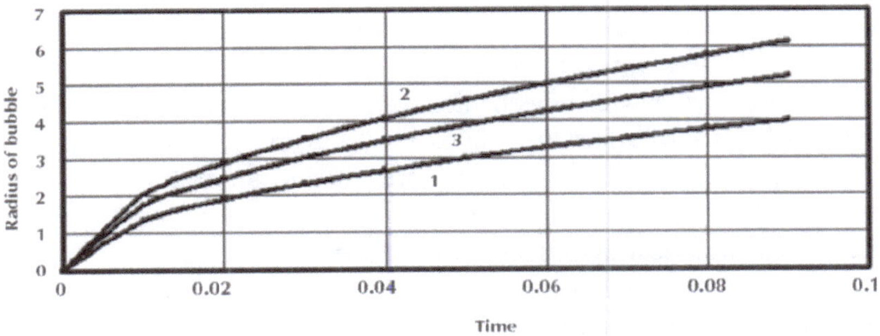

FIGURE 10 Dependence from time of vapor bubble radius. 1—water, 2—ethyl spirit, 3—water mixtures of ethyl spirit.

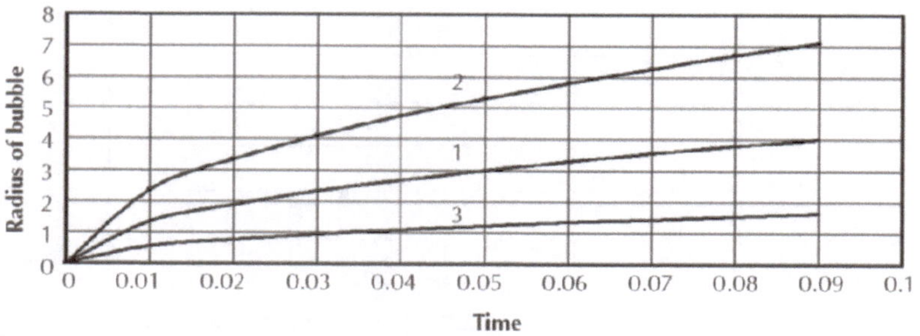

FIGURE 11 Dependence from time of vapor bubble radius. 1—water, 2—ethylene glycol, 3—water mixtures of ethylene glycol.

The solution of the reduced system of equations revealed an interesting effect. The parameters were characterized the dynamics of bubbles in aqueous ethyl alcohol in the field of variable pressure. They lied between the limiting values of relevant parameters for the pure components. It was for the case, which pressure drops and consequently the role of diffusion was unimportant.

A completely different situation is observed during the growth and collapse of vapor bubble in aqueous solutions of ethylene glycol. The effect of diffusion resistance, leads to inhibition of the rate of phase transformations. For pure components of the solution, the growth and the collapse rate of the bubble is much smaller than the corresponding values.

4.3.3 THE STRUCTURE OF THE PRESSURE WAVE FOR A SIMPLE PIPELINE SYSTEM

The wave dynamics of dispersed two-phase mixtures, in contrast to homogeneous media, is determined by processes of interaction between phases. The essential difference between mechanical and physical-chemical properties of the phases leads to the fact that the external disturbance has on the carrier and the dispersed phase of different actions. As a result of the wave front of finite intensity phase is no longer in equilibrium and the resulting relaxation processes can significantly influence the course.

The rates of relaxation processes determine the structure of individual relaxation zones for elementary waves, and the whole flow as a whole.

For example, the rapid mechanical fragmentation of drops for shock and detonation waves leads to the formation of a large number of secondary droplets, the surface area that is several orders of magnitude higher than those for the source of aerosol.

That leads to rapid evaporation of the liquid, mixing the vapor on gaseous oxidizer and the formation of a homogeneous air-fuel mixture [42]. The basis for studying the dynamics of disperses mixtures is a description of the mechanical motion of the mixture "in general" under review at the scale of the whole problem. Extensive interactions of the phases (including those caused by the deformation, fragmentation, evaporation and mixing), going at the scale of individual particles, determined by this macro scale motion, providing, in turn, a significant inverse effect on him.

Currently, conventional mathematical hydrodynamics model of heterogeneous environments is a model multispeed continuum, the most complete exposition in the works [43].

Investigate the structure of stationary shock waves in binary mixtures, propagating with the velocity of pressure wave. This speed is a fundamental parameter for modeling hydraulic transients. Consider the effect of the dynamics of vapor bubbles in gas-liquid two component mixtures on the propagation of shock waves, the effect of unsteady forces on transient flows, division of water columns in the vapor or steam bubbles, control steam pressure in cavities to pump up the direction of the pipeline. Solving this problem is extremely important to study all the conditions under which the piping system having adverse transients, in particular in the pumps and valves. In addition, it is important to develop methods of protection and devices to be used during the design and construction of separate parts of the system, as well as to identify their practical shortcomings.

In solving problems for the management of transients caused by abrupt change in pressure, suggests two possible strategies. The first strategy is to minimize the possibility of transient conditions during the design, determine the appropriate methods of flow control to eliminate the possibility of emergency and unusual situations in the system. The second strategy is to provide for the establishment of security device to control the possible transients due to events beyond control, such as equipment failures and power supply [44–45].

The Figs. 12 and 13 shows the curves of pressure of time, received both experimental and theoretical methods. It is seen that the results of theoretical calculations agree well with experimental data [46].

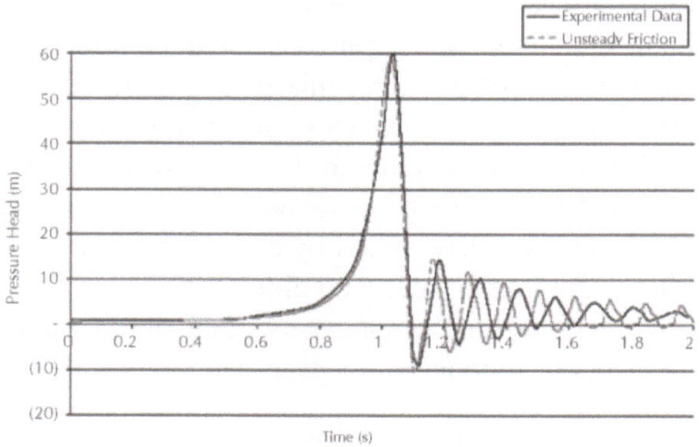

FIGURE 12 The structure of the pressure wave for a simple pipeline system using the model of equilibrium and no equilibrium friction.

In the experiments, the pipeline was equipped with a valve at the end of the main pipes, joined by timers to record time of closing. The characteristics of water hydraulic impact measured and recorded by extensometers in the computer memory. Pumping of water to the system was carried out by the pool, which allowed the pressure to stabilize the inlet.

The experiments were performed for three cases:

1. A simple positive hydraulic shock for a straight pipe of constant diameter. The measured characteristics were the basis for assessing the impact of changes in diameter and the local diversion to the distribution of water hammer.
2. Water hammer in pipelines with diameters change: contraction and expansion.
3. Water hammer in pipelines with a local diversion in two scenarios: the outflow from the brain to super compression pool and a free outflow from the brain (to atmospheric pressure, with the ability to absorb air in the negative phase). This was the main reason for the air intake in the negative phase.

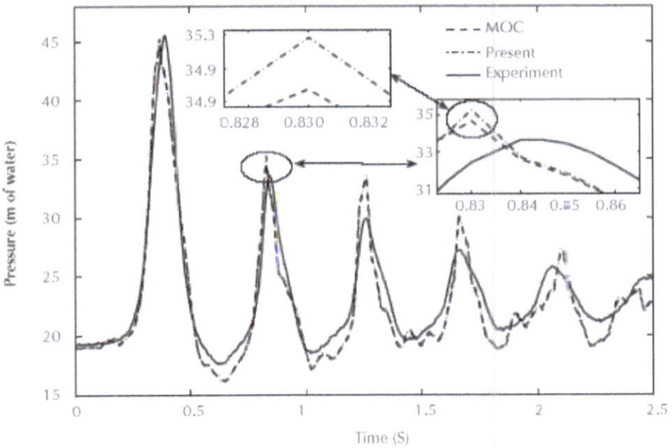

FIGURE 13 Comparison the results with experimental data.

Separation columns for pump power off on pipeline were carried out for two cases:

(a) With the surge tank and the assumption of local diversion: In this case, the air was sucked into the pipeline, and

(b) Without the surge tank, provided the local diversion.

It can be recorded pressure wave and the wave velocity in fast transients up to 5 ms (in this study to 1 s). The assessment procedure was used to analyze Curve data, which were obtained on real systems. Software regression curve providing customized features and provided a regression analysis. Thus, the regression model (first model) was used in the final procedure. This model was compared with the results of the model performance (second model). The calculation results allowed us to develop a technical solution for the management of transition processes in the pipeline system [47]. Figure 14 shows a scheme of how the developed device.

The surge tank with double bottom manages transitions, converting accumulated potential energy of water to kinetic energy. In critical situations during periods of rapid change in the nature of the flow of water from the reservoir flows into the system piping. The tank is usually located in the pumping station or highest point in the water system profile. In the vertical jumpers reservoir can be drilled various holes to control the supply of water from the system into the reservoir. But this design results in very little loss of flux, spilling from the tank. If there is overflow and leak, the tank can also act as a protective device against unintentional pressure increase

Thus, the device may serve as a way of protection from the overflow. There is another important problem. The problem is related to stability in the reservoir. It is necessary for avoiding rapid rising or lowering the water level in the reservoir.

Therefore, the surface area of the tank must be much greater cross section of the pipeline [48].

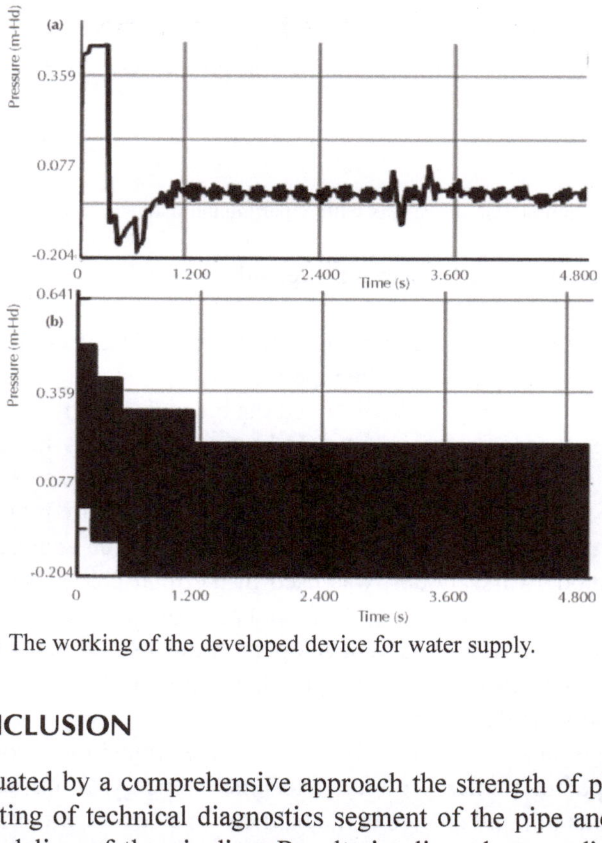

FIGURE 14 The working of the developed device for water supply.

4.4 CONCLUSION

It was evaluated by a comprehensive approach the strength of pipeline systems consisting of technical diagnostics segment of the pipe and the mathematical modeling of the pipeline. Results implicated on prediction of the selected model.

It was used a mathematical model for determining the nature of pulse propagation of pressure in pipelines. It was concluded the presence of leaks, bends, and local resistance. The model provides a basis for developing a new method for determining leakage and unauthorized connections. It gives a clear idea about the nature of pulse propagation of pressure in pipelines.

The experimental setup and performed on studies confirmed the validity of the proposed mathematical model, and comparing the results of calculations and experiments showed satisfactory agreement.

Industrial testing method for detection of leaks and to combat water hammer for the water pipeline showed the effectiveness of the proposed methods.

It was confirmed the two temperature model theoretically on anomalous braking speed of phase transformations in boiling of binary mixtures.

It was investigated a comparison on accuracy of the numerical methods, regression model and the method of characteristics for the analysis of transient flows. It was shown that the method of characteristics is computationally more efficient for the analysis of large water pipelines.

KEYWORDS

- **bubble binary solution**
- **hydraulic transition**
- **Nusselt parameter**
- **vapor bubble**
- **vena contraction**
- **water hammer**

CHAPTER 5

VAPOR PRESSURE AND SATURATION TEMPERATURE

CONTENTS

5.1 INTRODUCTION

Wood is a hygroscopic, porous, anisotropic and nonhomogeneous material. After log sawing, the lumber contains liquid water in fiber cavities (capillary water) and bound water inside the fiber wall (hygroscopic water). Porosity refers to volume fraction of void space. This void space can be actual space filled with air or space filled with both water and air. Capillary-porous materials are sometimes defined as those having pore diameter less than 10^{-7}m. Capillary porous materials were defined as those having a clearly recognizable pore space. In capillary porous material, transport of water is a more complex phenomena. In addition to molecular diffusion, water transport can be due to vapor diffusion, surface diffusion, Knudsen diffusion, capillary flow, and purely hydrodynamic flow. In hygroscopic materials, there is large amount of physically bound water and the material often shrinks during heating.

5.2 MATERIALS AND METHODS

In hygroscopic materials there is a level of moisture saturation below which the internal vapor pressure is a function of saturation and temperature. These relationships are called equilibrium moisture isotherms. Above this moisture saturation, the vapor pressure is a function of temperature only (as expressed by the Clapeyron equation) and is independent of the moisture level. Thus, above certain moisture level, all materials behave nonhygroscopic [1].

5.2.1 COMPUTER MODELS FOR HEAT FLOW IN NON-HOMOGENEOUS MATERIAL

Green wood contains a lot of water. In the outer parts of the stem, in the sapwood, spruce and pine have average moisture content of about 130%, and in the inner parts, in the heartwood, the average moisture content is about 35%. Wood drying is the art of getting rid of that surplus water under controlled forms. It will dry to an equilibrium moisture content of 8–16% fluid content when left in air which improves its stability, reduces its weight for transport, prepares it for chemical treatment or painting and improves its mechanical strength.

Water in wood is found in the cell cavities and cell walls. All void spaces in wood can be filled with liquid water called free water. Free water is held by adhesion and surface tension forces. Water in the cell walls is called bound water. Bound water is held by forces at the molecular level. Water molecules attach themselves to sites on the cellulose chain molecules. It is an intimate part of the cell wall but does not alter the chemical properties of wood. Hydrogen bonding is the predominant fixing mechanism. If wood is allowed to dry, the first water to be removed is free water. No bound water is evaporated until all free water has been removed. During removal of water, molecular energy is expended. Energy requirement for vaporization of bound water is higher than free water. Moisture content at which only the cell walls are completely saturated (all bound water) but no free water exists in all lumens is called the fiber saturation point (F.S.P). Typically the F.S.P of wood is within the range of 20–40% moisture content depending on temperature and wood species. Water in wood normally moves from high to low zones of moisture content. The surface of the wood must be drier than the interior if moisture is to be removed. Drying can be divided into two phases: movement of water from the interior to the surface of the wood, and removal of water from the surface. Water moves through the interior of the wood as a liquid or water vapor through various air passageways in the cellular structure of wood and through the cell walls [2].

Drying is a process of simultaneous heat and moisture transfer with a transient nature. The evolution process of the temperature and moisture with time must be predicted and actively controlled in order to ensure an effective and efficient drying operation. Lumber drying can be understood as the balance between heat transfer from airflow to wood surface and water transport from the wood surface to the airflow. Reduction in drying time and energy consumption offers the wood industries a great potential for economic benefit. In hygroscopic porous material like wood, mathematical models describing moisture and heat movements may be used to facilitate experimental testing and to explain the physical mechanisms underlying such mass transfer processes. The process of wood drying can be interpreted as simultaneous heat and moisture transfer with local thermodynamic equilibrium at each point within the timber. Drying of wood is in its nature an unsteady-state nonisothermal diffusion of heat and moisture, where temperature gradients may counteract with the moisture gradient [3].

5.2.2 SOME ASPECTS OF HEAT FLOW DURING DRYING PROCESS

5.2.2.1 STAGES OF DRYING

First stage: When both surface and core MC are greater than the F.S.P. Moisture movement is by capillary flow. Drying rate is evaporation controlled.

Second stage: When surface MC is less than the FSP and core MC is greater than the F.S.P. Drying is by capillary flow in the core and by bound water diffusion near the surface as fiber saturation line recedes into wood, resistance to drying increases. Drying rate is controlled by bound water diffusion.

Third stage: When both surface and core MC are less than the F.S.P. Drying is entirely by diffusion. As the MC gradient between surface and core becomes less, resistance to drying increases and drying rate decreases.

5.2.2.2 CAPILLARY

Capillary pressure is a driving force in convective wood drying at mild conditions. The temperature is higher outside than inside. The moisture profile during convective drying is in the opposite direction, namely, the drier part is toward the exposed surface of wood. This opposite pattern of moisture and temperature profiles lead to the concept of the wet front that separates the outer area, where the water is bound to the cell wall, from the inner area, where free water exists in liquid and vapor form. A wet front that moves slowly from the surface toward the center of a board during convective drying leads to subsequent enhancement of the capillary transportation. Capillary transportation can then be justified due to the moisture gradients developed around that area. When the drying conditions are mild, the drying period is longer so the relative portion of the total moisture removal, due to the capillary phenomena, is high, and it seems that this is the most important mass transfer mechanism [4].

5.2.2.3 BOUND WATER DIFFUSION

Credible data on the bound water diffusion coefficient in wood and the boundary condition for the interface between moist air and wood surface are very important for accurate description of timber drying as well as for the proper design and use of products, structures and buildings made of wood already dried below the fiber saturation point. During the last century, two groups of methods for measuring the bound water diffusion coefficient in wood were developed. The first one, traditionally called the cup method, uses data from the steady-state experiments of bound water transfer and is based on Fick's first law of diffusion. Unfortunately, the method is not valid for the bound water diffusion coefficient determination in wood because it cannot satisfy the requirements of the boundary condition of the first kind and the constant value of the diffusion coefficient. The second group of methods is based on the unsteady-state experiments and Fick's second law of diffusion. The common name of this group is the sorption method and it was developed to overcome the disadvantages of the cup technique [5].

5.2.2.4 DIFFUSION

In solving the diffusion equation for moisture variations in wood, some authors have assumed that the diffusion coefficient depends strongly on moisture content, while others have taken the diffusion coefficient as constant. It has been reported that the diffusion coefficient is influenced by the drying temperature, density and moisture content of timber. The diffusion coefficient of water in cellophane and wood substance was shown to increase with temperature in proportion to the increase in vapor pressure of water. It is also observed that the diffusion coefficient decreased with increasing wood density. Other factors affecting the diffusion coefficient that are yet to be quantified are the species (specific gravity) and the growth ring orientation. Literature has suggested that the ratios of radial and tangential diffusion coefficients vary for different tree species. The radial diffusion coefficient of New Zealand Pinus radiate has been estimated to be approximately 1.4 times the tangential diffusion coefficient. It is observed that for northern red oak, the diffusion coefficient is a function of moisture content only. It increases dramatically at low moisture content and tends to level off as the fiber saturation point is approached.

In a one-dimensional formulation with moisture moving in the direction normal to a specimen of a slice of wood of thickness 2a, the diffusion equation can be written as:

$$\frac{\partial(MC)}{\partial t}=\frac{\partial}{\partial X}\left(D\frac{\partial(MC)}{\partial X}\right)(0<X<a,t>0) \quad (1)$$

Where MC is moisture content, t is time, D is diffusion coefficient, and X is space coordinate measured from the center of the specimen.

The moisture content influences on the coefficient D only if the moisture content is below the fiber saturation point (F.S.P.)(typically 20%–30% for softwoods):

$$D(u)=\begin{cases} f_D(u) & ,u<u_{fsp} \\ f_D(u_{fsp}) & ,u\geq u_{fsp} \end{cases} \tag{2}$$

Where u_{fsp} denotes the F.S.P. and $f_D(u)$ is a function which expresses diffusion coefficient in moisture content, temperature and may be some other parameters of ambient air climate. The expression of $f_D(u)$ depends on variety of wood.

It was assumed that the diffusion coefficient bellow F.S.P. can be represented by:

$$f_D(u)=A.e^{-\frac{5280}{T}}.e^{\frac{B.u}{100}} \tag{3}$$

Where T is the temperature in Kelvin, u is percent moisture content, A and B can be experimentally determined.

The regression equation of diffusion coefficient of Pinus radiatatimber using the dry bulb temperature and the density is:

$$D(10^{-9})=1.89+0127\times T_{DB}-0.00213\times\rho_S \; (R^2=0.\,499) \tag{4}$$

The regression equations of diffusion coefficients below of Masson's pine during high temperature drying are:

TANGENTIAL DIFFUSION

$$D=0.0046MC^2+0.1753MC+4.2850\left(R^2=0.9391\right) \tag{5}$$

Radial diffusion

$$D=0.0092MC^2+0.3065MC+4.9243\left(R^2=0.9284\right) \tag{6}$$

The transverse diffusion coefficient D can be expressed by the porosity of wood v, the transverse bound water diffusion coefficient D_{vt} of wood and the vapor diffusion coefficient D_v in the lumens:

$$D=\frac{\sqrt{v}D_{bt}D_v}{(1-v)\left(\sqrt{v}D_{bt}+\left(1-\sqrt{v}\right)D_v\right)} \tag{7}$$

The vapor diffusion coefficient D_v in the lumens can be expressed as:

$$D_v = \frac{M_w D_a P_s}{SG_d \rho_w RT_k} \cdot \frac{d\varphi}{du} \tag{8}$$

where M_w (kg/kmol) is the molecular weight of water.

$$D_a = \frac{9.2.10^{-9} T_k^{2.5}}{(T_k + 245.18)} \tag{9}$$

is the inter diffusion coefficient of vapor in air,

$$SG_d = \frac{1.54}{(1+1.54u)} \tag{10}$$

is the nominal specific gravity of wood substance at the given bound water content. ρ_w= 103 kg/m^3 is the density of water, R = 8314 3 kmol, K is the gas constant, T_k is the Kelvin temperature, ψ is the relative humidity (%/100), and P_{sat} is saturated vapor pressure given by:

$$p_{sat} = 3390 \exp\left(-1.74 + 0.0759T_C - 0.000424T_C^2 + 2.44.10^{-6} T_C^3\right) \tag{11}$$

The derivative of air relative humidity ψ with respect to moisture content MC is given as:

$$MC = \frac{18}{w}\left(\frac{k_1 k_2 \psi}{1 + k_1 k_2 \psi} + \frac{k_2 \psi}{1 - k_2 \psi}\right) \tag{12}$$

where:

$$k_1 = 4.737 + 0.04773T_C - 0.00050012T_C^2 \tag{13}$$

$$k_2 = 0.7059 + 0.001695T_C + -0.000005638T_C^2 \tag{14}$$

$$W = 223.4 + .6942T_C + 0.01853T_C^2 \tag{15}$$

The diffusion coefficient D_{bt} of bound water in cell walls is defined according to the Arrhenius equation as:

$$D_{bt} = 7.10^{-6} \exp\left(-E_b / RT_k\right) \tag{16}$$

where:

$$E_b = \left(40.195 - 71.179Mc + 291Mc^2 - 669.92Mc^3\right).10^6 \tag{17}$$

is the activation energy.

The porosity of wood is expressed as:

$$\nu = 1 - SG(0.667 + Mc) \tag{18}$$

where specific gravity of wood *SG* at the given moisture content *u* is defined as:

$$SG = \frac{\rho_S}{\rho_W (1 + Mc)} = \frac{\rho_0}{\rho_W + 0.883\rho_0 Mc} \tag{19}$$

where ρ_s is density of wood, ρ_0 is density of oven-dry wood (density of wood that has been dried in a ventilated oven at approximately 104°C until there is no additional loss in weight).

Wood thermal conductivity (K_{wood}) is the ratio of the heat flux to the temperature gradient through a wood sample. Wood has a relatively low thermal conductivity due to its porous structure, and cell wall properties. The density, moisture content, and temperature dependence of thermal conductivity of wood and wood-based composites were demonstrated by several researchers. The transverse thermal conductivity can be expressed as:

$$K_{wood} = \left[SG \times (4.8 + 0.09 \times MC) + 0.57\right] \times 10^{-4} \frac{cal}{cm * Cs} \tag{20}$$

When moisture content of wood is below 40%.

$$K_{wood} = \left[SG \times (4.8 + 0.125 \times MC) + 0.57\right] \times 10^{-4} \frac{cal}{cm * Cs} \tag{21}$$

When moisture content of wood is above 40%.

The specific gravity and moisture content dependence of the solid wood thermal conductivity in the transverse (radial and tangential) direction is given by:

$$K_T = SG(K_{cw} + K_w.Mc) + K_a \nu \tag{22}$$

where:SG= specific gravity of wood, K_{cw} = Conductivity of cell wall substance (0.217 J /m/s/K), K_w= conductivity of water (0.4 J/m/s/K), K_a= conductivity of air (0.024 J/m/s/K), *Mc* = moisture content of wood (fraction), ν = porosity of wood.

The thermal conductivity of wood is affected by a number of basic factors: density, moisture content, extractive content, grain direction, structural irregularities such as checks and knots, fibril angle, and temperature. Thermal conductivity increases as density, moisture content, temperature, or extractive content of the wood increases. Thermal conductivity is nearly the same in the radial and tangential directions with respect to the growth rings.

The longitudinal thermal conductivity of solid wood is approximately 2.5 times higher than the transverse conductivity:

$$K_L = 2.5K_T \tag{23}$$

For moisture content levels below 25%, approximate thermal conductivity Kacross the grain can be calculated with a linear equation of the form:

$$K_{wood} = G(B + CM) + A \tag{24}$$

Where S*G* is specific gravity based on oven dry weight and volume at a given moisture content *MC* (%) and *A, B*, and *C* are constants. For specific gravity >0.3, temperatures around 24 °C, and moisture content values <25%, A = 0.01864, B = 0.1941, and C = 0.004064 (with k in W/(m·K)). Equation (24) was derived from measurements made by several researchers on a variety of species.

During the early stages of drying the material consists of so much water that liquid surfaces exist and drying proceeds at a constant rate. Constant drying rates are achieved when surface free water is maintained and the only resistance to mass transfer is external to the wood. The liquid water moves by capillary forces to the surface in same proportion of moisture evaporation. Moisture movement across the lumber will depend on the wood permeability and the drying rate itself is controlled by external conditions in this period. Part of energy received by the surface increase temperature in this region, and the heat transfer to the inner part of lumber starts. Since the moisture source for the surface is internal moisture, constant drying rates can only be maintained if there is sufficient moisture transport to keep the surface moisture content above the F.S.P. If this level is not maintained then some of the resistance to mass transfer becomes internal and neither the drying rate nor the surface temperature remains constant and drying proceeds to the falling rate period. As the lumber dries, the liquid water or wet line recedes into wood and the internal moisture movement involves the liquid flow and diffusion of water vapor and hygroscopic water. The effect of internal resistance on the drying rate increases. In the last phase (second falling rate period) there is no more liquid water in the lumber, and the drying rate is controlled only by internal resistance (material characteristics) until an equilibrium moisture content is reached [5–25].

5.3 RESULTS

A typical drying curve showing three stages of drying characteristic is illustrated in Fig. 1.

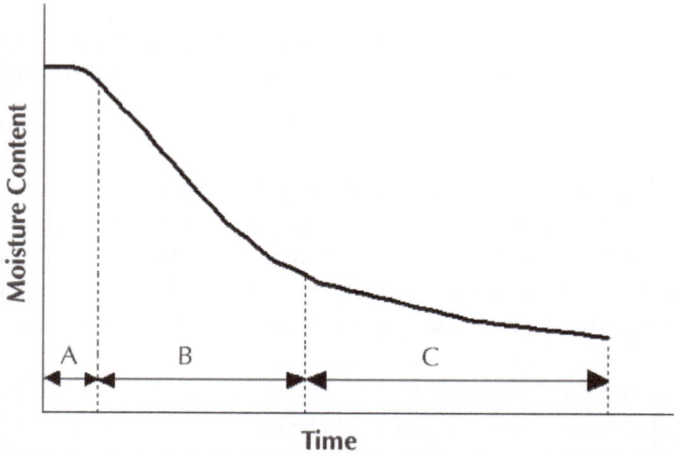

FIGURE 1 Drying characteristic of porous media: (a) constant rate region;(b) first falling rate region; (c) second falling rate region.

Pang et al. proposed that the three drying periods (constant rate, first falling rate and second falling rate)based on simulated drying of veneer be expressed by the following equations:

$$-\frac{d(MC)}{dt} = j_0 \text{ For MC>} M_{Cr1} \tag{25}$$

$$-\frac{d(MC)}{dt} = A + B * MC \text{ For } M_{cr1} > MC > M_{cr2} \tag{26}$$

$$-\frac{d(MC)}{dt} = \frac{A + B * M_{cr2}}{M_{cr2} - M_e} * \left(MC - M_e\right) \text{ For MC<} M_{cr2} \tag{27}$$

where: j_0 is constant drying rate, M_{Cr1} is the first critical moisture content, M_{cr2} is the second critical moisture content, constants A and B also vary with wood thickness, wood density, and drying conditions.

Moisture content of wood is defined as the weight of water in wood expressed as a fraction, usually a percentage, of the weight of oven dry wood. Moisture exists in wood as bound water within the cell wall, capillary water in liquid form and water vapor in gas form in the voids of wood. Capillary water bulk flow refers to the flow of liquid through the interconnected voids and over the surface of a solid due to molecular attraction between the liquid and the solid. Moisture content varies widely between species and within species of wood. It varies particularly between heartwood and sapwood. The amount of moisture in the cell wall may decrease as a result of extractive deposition when a tree undergoes change from sapwood to heartwood. The butt logs of trees may contain more water than the top logs. Variability of moisture content exists even within individual boards cut from the same tree. Green wood is often defined as freshly sawn wood in which the cell walls are completely saturated with water. Usually green wood contains additional water in the lumens. Moisture content at which both the cell lumens and cell walls are completely saturated with water is the maximum moisture content. An average green moisture content value taken from the Wood Handbook (Forest products society, 1999) of southern yellow pine (loblolly) is 33 and 110% for heartwood and sapwood, respectively. Sweet gum is 79 and 137% while yellow-poplar is 83 and 106% for heartwood and sapwood, respectively.

Permeability refers to the capability of a solid substance to allow the passage of gases or liquids under pressure. Permeability assumes the mass movement of molecules in which the pressure or driving force may be supplied by such sources as mechanically applied pressure, vacuum, thermal expansion, gravity, or surface tension. Under this condition, the permeability of wood is the dominant factor controlling moisture movement.

Fluid movement in wood is a very important process in wood products industries. An understanding of wood permeability is essential for determining lumber drying schedules for treating lumber and for producing high-quality wood products. The flow of gas inside the wood particle is limited due to the fact that wood consists of a large number of clustered small pores. The pore walls act as barriers largely preventing convective flow between adjacent pores. The wood annular rings also act as barriers for flow in the radial direction which makes flow in the axial direction more favorable and giving a lower permeability in the radial direction than in the axial direction where the axial flow is regarded as flow parallel to the wood fiber grains and the radial flow as flow perpendicular to the wood grains. The permeability in the wood cylinder is therefore an important parameter for the velocity field in the wood. The dry wood radial permeability is 10,000 times lower than the dry wood axial permeability. The chemical composition of the wood/char structure also affects the permeability, where the permeability in char is in order of 1,000 times larger than for wood.

Longitudinal flow becomes important, particularly in specimens having a low ratio of length to diameter, because of the high ratio of longitudinal to transverse permeability. Longitudinal permeability was found to be dependent upon specimen length in the flow direction, i.e., the decrease of specimen length appears result in greater permeability in less permeable species.

The effect of drying conditions on gas permeability and preservative treatability was assessed on western hemlock lumber. Although there were no differences in gas permeability between lumber dried at conventional and high temperatures, there were differences in preservative penetration. High temperature drying significantly reduced drying time, but did not appear to affect permeability or shell-to-core MC differences compared with drying at conventional temperature. Pits have a major influence on softwood permeability. Across pits can be impeded by aspiration or occlusion by deposition extractives on the membrane. Drying conditions can significantly affect pit condition, sometimes inducing aspiration that blocks both air and fluid flow. Pressure treatment is presumed to enhance preservative uptake and flow across pits, but the exact impact of pit condition (i.e., open or aspirated) is unknown. Drying conditions may also alter the state of materials deposited on pits, thereby altering the effects of pressure and perhaps the nature of preservative wood interactions. The latter effect may be especially important, since changes in wood chemistry could affect the rates of preservative fixation, which could produce more rapid preservative deposition on pit membranes that would slow further fluid ingress. The longitudinal permeability of the outer heartwood of each species also was determined to evaluate the effect of growth rate on the decrease in longitudinal permeability following sapwood conversion to heartwood. Faster diameter growth produced higher longitudinal permeability in the sapwood of yellow-poplar, but not in the sapwood of northern red oak or black walnut. Growth rate had no effect on either vessel lumen area percentage or decrease in longitudinal permeability in newly formed heartwood for all three species. Table 1 represents typical values for gas permeability. Values are given in orders of magnitude.

Darcy's law for liquid flow:

$$k = \frac{flux}{gradient} = \frac{V/(t \times A)}{\Delta P/L} = \frac{V \times L}{t \times A \times \Delta P} \quad (28)$$

where: k = Permeability [cm^3 (liquid)/ (cm atm sec)], V = Volume of liquid flowing through the specimen (cm^3), t = Time of flow (sec), A = Cross-sectional area of the specimen perpendicular to the direction of flow (cm^2), ΔP = Pressure difference between ends of the specimen (atm), L = Length of specimen parallel to the direction of flow (cm).

Darcy's law for gaseous flow:

$$K_g = \frac{V \times L \times P}{t \times A \times \Delta P \times \overline{P}} \quad (29)$$

where: K_g = Superficial gas permeability [cm^3(gas)/ (cm atm sec)], V = Volume of gas flowing through the specimen (cm^3 (gas)), P = Pressure at which V is measured (atm), t = Time of flow (sec), A = Cross-sectional area of the specimen perpendicular to the direction of flow (cm^2), ΔP = Pressure difference between ends of the specimen (atm), L = Length of specimen parallel to the direction of flow (cm), $\overline{P}$ = Average pressure across the specimen (atm).

TABLE 1 Typical values for gas permeability.

Type of sample	Longitudinal gas permeability [cm^3(gas)/(cm at sec)]
Red oak (R = 150 micrometers)	10,000
Basswood (R = 20 micrometers)	1,000
Maple, Pine sapwood, Coast Douglas-fir sapwood	100
Yellow-poplar sapwood, Spruce sapwood, Cedar sapwood	10
Coast Douglas-fir heartwood	1
White oak heartwood, Beech heartwood	0.1
Yellow-poplar heartwood, Cedar heartwood, Inland Douglasfir heartwood	0.01
Transverse Permeabilities (In approx. same species order as longitudinal)	0.001–0.0001

To simulate the heat and mass transport in drying, conservation equations for general nonhygroscopic porous media have been developed by Whitaker based on averaging procedures of all of the variables. These equations were further employed and modified for wood drying. Mass conservation equations for the three phases of moisture in local form are summarized in equations (30–32).

Water vapor:

$$\frac{\partial}{\partial t}\left(\varphi_g \rho_V\right) = -div\left(\rho_V V_V\right) + \dot{m}_{WV} + \dot{m}_{bV} \tag{30}$$

Bound water:

$$\frac{\partial}{\partial t}\left(\varphi_s \rho_b\right) = -div\left(\rho_b V_b\right) + \dot{m}_{bV} + \dot{m}_{wb} \tag{31}$$

Free water:

$$\frac{\partial}{\partial t}\left(\varphi_w \rho_w\right) = -div\left(\rho_w V_w\right) - \dot{m}_{wv} - \dot{m}_{wb} \tag{32}$$

Where the velocity of the transported quantity is denoted by V_i, ρ_i is the density, and m_{ij} denotes the transition from phases i and j. From here on, the subscripts w, b, v, and$s$ refer, respectively, to free water, bound water, water vapor, and the solid skeleton of wood. Denoting the total volume by V and the volume of the phase i by V_i, the volumetric fraction of this phase is:

$$\varphi_i = \frac{V_i}{V} \tag{33}$$

with the geometrical constraint:

$$\varphi_g + \varphi_s + \varphi_w = 1 \tag{34}$$

Darcy's law, by using relative permeabilities, provides expressions for the free liquid and gas phase velocities as follows:

$$\mathrm{v}_l = -\frac{K_l K_{rl}}{\mu_l} \nabla P_l \tag{35}$$

and

$$\mathrm{v}_v = -\frac{K_v K_{rv}}{\mu_v} \nabla P_v \tag{36}$$

Where K is the intrinsic permeability (m^2), K_r is the relative permeability, P is the pressure (Pa), and μ is the viscosity (Pa.s).

The heat flux (q) and the moisture flux (N_v) are estimated by:

$$q = h\left(T_G - T_{surf}\right) \tag{37}$$

$$N_v = \psi K_0\left(Y_{surf} - Y_G\right) = \beta\left(p_G^v - p_{ats}^v\right) \tag{38}$$

In which T_{surf}, $_{surf}$ and p_s^n are the wood temperature, the air humidity and the vapor partial pressure, respectively, at the wood surface and, T_G, $_G$ and p_G^n are the corresponding parameters in the air stream. The heat-transfer coefficient is represented by h. The mass-transfer coefficient is β when vapor partial pressure difference is taken as driving force and is k_0 when humidity difference is taken as the driving force with ψ being the humidity factor. The mass-transfer coefficient related to humidity difference is a function of distance along the airflow direction from the inlet side. The heat-transfer coefficient is correlated to the mass-transfer coefficient, as shown by and can be calculated from it. The humidity coefficient φ has been found to vary from 0.70 to 0.76, depending on the drying schedules and board thickness.

For the moisture mass transfer and balance, the moisture loss from wood equals the moisture gain by the hot air, and the moisture transfer rate from the board is described by mass transfer coefficient multiplied by driving force (humidity difference, for example). These considerations yield:

$$-\frac{\partial}{\partial \tau}\left[MC.\rho_s.(1-\varepsilon)\right] = G.\frac{\partial Y}{\partial X} = \begin{cases} -\psi K_0.a.\left(Y_{surf} - Y_G\right)(condensation) \\ \psi K_0.a.f.\left(Y_{sat} - Y_G\right)(evaporation) \end{cases} \tag{39}$$

Where MC is the wood moisture content, ρ_s is the wood basic density, ε is the void fraction in the lumber stack, a is the exposed area per unit volume of the stack and G is the dry air mass flow rate. In order to solve the above equations, the relative drying rate (f) needs to be defined which is a function of moisture content.

For the heat transfer and balance, the energy loss from the hot air equals the heat gain by the moist wood. The convective heat transfer is described by product of heat transfer coefficient and the temperature difference between the hot air and the wood surface. The resultant relationships are as follows:

$$\frac{\partial T_{wood}}{\partial \tau}=\frac{(1+\alpha_R-\alpha_{LS})}{\rho_s.(1-\varepsilon).C_{Pwood}}.\left[h.a.(T_G-T_{wood})-G.\Delta H_{wv}.\frac{\partial Y_G}{\partial X}\right]$$

$$\frac{\partial T_G}{\partial X}=\frac{\left(h.a+G.C_{Pv}\frac{\partial Y_G}{\partial Z}\right).(T_G-T_{wood})}{G.(C_{Pv}+Y_G.C_{Pv})} \tag{40}$$

In the above equations, T_{wood} is the wood temperature, α_R and α_{LS} are coefficients to reflect effects of heat radiation and heat loss, C_{Pwood} is the specific heat of wood, and ΔH_{wv} is theof water evaporation. These equations have been solved to determine the changes of air temperature and wood temperature along the airflow direction and with time.

The energy rate balance (kW) of a drying air adjacent to the wood throughout the wood board can be represented as follows:

$$\frac{1}{2}V_a\rho_{a,mt}cp_{a,mt}\frac{dT_a}{dt}=\frac{1}{2}vA_{cs}cp_{a,mt}\left(T_{a,in-}T_{a,ex}\right)+\dot{Q}_{evap}-\dot{Q}_{conv} \tag{41}$$

Where $\dot{Q}_{evap}$ and $\dot{Q}_{conv}$ (kW) are the evaporation and convection heat transfer rates between the drying air and wood, which can be calculated as follows:

$$\dot{Q}_{evap}=r\dot{m}_{wv,s}A_{surf} \tag{42}$$

$$\dot{Q}_{conv}=hA\left(T_a-T_{SO}\right) \tag{43}$$

The specific water vapor mass flow rate ($\dot{m}_{wv,surf}$)(kg/m^2 s) to the drying air can be calculated as follows:

$$\dot{m}_{wv,surf}=\frac{h_D}{R_{wv}T_{SO}}\left(P_{wv,surf}-P_{wv,a}\right) \tag{44}$$

The vapor pressure on the wood surface can be determined from the sorption isotherms of wood. The mass transfer coefficient (h_D)(m/s) can be calculated from the convection heat transfer coefficient (h)(kW/m^2 K) as follows:

$$h_D=h\frac{1}{\rho_{a,mt}cp_{a,mt}Le^{0.58}}/\left(1-\frac{\rho_{wv,m}}{P}\right) \tag{45}$$

Water transfer in wood involves liquid free water and water vapor flow while MC of lumber is above the F.S.P.

According to Darcy's law the liquid free water flux is in proportion to pressure gradient and permeability. So Darcy's law for liquid free water may be written as:

$$J_f = \frac{K_l \rho_l}{\mu_l} \cdot \frac{\partial P_c}{\partial \chi} \tag{46}$$

where: J_f=liquid free water flow flux, kg/ m^2 ·s, K_l=specific permeability of liquid water, $m^3(liquid)/m$, ρ_l =density of liquid water, kg/m^3, μ_l = viscosity of liquid water, $p_a \cdot s$, P_c = capillary pressure, p_a , χ = water transfer distance, m, $\partial p_c / \partial \chi$ = capillary pressure gradient, p_a / m.

The water vapor flow flux is also proportional to pressure gradient and permeability as follows:

$$J_{vf} = \frac{K_V \rho_v}{\mu_V} \cdot \frac{\partial P_V}{\partial \chi} \tag{47}$$

where: J_{vf} = water vapor flow flux, kg/m^2·s, K_v = specific permeability of water vapor, $m^3(vapor)/m$, ρ_v,μ_v = density and viscosity of water vapor, respectively, kg/m^3 and $p_a \cdot s$, $\partial p_V / \partial \chi$ =vapor partial pressure gradient, p_a / m .

Therefore, the water transfer equation above F.S.P. during high temperature drying can be written as:

$$\rho_s \frac{\partial (MC)}{\partial t} = \frac{\partial}{\partial x}\left(J_f + J_{vf}\right) \tag{48}$$

where: ρ_S = basic density of wood, kg/m^3 , MC = moisture content of wood, %, t = time, s, $\partial(MC)/\partial t$ = the rate of moisture content change, %/s, x = water transfer distance, m.

5.3.1 WATER TRANSFER MODEL BELOW F.S.P.

Water transfer in wood below F.S.P. involves bound water diffusion and water vapor diffusion. The bound water diffusion in lumber usually is unsteady diffusion; the diffusion equation follows Fick's second law as follows:

$$\frac{\partial (MC)}{\partial t} = \frac{\partial}{\partial x}\left(D_b \frac{\partial (MC)}{\partial x}\right) \tag{49}$$

where D_b is bound water diffusion coefficient, m^2/s, $\partial(MC)/\partial x$ is MC gradient of lumber, %/m. The bound water diffusion flux J_b can be expressed as:

$$J_b = D_b \rho_s \frac{\partial(MC)}{\partial x} \tag{50}$$

where: ρ_S is basic density of wood, kg/m^3.

The water vapor diffusion equation is similar to bound water diffusion equation as follows

$$\frac{\partial(MC)}{\partial t} = \frac{\partial}{\partial(MC)}\left(D_V \frac{\partial(MC)}{\partial x}\right) \tag{51}$$

where D_V is water vapor diffusion coefficient, m^2/s. The water vapor diffusion flux can be expressed as:

$$J_V = D_V \rho_s \frac{\partial(MC)}{\partial x} \tag{52}$$

Therefore, the water transfer equation below F.S.P. during high temperature drying can be expressed as:

$$\rho_s \frac{\partial(MC)}{\partial t} = \frac{\partial}{\partial x}(J_b + J_V) \tag{53}$$

Two types of wood samples (namely; Guilan spruce and pine) were selected for drying investigation. Natural defects such as knots, checks, splits, etc., which would reduce strength of wood are avoided. All wood samples were dried to a moisture content of approximately 30%.The effect of drying temperature and drying modes on the surface roughness, hardness and color development of wood samples are evaluated.

The average roughness is the area between the roughness profile and its mean line, or the integral of the absolute value of the roughness profile height over the evaluation length:

$$R_a = \frac{1}{L}\int_0^L |r(x)dx| \tag{54}$$

When evaluated from digital data, the integral is normally approximated by a trapezoidal rule:

$$R_a = \frac{1}{N}\sum_{n=1}^{N} |r_n| \tag{55}$$

The root-mean-square (rms) average roughness of a surface is calculated from another integral of the roughness profile:

$$R_q = \sqrt{\frac{1}{L}\int_0^L r^2(x)dx} \tag{56}$$

The digital equivalent normally used is:

$$R_q = \sqrt{\frac{1}{N}\sum_{n=1}^{N} r_n^2} \tag{57}$$

R_z (ISO) is a parameter that averages the height of the five highest peaks plus the depth of the five deepest valleys over the evaluation length. These parameters, which are characterized by ISO 4287 were employed to evaluate influence of drying methods on the surface roughness of the samples.

We investigated the influence of drying temperatures on the surface roughness characteristics of veneer samples as well. The results showed that the effect of drying temperatures used in practice is not remarkable on surface roughness of the sliced veneer and maximum drying temperature (130 °C) applied to sliced veneers did not affect significantly surface roughness of the veneers. Veneer sheets were classified into four groups and dried at 20, 110, 150, and 180 °C. According to the results, the smoothest surfaces were obtained for 20 °C drying temperature while the highest values of surface roughness were obtained for 180 °C. Because some surface checks may develop in the oven-drying process. It was also found in a study that the surface roughness values of beech veneers dried at 110 °C was higher than that of dried at 20 °C.

In another experimental study, veneer sheets were oven-dried in a veneer dryer at 110 °C(normal drying temperature) and 180° C (high drying temperature) after peeling process. The surfaces of some veneers were then exposed at indoor laboratory conditions to obtain inactive wood surfaces for glue bonds, and some veneers were treated with borax, boric acid and ammonium acetate solutions. After these treatments, surface roughness measurements were made on veneer surfaces. Alder veneers were found to be smoother than beech veneers. It was concluded that the values mean roughness profile (R_a) decreased slightly or no clear changes were obtained in R_a values after the natural inactivation process. However, little increases were obtained for surface roughness parameters, no clear changes were found especially for beech veneers.

The changes created by weathering on impregnated wood with several different wood preservatives were investigated. The study was performed on the accelerated weathering test cycle, using UV irradiation and water spray in order to simulate natural weathering. Wood samples were treated with ammonium copper quat (ACQ 1900 and ACQ 2200), chromated copper arsenate (CCA), Tanalith E 3491 and Wolmanit CX-8 in accelerated weathering experiment. The changes on the surface of the weathered samples were characterized by roughness measurements on the samples with 0, 200, 400 and 600 h of total weathering. Generally the surface values of alder wood treated with copper-containing preservatives decreased with over the irradiation time except for treated Wolmanit CX-82% when comparing unweathered values. Surface values of pine treated samples generally increased with increasing irradiation time except for ACQ-1900 groups.

Because the stylus of detector was so sensitive first each sample was smoothened with emery paper then measurement test was performed before and after drying. The Mitutoyo Surface roughness tester SJ-201P instrument was employed for surface roughness measurements. Cut-off length was 2.5 mm, sampling length was 12.5 mm and detector tip radius was 5 μm in the surface roughness measurements. Tables2 and 3displays the changes in surface roughness parameters (R_a, R_z and R_q) of the Pine and Guilan spruce at varying drying methods. In both cases the surface roughness becomes higher during microwave and infrared heating while surface smoothness of both pine and Guilan spruce increased during convection and combined drying. However, the roughness of wood is a complex phenomenon because wood is an anisotropic and heterogeneous material. Several factors such as anatomical differences, growing characteristics, machining properties of wood, pretreatments (e.g., steaming, drying, etc.) applied to wood before machining should be considered for the evaluation of the surface roughness of wood.

TABLE 2 Surface roughness (μm) for pine.

Drying methods	Drying conditions	R_a	R_z	R_q
Microwave	Before drying	4.52	24.68	5.39
	After drying	5.46	30.21	6.62
Infrared	Before drying	4.42	25.52	5.43
	After drying	4.87	26.55	5.69
Convection	Before drying	4.66	26.87	5.86
	After drying	4.08	24.64	5.12
Combined	Before drying	5.23	32.59	6.42
	After drying	3.41	21.7	4.27

TABLE 3 Surface roughness (μm) for Guilan spruce.

Drying methods	Drying conditions	R_a	R_z	R_q
Microwave	Before drying	6.44	34.18	7.85
	After drying	7.77	44.3	9.82
Infrared	Before drying	4.92	30.61	6.30
	After drying	6.42	38.93	8.17
Convection	Before drying	4.97	32.41	6.5
	After drying	4.78	32.27	6.34
Combined	Before drying	10.41	59.5	13.37
	After drying	9.11	54.31	11.5

Hardness represents the resistance of wood to indentation and marring. In order to measure the hardness of wood samples, the Brinell hardness method was applied. In this method a steel hemisphere of diameter 10 mm was forced into the surface under test. The Brinell method measures the diameter of the mark caused by the steel ball in the specimens. The specimens were loaded parallel and perpendicular to the direction of wood grains. After applying the force the steel ball was kept on the surface for about 30 seconds. The values of hardness are shown in Figs. 2and 3,respectively. In both type of samples the hardness measured in longitudinal direction is reported to be higher than tangential. The amount of fibers and its stiffness carrying the load are expected to be lower when the load direction is angled to the grain. Results showed that hardness of wood increased in combined drying. The hardness of wood is proportional to its density. The hardness of wood varies, depending on the position of the measurement. Latewood is harder than early wood and the lower part of a stem is harder than the upper part. Increase in moisture content decreases the hardness of wood. It was observed the effect of different drying temperatures during air circulation drying. The result indicates no significant influence of temperature on hardness; still the specimens dried at higher temperature gave a hard and brittle impression. It was also investigated whether wood hardness is affected by temperature level during microwave drying and whether the response is different from that of conventionally dried wood. It was concluded that there is a significant difference in wood hardness parallel to the grain between methods when drying progresses to relatively lower level of moisture content, that is, wood hardness becomes higher during microwave drying. Variables such as density and moisture content have a greater influence on wood hardness than does the drying method or the drying temperature.

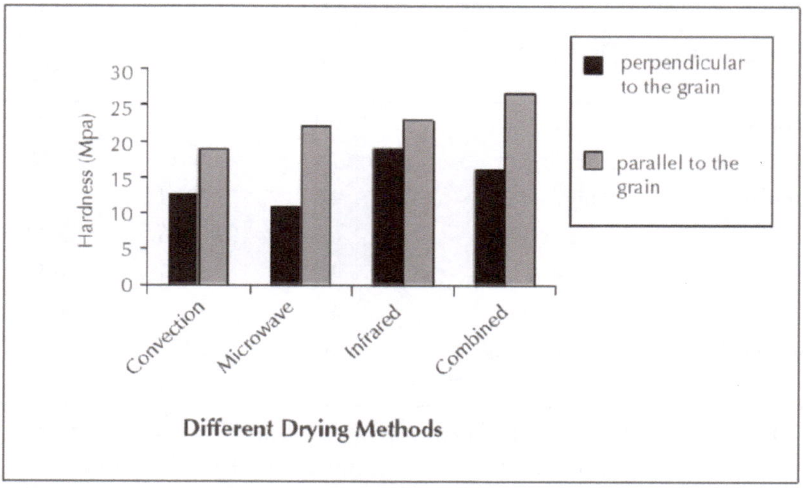

FIGURE 2 Brinell hardness for Guilan spruce.

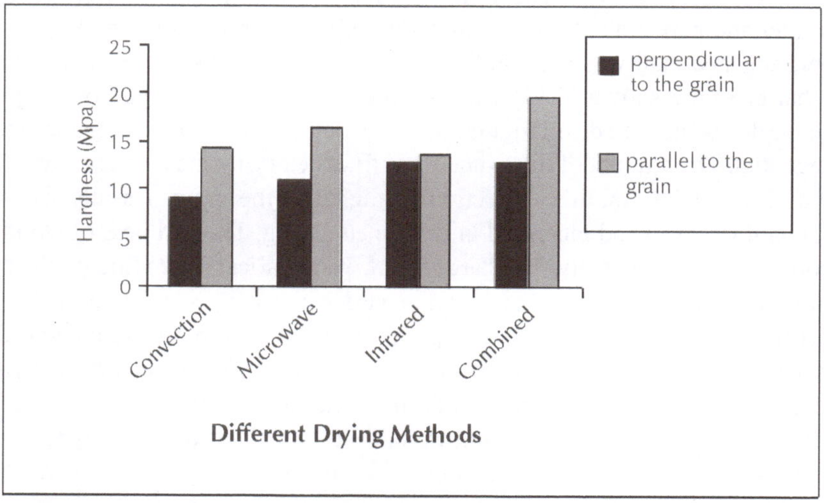

FIGURE 3 Brinell hardness for pine.

Color development of wood surfaces can be measured by using optical devices such as spectrophotometers. With optical measurement methods, the uniformity of color can be objectively evaluated and presented as L*, a* and b* coordinates named by CIEL*a*b* color space values. Measurements were made both on fresh and dried boards and always from the freshly planed surface. Three measurements in each sample board were made avoiding knots and other defects and averaged to one recording. The spectrum of reflected light in the visible region (400–750 nm) was measured and transformed to the CIEL*a*b* color scale using a 10° standard observer and D65 standard illuminant.

These color space values were used to calculate the total color change (ΔE^*) applied to samples according to the following equations:

$$\begin{aligned} \Delta L^* &= L_f^* - L_i^* \\ \Delta a^* &= a_f^* - a_i^* \\ \Delta b^* &= b_f^* - b_i^* \\ \Delta E^* &= \sqrt{(\Delta L^*)^2 + (\Delta a^*)^2 + (\Delta b^*)^2} \end{aligned} \tag{58}$$

f and *i* are subscripts after and before drying, respectively.

In this three dimensional coordinates, L* axis represents nonchromatic changes in lightness from an L* value of 0 (black) to an L* value of 100 (white), +a* represents red, –a* represents green, +b* represents yellow and –b* represents blue.

As can be seen from Figs. 4 and 5 color space values of both pine and Guilan spruce changed after drying [20–25].

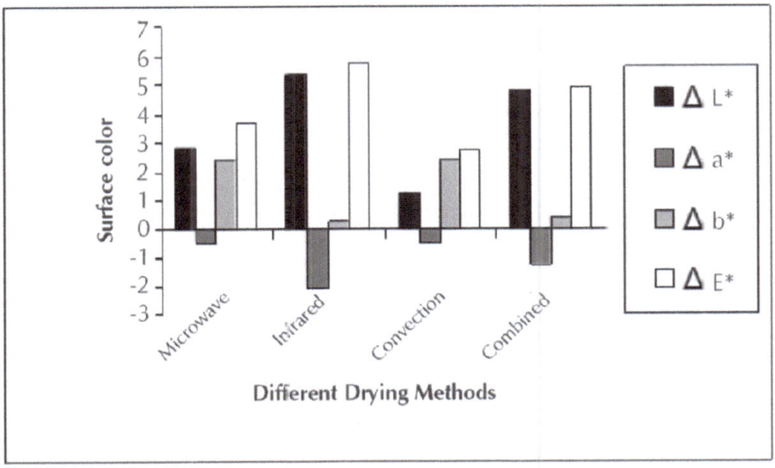

FIGURE 4 Surface color of Pine.

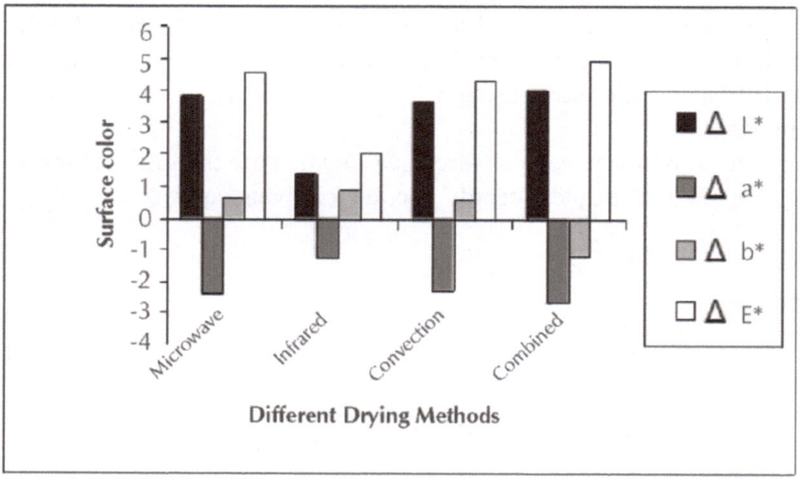

FIGURE 5 Surface color of Guilan spruce.

5.4 CONCLUSION

Results shows that Δa^* generally decreased but Δb^* increased for both pine and Guilan spruce wood samples except for Guilan spruce during combined heating. The lightness values ΔL^* increased during drying. The L^* of wood species such as tropical woods, which originally have dark color increases by exposure to light. This is due to the special species and climate condition of Guilan spruce and pines wood. Positive values of Δb^* indicate an increment of yellow color and negative values an increase of blue color. Negative values of Δa^* indicate a tendency of wood surface to greenish. A low ΔE^* corresponds to a low color change or a stable color. The biggest changes in color appeared in ΔE^* values of pine samples during infrared drying while for Guilan spruce it was reversed. Due to differences in composition of wood components, the color of fresh, untreated wood varies between different species, between different trees of the same species and even within a tree. Within a species wood color can vary due to the genetic factors and environmental conditions. In discoloration, chemical reactions take place in wood, changing the number and type of chromophores.

Discolorations caused by the drying process are those that actually occur during drying and are mainly caused by nonmicrobial factors. Many environmental factors such as solar radiation, moisture and temperature cause weathering or oxidative degradation of wooden products during their normal use; these ambient phenomena can eventually change the chemical, physical, optical and mechanical properties of wood surfaces.

A number of studies have been conducted that have attempted to find a solution to kiln brown stain, the majority of them being pretreatment processes. Biological treatment, compression rolling, sap displacement and chemical inhibitors have all been used as pretreatments. In all cases these processes were successful in reducing or eliminating stain but were not considered economically viable. Vacuum drying and modified schedules have been tried as modified drying processes with only limited success. Within industry various schedules have been developed, though these are generally kept secret and it is difficult to gauge their success. Generally it seems that industry has adopted a post-drying process involving the mechanical removal of the kiln brown stain layer.

Microwave processing of materials is a relatively new technology that provides new approaches to improve the physical properties of materials. Microwave drying generate heat from within the grains by rapid movement of polar molecules causing molecular friction and help in faster and more uniform heating than does conventional heating. If wood is exposed to an electromagnetic field with such high frequency as is characteristic for microwaves, the water molecules, which are dipoles, begin to turn at the same frequency as the electromagnetic field. Wood is a complex composite material, which consists mainly of cellulose (40–45%), hemicelluloses (20–30%) and lignin (20–30%). These polymers are also polar molecules and therefore even they are likely to be affected by the electromagnetic field. This could possibly cause degradation in terms wood hardness. For Guilan spruce the average of hardness is shown to be much higher than pine. From the experimental results it can be observed that in combined microwave dryer, the hardness was relatively improved in comparison to the other drying methods. Microwave and infrared drying can increase wood surface roughness while the smoothness of wood increases during convection and combined drying. The effect varies with the wood species. Thus this work suggests keeping the core temperature below the critical value until the wood has dried below fiber saturation as one way of ensuring that the dried wood is acceptably bright and light in color.

KEYWORDS

- **electromagnetic field**
- **Eulerian based model**
- **high frequency**
- **microwave**
- **Navier-Stokes equations**

CHAPTER 6

FINITE DIFFERENCE AND METHOD OF CHARACTERISTICS FOR TRANSITIONAL FLOW

CONTENTS

6.1 INTRODUCTION

This chapter discuss on Eulerian based model for water hammer. This model was defined by the method of characteristics (MOC), finite difference form. The method was encoded into an existing hydraulic simulation model. The surge wave was assumed as a failure factor in an elastic case of water pipeline with free water bubble. The results were compared by regression analysis. It indicated that the accuracy of the Eulerian based model for water transmission line.

6.2 MATERIALS AND METHODS

Water hammer as a fluid dynamics phenomenon is an important case study for designer engineers. This phenomenon has a complex mathematical behavior. Today high technology provides a suitable condition for finding new methods in order to reduction of water hammer disaster. Water hammer disaster can be happened at earthquake or tsunami. For example, at these critical conditions, water transmission pipeline control at power plants, water treatment plants, water transmission and distribution plants will be at high risk due to damage or failure hazard. As a side effect of water hammer phenomenon, this situation increases the probability of surge wave generation. Surge pressure and velocity of surge wave acts at fast transients, down to 5 ms. Therefore it must be detected on actual systems (by field tests) by high technology and high speed detectors. Also, besides the flow and pressure, it must be computed by computational model. The recording of fast transients needs to use the high technology and online data intercommunication. Water transmission failure sometimes happens due to unusual factors, which can suddenly change in the boundaries of the system. High surge pressure at earth quake, pump power off and inlet air by the air inlet valve, high discharge rate due to connections and consumers are some of the unusual factors which suddenly change in the boundaries of the system. Most of the transients in water and wastewater systems are the result of changes in the properties and boundaries of the system[1].It can generate the spread of the surge wave and changes at liquids properties in pipes and channels. It causes the formation and collapse of vapor bubbles or cavitations and air leakage Hariri Asli, et al. [2]. The study of hydraulic transients began with the work of Zhukovski [1]. Many researchers have made significant contributions in this area, including Parmakian [3], Wood [4] who popularized and perfected the graphical method of calculation. Wylie and Streeter [5]MOC combined with computer modeling. Subject of transients in liquids are still growing fast around the world. Brunone et al. [6], Koelle and Luvizotto [7], Filion and Karney [8], Hamam and Mc Corquodale [9], Savic and Walters [10], Walski and Lutes [11], Lee and Pejovic [12], have been developed various methods of investigation of transient pipe flow. These ranges of methods are included by approximate equations to numerical solutions of the nonlinear Navier–Stokes equations. Various methods have been developed to solve transient flow in pipes. These ranges have been formed from approximate equations to numerical solutions of the nonlinear Navier–Stokes equations. Elastic theory describes the unsteady flow of a compressible liquid in an elastic system. Transient theory stem from the two governing equations. The continuity equation and the momentum equation are needed to determine velocity and surge pressure in a one-dimensional flow system. Solving these two equations produces a theoretical result that usually corresponds quite closely to actual system measurements, if the data and assumptions used to build the numerical model are valid. Among the approaches proposed to solve.

The single-phase (pure liquid) water hammer equations are the MOC finite differences (FD), wave characteristic method (WCM), finite elements (FE), and finite volume (FV). One difficulty that commonly arises relates to the selection of an appropriate level of time step to use for the analysis. The obvious trade off is between computational speed and accuracy. In general, for the smaller the time step, there is the longer the run time but the greater the numerical accuracy.

An evaluation of surge or pressure wave in an elastic case with the free water bubble. It started with the solving of approximate equations by numerical solutions of the nonlinear Navier-Stokes equations based on the MOC. Then it derived the Zhukouski formula and velocity of surge or pressure wave in an elastic case with the high value of free water bubble. So the numerical modeling and simulation, which was defined by "MOC" provided a set of results. Basically the "MOC" approach transforms the water hammer partial differential equations into the ordinary differential equations along the characteristic lines[13–44].

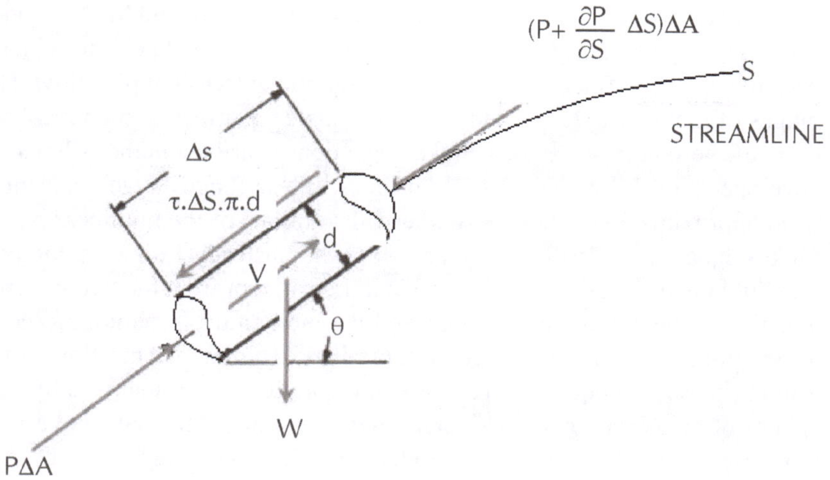

FIGURE 1 Newton second law(conservation of momentum equation)for fluid element.

It is defined as the combination of momentum equation (Fig. 1) and continuity equation (Fig. 2) for determining the velocity and pressure in a one-dimensional flow system. The solving of these equations produces a theoretical result that usually corresponds quite closely to actual system measurements.

$$P\Delta A-(P+\frac{\partial P}{\partial S}.\Delta S)\Delta A-W.\sin\theta-\tau.\Delta S.\pi.d=\frac{W}{g}.\frac{dV}{dt}$$

Both sides are divided by m and with assumption:

$$\frac{\partial Z}{\partial S}=+\sin\theta-\frac{1}{\partial}.\frac{\partial P}{\partial S}-\frac{\partial Z}{\partial S}-\frac{4\tau}{\gamma D}=\frac{1}{g}.\frac{dV}{dt} \quad (1)$$

$$\Delta A=\frac{\Pi.D^2}{4}$$

If fluid diameter assumed equal to pipe diameter then:

$$\frac{-1}{\gamma}.\frac{\partial P}{\partial S}-\frac{\partial Z}{\partial S}-\frac{4\tau_\circ}{\gamma.D} \quad (2)$$

$$\tau_\circ=\frac{1}{8}\rho.f.V^2$$

$$-\frac{1}{\gamma}\cdot\frac{\partial P}{\partial S}-\frac{\partial Z}{\partial S}-\frac{f}{D}\cdot\frac{V^2}{2g}=\frac{1}{g}\cdot\frac{dV}{dt} \tag{3}$$

$$V^2=V\,|\,V\,|,\frac{dV}{dt}+\frac{1}{\rho}\cdot\frac{\partial P}{\partial S}+g\frac{dZ}{dS}+\frac{f}{2D}V|V|=0 \text{ (Euler equation)} \tag{4}$$

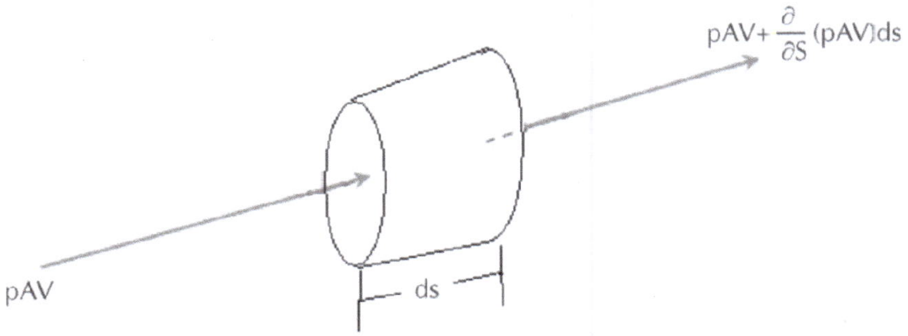

FIGURE 2 Continuity equation (conservation of mass) for fluid element.

For finding (V) and (P) we need to conservation of mass law (Fig. 2)

$$\rho AV-\left[\rho AV-\frac{\partial}{\partial S}(\rho AV)dS\right]=\frac{\partial}{\partial t}(\rho AdS)-\frac{\partial}{\partial S}(\rho AV)dS=\frac{\partial}{\partial t}(\rho AdS) \tag{5}$$

$$-\left(\rho A\frac{\partial V}{\partial S}dS+\rho V\frac{\partial A}{\partial S}dS+AV\frac{\partial \rho}{\partial S}dS\right)=\rho A\frac{\partial}{\partial t}(dS)+\rho dS\frac{\partial A}{\partial t}+AdS\frac{\partial p}{\partial t}, \tag{6}$$

$$\frac{1}{\rho}\left(\frac{\partial \rho}{\partial t}+V\frac{\partial \rho}{\partial S}\right)+\frac{1}{A}\left(\frac{\partial A}{\partial t}+V\frac{\partial A}{\partial S}\right)+\frac{1}{dS}\cdot\frac{\partial}{\partial t}(dS)+\frac{\partial V}{\partial S}=\circ$$

With $\frac{\partial \rho}{\partial t}+V\frac{\partial \rho}{\partial S}=\frac{d\rho}{dt}$ and $\frac{\partial A}{\partial t}+V\frac{\partial A}{\partial S}=\frac{dA}{dt}$

$$\frac{1}{\rho}\cdot\frac{d\rho}{dt}+\frac{1}{A}\cdot\frac{dA}{dt}+\frac{\partial V}{\partial S}+\frac{1}{dS}\cdot\frac{1}{dt}(dS)=\circ, \tag{7}$$

$K = \left(\dfrac{d\rho}{\left(\dfrac{d\rho}{\rho} \right)} \right)$ (Fluid module of elasticity) then:

$$\frac{1}{\rho}.\frac{d\rho}{dt} = \frac{1}{k}.\frac{d\rho}{dt} \tag{8}$$

Put Eq. (4.2.7) into Eq. (4.2.8), Then:

$$\frac{\partial V}{\partial S} + \frac{1}{k}.\frac{d\rho}{dt} + \frac{1}{A}\cdot\frac{dA}{dt} + \frac{1}{dS}.\frac{d}{dt}(dS) = \circ$$

$$\rho\frac{\partial V}{\partial S} + \frac{d\rho}{dt}\rho\left[\frac{1}{k} + \frac{1}{A}.\frac{dA}{d\rho} + \frac{1}{dS}\cdot\frac{d}{d\rho}(dS)\right] = \circ \tag{9}$$

$$\rho\left[\frac{1}{k} + \frac{1}{A}.\frac{dA}{dt} + \frac{1}{dS}\cdot\frac{d}{d\rho}(dS)\right] = \frac{1}{C^2}$$

Then $$C^2\frac{\partial V}{\partial S} + \frac{1}{\rho}.\frac{d\rho}{dt} = \circ \text{ (Continuity equation)} \tag{10}$$

Partial differential Eqs.(4)and(10) are solved by MOC:

$$\frac{dp}{dt} = \frac{\partial p}{\partial t} + \frac{\partial p}{\partial S}\cdot\frac{dS}{dt} \tag{11}$$

$$\frac{dV}{dt} = \frac{\partial V}{dt} + \frac{\partial V}{\partial S}\cdot\frac{dS}{dt} \tag{12}$$

Then

$$\left| \begin{array}{l} \dfrac{\partial V}{\partial t} + \dfrac{1}{\rho}\dfrac{\partial p}{\partial S} + g\dfrac{dz}{dS} + \dfrac{f}{2D}V|V| = \circ, \\ C^2\dfrac{\partial V}{\partial S} + \dfrac{1}{\rho}\dfrac{\partial P}{\partial t} = \circ, \end{array} \right. \tag{13) and (14}$$

By Linear combination of Eqs.(13) and (14)

$$\lambda\left(\frac{\partial V}{\partial t}+\frac{1}{\rho}\frac{\partial p}{\partial S}+g.\frac{dz}{dS}+\frac{f}{2D}V|V|\right)+C^{2}\frac{\partial V}{\partial S}+\frac{1}{\rho}\frac{\partial p}{\partial t}=\circ \tag{15}$$

$$(\lambda\frac{\partial V}{\partial t}+C^2\frac{\partial V}{\partial S})+(\frac{1}{\rho}.\frac{\partial \rho}{\partial t}+\frac{\lambda}{\rho}.\frac{\partial P}{\partial S})+\lambda.g.\frac{dz}{dS}+\frac{\lambda.f}{2D}V|V|=\circ \tag{16}$$

$$\lambda\frac{\partial V}{\partial t}+C^2\frac{\partial V}{\partial S}=\lambda\frac{dV}{dt}\Rightarrow\lambda\frac{dS}{dt}=C^2 \tag{17}$$

$$\frac{1}{\rho}.\frac{\partial p}{\partial t}+\frac{\lambda}{\rho}.\frac{\partial \rho}{\partial S}=\frac{1}{\rho}.\frac{d\rho}{dt}\Rightarrow$$

$$\frac{\lambda}{\rho}=\frac{1}{\rho}.\frac{dS}{dt} \tag{18}$$

$$\left|\frac{C^2}{\lambda}=\lambda \text{ (By removing } \frac{dS}{dt}\text{)}, \ \lambda=\pm C\right.$$

For $\lambda=\pm C$ From Eq.(18) we have:

$$C\frac{dV}{dt}+\frac{1}{\rho}.\frac{dp}{dt}+C.g.\frac{dz}{dS}+C.\frac{f}{2D}V|V|=\circ$$

With dividing both sides by "C":

$$\frac{dV}{dt}+\frac{1}{c.\rho}\frac{dP}{dt}+g.\frac{dz}{dS}+\frac{f}{2D}V|V|=\circ \tag{19}$$

For $\lambda=-C$ by Eq.(16):

$$\frac{dV}{dt}-\frac{1}{c.\rho}\frac{dp}{dt}+g\frac{dZ}{dS}+\frac{f}{2D}V|V|=\circ \tag{20}$$

If $\rho=\rho.g(H-Z)$

From Eqs. (9) and (10):

$$\left|\begin{array}{l}\dfrac{dV}{dt}+\dfrac{g}{c}\cdot\dfrac{dH}{dt}+\dfrac{f}{2D}V|V|=\circ \\ if:\dfrac{dS}{dt}=C,\end{array}\right. , \qquad \text{(21) and (22)}$$

$$\left| \begin{array}{l} \frac{dV}{dt} + \frac{g}{c} \cdot \frac{dH}{dt} + \frac{f}{2D} V|V| = \circ, \\ if: \frac{dS}{dt} = -C, \end{array} \right. \quad (23) \text{ and } (24)$$

The MOC is a finite difference technique which pressures were computed along the pipe for each time step (1–35).

The calculation automatically subdivided the pipe into sections (intervals) and selected a time interval for computations Eqs.(22)and(24) are the characteristic equation of Eqs. (21)and(23).

If $f = 0$ Then Eq. (23) will be (Fig. 3),

$$\frac{dV}{dt} - \frac{g}{c} \cdot \frac{dH}{dt} = \circ \text{ or}$$

$$dH = \left(\frac{C}{g} \right) dV, (Zhukousky), \quad (25)$$

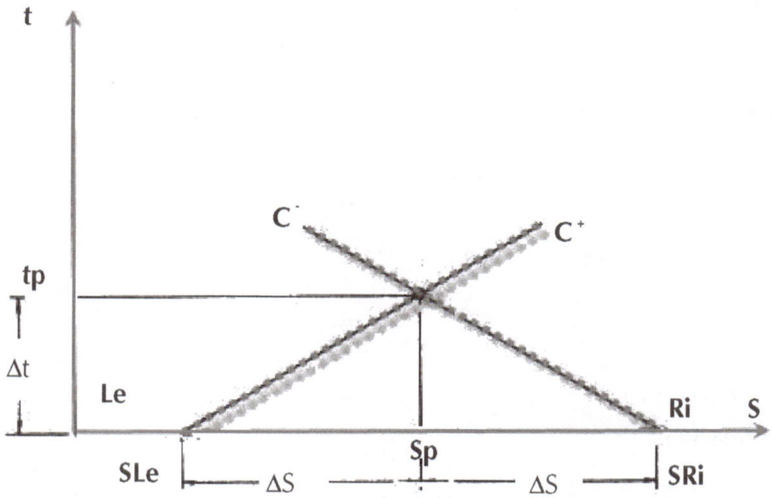

FIGURE 3 Intersection of characteristic lines with positive and negative slope.

If the pressure at the inlet of the pipe and along its length is equal to p_0, then slugging pressure undergoes a sharp increase:

$$\Delta p : p = p_0 + \Delta p$$

The Zhukouski formula is as flowing:

$$\Delta p = \left(\frac{C.\Delta V}{g} \right) \tag{26}$$

The speed of the shock wave is calculated by the formula:

$$C = \sqrt{\frac{g.\frac{E_W}{\rho}}{1 + \frac{d}{t_W} \cdot \frac{E_W}{E}}}, \tag{27}$$

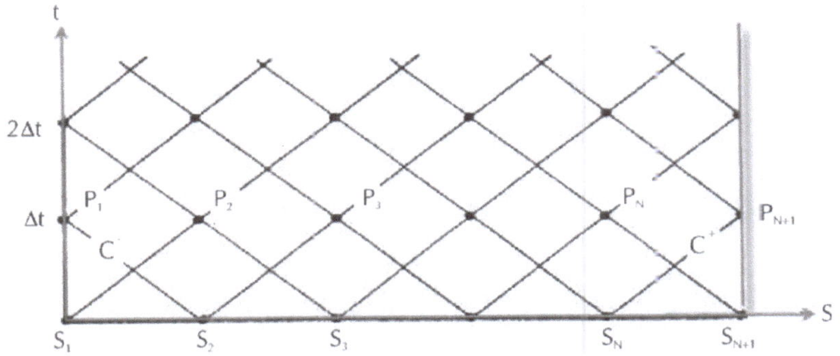

FIGURE 4 Set of characteristic lines intersection for assumed pipe.

By finite difference method of water hammer:

$$T_p - 0 = \Delta t$$

$$c^+ : (V_p - V_{Le})/(T_p - \circ) + (\frac{g}{c})(H_p - H_{Le})/(T_p - \circ) + fV_{Le}|V_{Le}|/2D) = \circ, \tag{28}$$

$$c^- : (V_p - V_{Ri})/(T_p - \circ) + (\frac{g}{c})(H_p - H_{Ri})/(T_p - \circ) + fV_{Ri}|V_{Ri}|/2D) = \circ, \tag{29}$$

$$c^+ : (V_p - V_{Le}) + (\frac{g}{c})(H_p - H_{Le}) + (f.\Delta t)(f.V_{Le}|V_{Le}|/2D) = \circ, \tag{30}$$

$$c^- : (V_p - V_{Ri}) + (\frac{g}{c})(H_p - H_{Ri}) + (f.\Delta t)(fV_{Ri}|V_{Ri}|/2D) = \circ, \tag{31}$$

$$V_P = \frac{1}{2}\left[(V_{Le}+V_{Ri})+\frac{g}{c}\left(H_{Le}-H_{Ri}\right)-(f.\Delta t/2D)(V_{Le}|V_{Le}|-V_{Ri}|V_{Ri}|)\right], \quad (32)$$

$$H_P = \frac{1}{2}\left[\frac{c}{g}(V_{Le}+V_{Ri})+(H_{Le}-H_{Ri})-\frac{c}{g}(f.\Delta t/2D)(V_{Le}|V_{Le}|-V_{Ri}|V_{Ri}|)\right], \quad (33)$$

$V_{Le}, V_{Ri}, H_{Le}, H_{Ri}, f, D$ are initial conditions parameters.

They are applied for solution at steady state condition. Water hammer equations calculation starts with pipe length "L" divided by "N" parts:

$$\Delta S = \frac{L}{N} \text{ and } \Delta t = \frac{\Delta s}{C}$$

Equations (4.2.28) and (4.2.29) are solved for the range P_2 through P_N, therefore, H and V are found for internal points. Hence:
At P_1 there is only one characteristic Line (c^-)
At P_{N+1} there is only one characteristic Line (c^+)
For finding H and V at P_1 and P_{N+1} the boundary conditions are used[45–124].

6.3 RESULTS

The challenge of selecting a time step is made difficult in pipeline systems by two conflicting constraints. They are defined by dynamic model for water hammer, first of all for calculating many boundary conditions, such as obtaining the head and discharge at the junction of two or more pipes. It is necessary that the time step be common to all pipes. The second constraint arises from the nature of the "MOC." If the adjective terms in the governing equations are neglected (as is almost always justified), the "MOC" requires that ratio of the distance x to the time step t be equal to the wave speed in each pipe. In other words, the Courant number should ideally be equal to one and must not exceed one by stability reasons.

TABLE 1 Eulerian based computational dynamic model.

Pipe	Length (m)	Diameter (mm)	Hazen-Williams Friction Coef.	Velocity (m/s)
P2	60.7	1200	90	2.21
P3	311	1200	90	2.21
P4	1	1200	90	2.65
P5	0.5	1200	91	2.65

TABLE 1 *(Continued)*

Pipe	Length (m)	Diameter (mm)	Hazen-Williams Friction Coef.	Velocity (m/s)
P6	108.7	1200	67	2.65
P7	21.5	1200	90	2.21
P8	15	1200	86	2.21
P9	340.7	1200	90	2.21
P10	207	1200	90	2.21
P11	339	1200	90	2.21
P12	328.6	1200	90	2.21
P13	47	1200	90	2.21
P14	590	1200	90	2.21
P15	49	1200	90	2.21
P16	224	1200	90	2.21
P17	18.4	1200	90	2.21
P18	14.6	1200	90	2.21
P19	12	1200	90	2.21
P20	499	1200	90	2.21
P21	243.4	1200	90	2.21
P22	156	1200	90	2.21
P23	22	1200	90	2.21
P24	82	1200	90	2.21
P25	35.6	1200	90	2.21
P0	0.5	1200	90	2.65
P1	0.5	1200	90	2.65

TABLE 2 Computational results of Eulerian based model.

Type of Point	Maximum Volume	Type of Point	Maximum Volume	Type of Point	Maximum Volume	Type of Point	Maximum Volume
P0:J1	Vapor	P12:J13	Vapor	P16:J16	Vapor	P21:J22	Vapor
P0:J2	Vapor	P13:33.33%	Vapor	P16:J17	Air	P22:10.00%	Vapor

TABLE 2 *(Continued)*

Type of Point	Maximum Volume	Type of Point	Maximum Volume	Type of Point	Maximum Volume	Type of Point	Maximum Volume
P1:J2	Vapor	P13:66.67%	Vapor	P17:J17	Air	P22:20.00%	Vapor
P1:J3	Vapor	P13:J13	Vapor	P17:J18	Vapor	P22:30.00%	Vapor
P10:15.38%	Vapor	P13:J14	Vapor	P18:J18	Vapor	P22:40.00%	Vapor
P10:23.08%	Vapor	P14:10.81%	Vapor	P18:J19	Vapor	P22:50.00%	Vapor
P10:30.77%	Vapor	P14:13.51%	Vapor	P19:J19	Vapor	P22:60.00%	Vapor
P10:38.46%	Vapor	P14:16.22%	Vapor	P19:J20	Air	P22:70.00%	Vapor
P10:46.15%	Vapor	P14:18.92%	Vapor	P2:25.00%	Vapor	P22:80.00%	Vapor
P10:53.85%	Vapor	P14:2.70%	Vapor	P2:50.00%	Vapor	P22:90.00%	Vapor
P10:61.54%	Vapor	P14:21.62%	Vapor	P2:75.00%	Vapor	P22:J22	Vapor
P10:69.23%	Vapor	P14:24.32%	Vapor	P2:J6	Vapor	P22:J23	Vapor
P10:7.69%	Vapor	P14:27.03%	Vapor	P2:J7	Vapor	P23:50.00%	Vapor
P10:76.92%	Vapor	P14:29.73%	Vapor	P20:12.90%	Vapor	P23:J23	Vapor
P10:84.62%	Vapor	P14:32.43%	Vapor	P20:16.13%	Vapor	P23:J24	Vapor
P10:92.31%	Vapor	P14:35.14%	Vapor	P20:19.35%	Vapor	P24:20.00%	Vapor
P10:J10	Vapor	P14:37.84%	Vapor	P20:22.58%	Vapor	P24:40.00%	Vapor
P10:J11	Vapor	P14:40.54%	Vapor	P20:25.81%	Vapor	P24:60.00%	Vapor
P11:14.29%	Vapor	P14:43.24%	Vapor	P20:29.03%	Vapor	P24:80.00%	Vapor
P11:19.05%	Vapor	P14:45.95%	Vapor	P20:3.23%	Vapor	P24:J24	Vapor
P11:23.81%	Vapor	P14:48.65%	Vapor	P20:32.26%	Vapor	P24:J28	Air
P11:28.57%	Vapor	P14:5.41%	Vapor	P20:35.48%	Vapor	P25:33.33%	Vapor
P11:33.33%	Vapor	P14:51.35%	Vapor	P20:38.71%	Vapor	P25:66.67%	Vapor
P11:38.10%	Vapor	P14:54.05%	Vapor	P20:41.94%	Vapor	P25:J28	Air
P11:4.76%	Vapor	P14:56.76%	Vapor	P20:45.16%	Vapor	P25:N1	Vapor
P11:42.86%	Vapor	P14:59.46%	Vapor	P20:48.39%	Vapor	P3:10.00%	Vapor
P11:47.62%	Vapor	P14:62.16%	Vapor	P20:51.61%	Vapor	P3:15.00%	Vapor
P11:52.38%	Vapor	P14:64.86%	Vapor	P20:54.84%	Vapor	P3:20.00%	Vapor
P11:57.14%	Vapor	P14:67.57%	Vapor	P20:58.06%	Vapor	P3:25.00%	Vapor
P11:61.90%	Vapor	P14:70.27%	Vapor	P20:6.45%	Vapor	P3:30.00%	Vapor

TABLE 2 *(Continued)*

Type of Point	Maximum Volume	Type of Point	Maximum Volume	Type of Point	Maximum Volume	Type of Point	Maximum Volume
P11:66.67%	Vapor	P14:72.97%	Vapor	P20:61.29%	Vapor	P3:35.00%	Vapor
P11:71.43%	Vapor	P14:75.68%	Vapor	P20:64.52%	Vapor	P3:40.00%	Vapor
P11:76.19%	Vapor	P14:78.38%	Vapor	P20:67.74%	Vapor	P3:45.00%	Vapor
P11:80.95%	Vapor	P14:8.11%	Vapor	P20:70.97%	Vapor	P3:5.00%	Vapor
P11:85.71%	Vapor	P14:81.08%	Vapor	P20:74.19%	Vapor	P3:50.00%	Vapor
P11:9.52%	Vapor	P14:83.78%	Vapor	P20:77.42%	Vapor	P3:55.00%	Vapor
P11:90.48%	Vapor	P14:86.49%	Vapor	P20:80.65%	Vapor	P3:60.00%	Vapor
P11:95.24%	Vapor	P14:89.19%	Vapor	P20:83.87%	Vapor	P3:65.00%	Vapor
P11:J11	Vapor	P14:91.89%	Vapor	P20:87.10%	Vapor	P3:70.00%	Vapor
P11:J12	Vapor	P14:94.59%	Vapor	P20:9.68%	Vapor	P3:75.00%	Vapor
P12:14.29%	Vapor	P14:97.30%	Vapor	P20:90.32%	Vapor	P3:80.00%	Vapor
P11:33.33%	Vapor	P14:51.35%	Vapor	P20:38.71%	Vapor	P25:66.67%	Vapor
P11:38.10%	Vapor	P14:54.05%	Vapor	P20:41.94%	Vapor	P25:J28	Air
P11:4.76%	Vapor	P14:56.76%	Vapor	P20:45.16%	Vapor	P25:N1	Vapor
P12:19.05%	Vapor	P14:J14	Vapor	P20:93.55%	Vapor	P3:85.00%	Vapor
P12:23.81%	Vapor	P14:J15	Air	P20:96.77%	Vapor	P3:90.00%	Vapor
P12:28.57%	Vapor	P15:33.33%	Vapor	P20:J20	Air	P3:95.00%	Vapor
P12:33.33%	Vapor	P15:66.67%	Vapor	P20:J21	Vapor	P3:J7	Vapor
P12:38.10%	Vapor	P15:J15	Air	P21:12.50%	Vapor	P3:J8	Vapor
P12:4.76%	Vapor	P15:J16	Vapor	P21:18.75%	Vapor	P4:J3	Vapor
P12:42.86%	Vapor	P16:14.29%	Vapor	P21:25.00%	Vapor	P4:J4	Vapor
P12:47.62%	Vapor	P16:21.43%	Vapor	P21:31.25%	Vapor	P5:J26	Air
P12:52.38%	Vapor	P16:28.57%	Vapor	P21:37.50%	Vapor	P5:J4	Vapor
P12:57.14%	Vapor	P16:35.71%	Vapor	P21:43.75%	Vapor	P6:14.29%	Vapor
P12:61.90%	Vapor	P16:42.86%	Vapor	P21:50.00%	Vapor	P6:28.57%	Vapor
P12:66.67%	Vapor	P16:50.00%	Vapor	P21:56.25%	Vapor	P6:42.86%	Vapor
P12:71.43%	Vapor	P16:57.14%	Vapor	P21:6.25%	Vapor	P6:57.14%	Vapor
P12:76.19%	Vapor	P16:64.29%	Vapor	P21:62.50%	Vapor	P6:71.43%	Vapor

TABLE 2 *(Continued)*

Type of Point	Maximum Volume	Type of Point	Maximum Volume	Type of Point	Maximum Volume	Type of Point	Maximum Volume
P12:80.95%	Vapor	P16:7.14%	Vapor	P21:68.75%	Vapor	P6:85.71%	Vapor
P12:85.71%	Vapor	P16:71.43%	Vapor	P21:75.00%	Vapor	P6:J26	Air
P12:9.52%	Vapor	P16:78.57%	Vapor	P21:81.25%	Vapor	P6:J27	Vapor
P12:90.48%	Vapor	P16:85.71%	Vapor	P21:87.50%	Vapor	P7:50.00%	Vapor
P12:95.24%	Vapor	P16:92.86%	Vapor	P21:93.75%	Vapor	P7:J27	Vapor
P12:J12	Vapor	—	—	P21:J21	Vapor	—	—
P12:19.05%	Vapor	P14:J14	Vapor	P20:93.55%	Vapor	P3:85.00%	Vapor
P12:23.81%	Vapor	P14:J15	Air	P20:96.77%	Vapor	P3:90.00%	Vapor
P12:28.57%	Vapor	P15:33.33%	Vapor	P20:J20	Air	P3:95.00%	Vapor
P12:33.33%	Vapor	P15:66.67%	Vapor	P20:J21	Vapor	P3:J7	Vapor
P12:38.10%	Vapor	P15:J15	Air	P21:12.50%	Vapor	P3:J8	Vapor
P12:4.76%	Vapor	P15:J16	Vapor	P21:18.75%	Vapor	P4:J3	Vapor
P12:42.86%	Vapor	P16:14.29%	Vapor	P21:25.00%	Vapor	P4:J4	Vapor
P12:47.62%	Vapor	P16:21.43%	Vapor	P21:31.25%	Vapor	P5:J26	Air
P12:52.38%	Vapor	P16:28.57%	Vapor	P21:37.50%	Vapor	P5:J4	Vapor
P12:57.14%	Vapor	P16:35.71%	Vapor	P21:43.75%	Vapor	P6:14.29%	Vapor
P12:61.90%	Vapor	P16:42.86%	Vapor	P21:50.00%	Vapor	P6:28.57%	Vapor
P12:66.67%	Vapor	P16:50.00%	Vapor	P21:56.25%	Vapor	P6:42.86%	Vapor
P12:71.43%	Vapor	P16:57.14%	Vapor	P21:6.25%	Vapor	P6:57.14%	Vapor
P12:76.19%	Vapor	P16:64.29%	Vapor	P21:62.50%	Vapor	P6:71.43%	Vapor
P12:80.95%	Vapor	P16:7.14%	Vapor	P21:68.75%	Vapor	P6:85.71%	Vapor
P12:85.71%	Vapor	P16:71.43%	Vapor	P21:75.00%	Vapor	P6:J26	Air
P12:9.52%	Vapor	P16:78.57%	Vapor	P21:81.25%	Vapor	P6:J27	Vapor
P12:90.48%	Vapor	P16:85.71%	Vapor	P21:87.50%	Vapor	P7:50.00%	Vapor
P12:95.24%	Vapor	P16:92.86%	Vapor	P21:93.75%	Vapor	P7:J27	Vapor
P12:J12	Vapor	—	—	P21:J21	Vapor	—	—

Faced with this challenge, this work tried for ways of relaxing the numerical constraints.

For the velocity of surge or pressure wave in an elastic case with free water bubble, the flowing equation would be valid:

$$C = \frac{1}{\left[\rho\left(\left(\frac{1}{E_W}\right)+\left(\frac{D}{E.t_W}\right)+\frac{n}{P}\right)\right]^{\frac{1}{2}}}, \tag{34}$$

For the velocity of surge or pressure wave in elastic case(Table 1) with the high value of free water bubble(Table 2) the flowing equation would be valid:

$$C = \left(\frac{g.h}{n}\right)^{1/2}, \tag{35}$$

Really stopping of a second layer of liquid exerts pressure on the following layers gradually caused high wave pressure. It acts directly at the valve extends to the rest of the pipeline against fluid flow speed C.

If the pressure at the beginning of the pipeline remains unchanged then after the shock of the initial section of the tube, it begins the reverse movement of the shock wave with the same velocity C and instantaneous changes.

Instantaneous changes in the rate of flow at the pipeline causes the surge wave. This phenomena occurs during water hammer are explained on the basis of compressibility of liquid drops.

This models used for laboratory; computational and field-tests experiments in order to presentation of water hammer phenomenon at the water pipeline.

Consistency for the observed values of maximal pressure the corresponding values were calculated according to Zhukouski's formula. In the final procedure it was compared the results of Eulerian based computational model(1) through (1)with the results of field test model(36)for transient flow.

TABLE 3 Field tests and computational results of Eulerian based model in the water pipeline.

Type of Point	Percent of air volume	Velocity of surge wave (m/s)	Head (m)
P10	15.38%	113.6	129
P11	14.29%	94.9	126.1
P12	14.29%	96.8	133.1
P13	33.33%	63.5	132.7
P14	10.81%	121	146.2
P15	33.33%	66.5	146
P16	14.29%	102.4	146.9
P2	25.00%	79.2	156.8
P20	12.90%	110.3	145.9
P21	12.50%	105	143.1
P22	10.00%	118.4	140.2
P23	50.00%	53	140.4
P24	20.00%	83	137.8
P25	33.33%	56.2	104.3
P3	10.00%	115.5	133.3
P6	14.29%	106	157.3
P7	50.00%	56	157.2

Curve estimation is the most appropriate when the relationship between the dependent variable and the independent variable is not necessarily linear.

Linear regression is used to model the value of a dependent scale variable based on its linear relationship to one or more predictors.

Non-linear regression is appropriate when the relationship between the dependent and independent variables is not intrinsically linear. Binary logistic regression is most useful in modeling of the event probability for a categorical response variable with two outcomes.

The auto-regression procedure is an extension of ordinary least-squares regression analysis specifically designed for time series.

One of the assumptions underlying ordinary least-squares regression is the absence of auto-correlation in the model residuals. Time series, however, often exhibit first-order auto-correlation of the residuals. In the presence of auto-correlated residuals, the linear regression procedure gives inaccurate estimates of how much of the series variability is accounted for by the chosen predictors. This can adversely affect the choice of predictors, and hence the validity of the model. The auto-regression procedure accounts for first-order auto-correlated residuals. It provides reliable estimates of both goodness of-fit measures and significant levels of chosen predictor variables.

One of the approaches of this work was the definition of a model by regression. It was defined based on the relationship between the dependent and independent data or variables for surge wave. It was showed in(Fig. 5) and Eq. (36). The variables are as follows: C – Velocity of surge wave $\left(m/_s\right)$ as a dependent variable with nomenclature "*Y*." The independent variable with nomenclature "*X*" such as: n – Percent of air volume $\left(m^3\right)$.

The curve estimation procedure allows quick estimating regression statistics, and producing related plots for different models. Hence the auto-regression procedure by regression software "SPSS 10.0.5" was selected for the curve estimation procedure in the present work. Therefore, the regression model was built based on the field test data.

Regression software "SPSS" fitted the function curve and provided regression analysis. So, the regression model(36) was found in the final procedure. By this model, laboratory and field test results were compared by computational model(35). The main practical aim of the present work was concentrated on the definition of a condition base maintenance (CM) method for all water transmission systems.

The data collection procedures were as follows:

At fast transients, down to 1 s, surge pressure and velocity of surge wave were recorded. These data were used for curve estimation procedure. They were detected on actual systems (field tests). Also, flow and pressure were computed by computational model(32) and (33). These data were compared by flow and pressure data which were collected from actual systems (field tests). The model was calibrated using one set of data, without changing parameter values. It was used to match a different set of results. Curve estimation procedure (Figs. 6–9) for surge wave velocity was formed by estimating regression statistics. Although related plots for the field test model were produced.

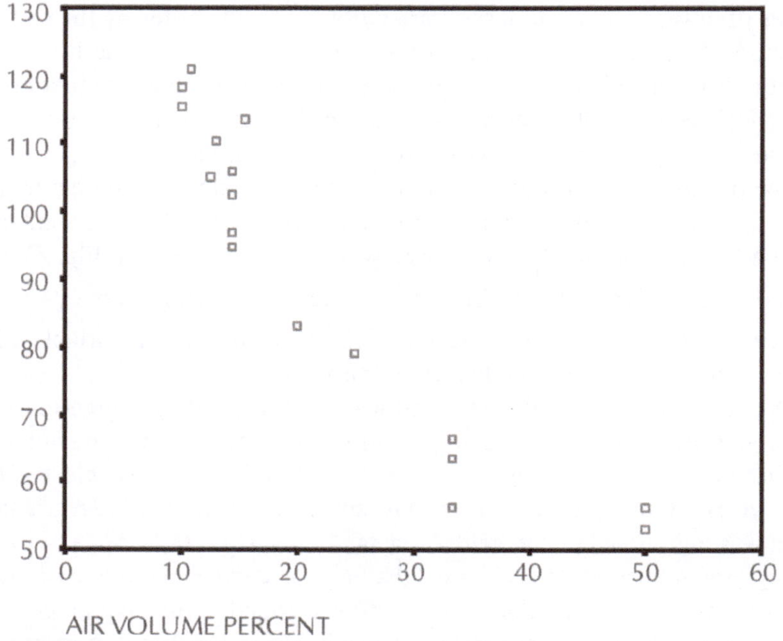

FIGURE 5. 2-D Scatter diagram of surge wave speed for water pipeline with free water bubble.

TABLE 4 Regression for model summary.

Model	R	R Square	Adjusted R Square	Std. Error of the Estimate
1	0.930[a]	0.865	0.856	9.08222

[a]Predictors: (Constant), AIR VOLUME PERCENT

TABLE 5 Model summary for function of regression model.

Model		Sum of Squares	df	Mean Square	F	Sig.
1	Regression	7934.64	1	7934.64	96.193	0.000[a]
	Residual	1237.3	15	82.487		0.000[b]
	Total	9171.94	16			

[a]Predictors: (Constant), AIR VOLUME PERCENT.
[b]Dependent Variable: SURGE VELOCITY(M/S).

TABLE 6 Variables and constant of function for Regression model.

Model		Unstandardized Coefficients		Standardized Coefficients	T	Sig.
		B	Std. Error	Beta		
1	(Constant)	127.451	4.35		29.301	0[a]
Air volume percent	−1.673	0.171	−0.93	−9.808	0	

[a]Dependent Variable: SURGE VELOCITY(M/S).

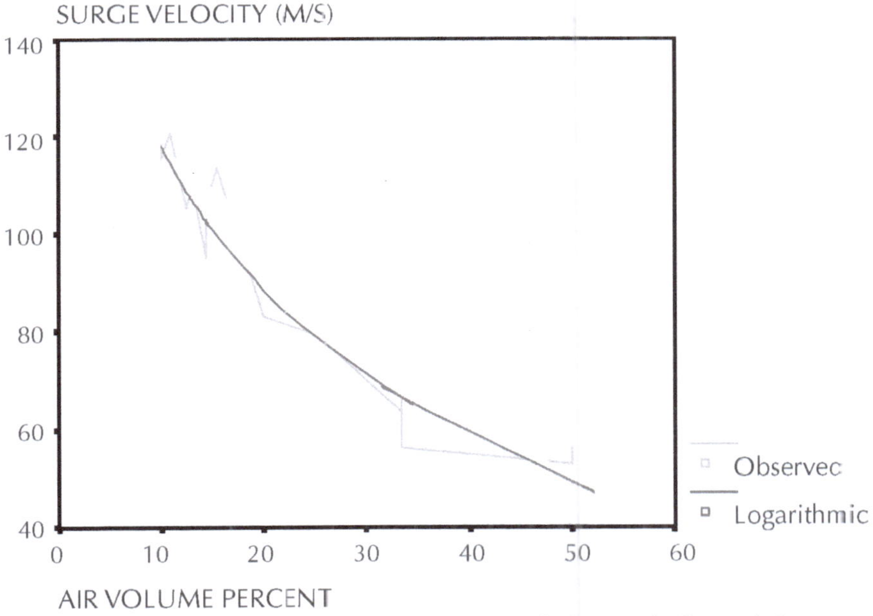

FIGURE 6 Logarithmic curve fit for surge wave speed of water pipeline with free water bubble.

TABLE 7 List wise deletion of missing data for curve fit by Logarithmic method.

Multiple R	0.96799
R Square	0.93701
Adjusted R Square	0.93281
Standard Error	6.20621

TABLE 8 Analysis of variance for curve fit by Logarithmic method.

	dF	Sum of Squares	Mean Square
Regression	1	8594.1825	8594.1825
Residuals	15	577.7563	38.5171

F = 223.12650 and Signif F = 0.0000.

TABLE 9 Variables of the equation for curve fit by Logarithmic method.

Variable	B	SE B	Beta	T Sig T
VAR00002	–42.606130	2.852309	–0.967992	–14.937
(Constant)	215.971673	8.522770	–25.341	

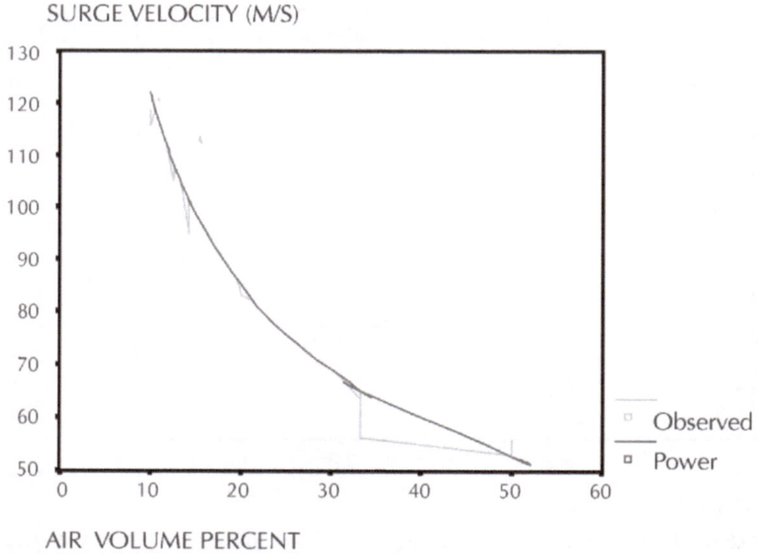

FIGURE 7 Power curve fit for surge wave speed of water pipeline with free water bubble.

TABLE 10 List wise deletion of missing data for curve fit by power method.

Multiple R	0.97433
R Square	0.94932
Adjusted R Square	0.94594
Standard Error	0.06739

This work presented regression software "SPSS10.0.5" for fitting the function curve and providing regression analysis. So, by comparison of 10 methods (i.e., logarithmic method with Power method), finally the Power method (.1) with standard error 0.06739 was selected as a regression model due to curve fitting (Fig. 6).

TABLE 11 Analysis of variance for curve fit by Power method.

	dF	Sum of Squares	Mean Square
Regression	1	1.2761075	1.2761075
Residuals	15	0.0681247	0.0045416

F = 280.97904 and Signif F = 0.0000.

TABLE 12 Curve fit by powermethod

Variable	B	SE B	Beta	T Sig T
VAR00002	–0.519175	0.030973	–0.974331	–16.762
(Constant)	402.197842	37.222066	—	10.805

TABLE 13 Variables of the equation for curve fit by Power method

Dependent Mth	Rsq	d.f.	F Sigf	b_0	b_1
VAR00001 POW	0.949	15	280.98	402.198	–0.5192

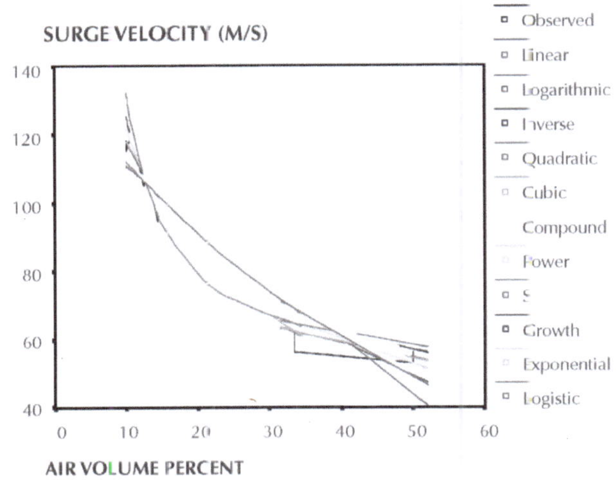

FIGURE 8 Comparison of curve fit for a series of functions due to surge wave in water pipeline.

In power functions, however, a variable base is raised to a fixed exponent. The parameter b_o serves as a simple scaling factor, moving the values of X^{b_1} up or down as b_o increases or decreases, respectively the parameter b_1, called either the exponent or the power, determines the function's rates of growth or decay. Depending on whether it is positive or negative, a whole number or a fraction, b_1 will also determine the function's overall shape and behavior. So the flowing equation derived by the fitting function curve and regression analysis.

$$f(X) = b_o X^{b_1}, \tag{36}$$

$$f(X) = (402.198) * X^{(-.5192)}$$

TABLE 14 Field tests and computational results of Eulerian based model.

Type of Point	Percent of air volume (cm)	Type of Point	Percent of air volume (cm)
P10	92.31%	P20:J20	Air
P11	95.24%	P21	93.75%
P12	95.24%	P22	90.00%
P13	66.67%	P23	50.00%
P14	97.30%	P24	80.00%
P14:J15	Air	P24:J28	Air
P15	66.67%	P25	66.67%
P15:J15	Air	P25:J28	Air
P16	92.86%	P3	95.00%
P16:J17	Air	P5:J26	Air
P17:J17	Air	P6	85.71%
P19:J20	Air	P6:J26	Air
P20	96.77%	P7	50.00%

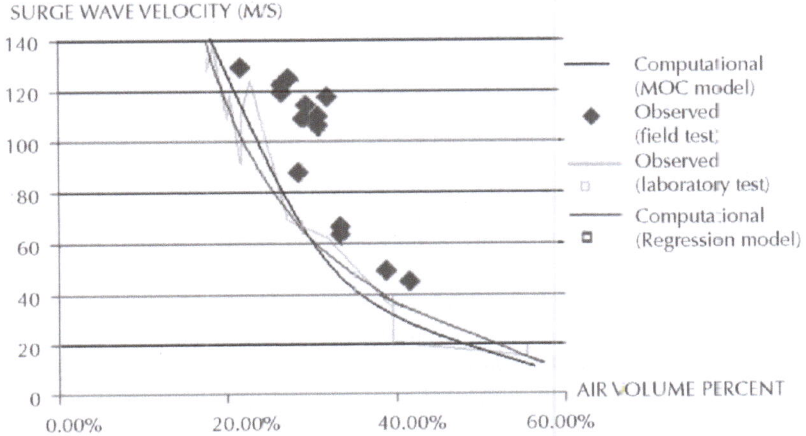

FIGURE 9 Comparison of computational results with laboratory and field test results.

This work selected the power function as a regression model based on the field test metering. Hence it was compared with the computational results[125–154].It was assumed the penetrated air volume in the water pipeline, which was showed in (Fig. 9).In present work the equation that was offered by regression is as the flowing:

$$SurgeVelocity = 402.198 * (AirVolume)^{(-.5192)}$$

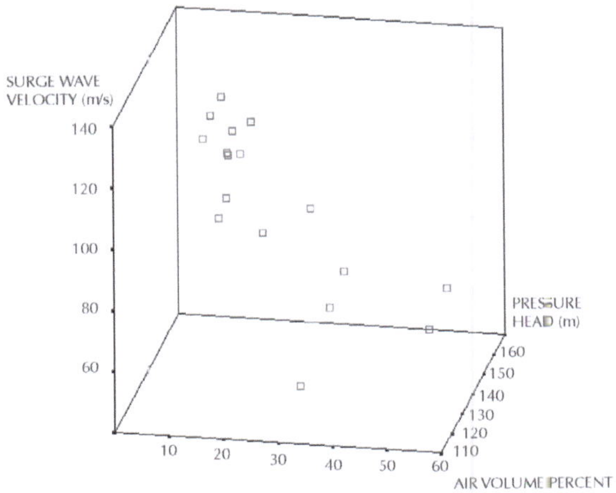

FIGURE 10 3-D Scatter diagram of surge wave speed for water pipeline with free water bubble.

Some of the reasons for changes at liquid properties in pipes and channels are due to the spread of the surge wave, the formation and collapse of vapor bubbles (cavitation) and air leak or disconnection of the system. The surge wave velocity has proportional changes against to changes in percent of the formation and collapse of vapor bubbles (Figs. 9 and 10).

The pipe system has a characteristic time period $T = \frac{2L}{a}$, where L is equal to 590 (m).It is the longest possible path through the system. The pressure wave speed a is equal to1084 $\left(m/s\right)$. The time period is equal to 1.1 (S). It is equal to the time for a pressure wave to travel the pipe system's greatest length two times. Generally it is recommended that the run duration equals or exceeds *T*, which run duration or total simulation time was equal to 5 (S). Run duration is measured either in seconds or as a number of time steps. Time steps typically range from a few hundredths of a second to a few seconds, depending on the system and the pressure wave speed. The run duration has a direct effect on the modeling computation time, along with the time step selected for the simulation. Another factor to consider definition of run duration is to allow enough time for friction to significantly dampen the transient energy at vapor pressure $-10.0(m)$.

The conclusions were drawn on the basis of experiments and calculations for the pipeline with a local leak. Hence, the most important effects that were observed are as follows:

The pressure wave speed generated by water hammer phenomenon was influenced by some additional factors. Therefore, the ratio of local leakage and discharge from the leak location was mentioned. The effect of total discharge from the pipeline and its effect on the values of wave oscillations period were studied. The outflow to the surge tank from the leak affected the value of wave celerity. The pipeline was equipped with the valve at the end of the main pipe, which was joined with the closure time register. The water hammer pressure characteristics were measured by extensometers, as similar as in the work of (Kodura and Weinerowska, 2005). Simultaneously it was recorded in the computer's memory. The supply of the water at the system was realized with the use of the reservoir, which enabled inlet pressure stabilization. In positive water hammer in the pipeline was introduced with local leaks in two scenarios; first; with the outflow from the leak to the overpressure reservoir, second; with free outflow from the leak to atmospheric pressure, with the possibility of sucking air in the negative phase. The bubbles in nonlinear dynamics fluid state acts as a separator gate for two distinct parts of the flow at upstream and downstream of separator gate. It causes high surge wave velocity at one of these distinct parts. Therefore, the compressed air builds high-pressure flow, which can destroy the water pipeline.

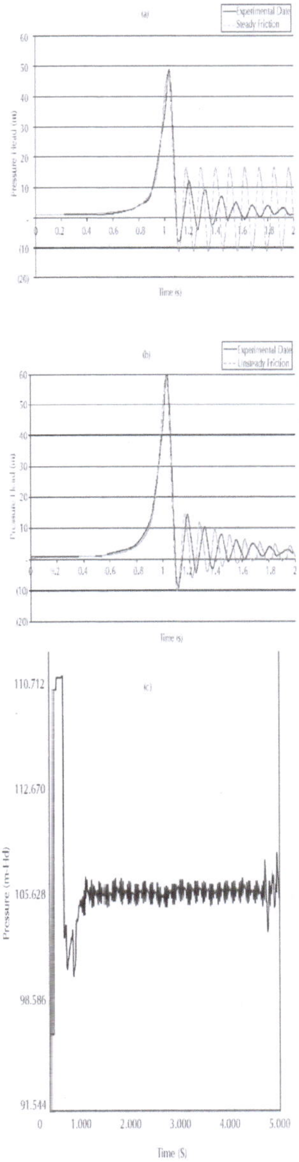

FIGURE 11 Pressure head histories for a single piping system using; (a) steady friction, (b) unsteady friction, Chaudhry, (c) steady friction, present computational results.

Chaudhry [13], Obtained pressure heads by the steady and unsteady friction model (Fig. 10). Comparison showed similarity in present work and work of Chaudhry [13], (Figs. 11(a) and 11(b)) results. Parmakian [14],Streeter and Wylie [15].

The MOC based scheme was most popular because this scheme provided the desirable attributes of accuracy, numerical efficiency and programming simplicity in the work of Parmakian [14], Wylie and Streeter [15].The water hammer software package was used in present chapter. The Numerical solution of the nonlinear Navier-Stokes equations and MOC was used by software package.

Leon [16] comparison showed similarity between present work results against the work of Leon [16].

The results are as following:

1. Numerical tests showed that the proposed second-order formulation at boundary conditions (achieved by using virtual cells) is second-order. In addition, the proposed formulation maintains the conservation property of FV schemes and introduced no unphysical perturbations into the computational domain.
2. Numerical tests were performed for smooth (i.e., flows that do not present discontinuities) and sharp transients.
3. The high efficiency of the proposed scheme was important for real-time control (RTC) of water hammer flows in large networks.

Changes at system boundaries (Sudden changes) created a transient pressure pulse. In this regard, model design needed to find the relation between many variables accordance to fluid transient. Therefore, a computational technique was presented and the results were compared by field tests. In present work after closing the valves on the horizontal pipe of constant diameter, which moves the liquid with an average speed V_0, a liquid layer, located directly at the gate, immediately stops. Then successively terminate movement of the liquid layers (turbulence, counter flows) to increase with time away from the gate. In this work the air was sucked into the pipeline. Pressure wave velocity was recorded in fast transient up to 5 ms(in this work 1 second). The assessment procedure was used to analysis the collected data, which were obtained at real system by Hariri et al. [17].

Water hammer numerical modeling and simulation processes were presented by three models:

1. Laboratory Model,
2. MOC Model, and
3. Regression Model.

Three dimensional heat transfer characteristics and pressure drop of water flow in a set of rectangular mictrotubes were numerically investigated. A FV solver was employed to predict the temperature field for the conjugate heat transfer problem in both the solid and liquid regions in the microchannels. The full Navier-Stoke's approach was examined for this kind of narrow tubes for the pressure drop evaluations. The complete form of the energy equation with the dissipation terms was also linked to the momentum equations. The computed thermal characteristics and pressure drop showed good agreements with the experimental data. The effects of flow rate and channel geometry on the heat transfer capability and pressure drop of the system were also predicted.

To design an effective microtubes heat sink, fundamental understanding of the characteristics of heat transfer and fluid flow in microchannels are necessary. At the early stages, the designs and relations of macroscale fluid flow and heat transfer were employed. However, many experimental observations in microchannels deviate significantly from those in macroscale channels.

These disagreements were first observed by Tuckerman and Pease[18]. They demonstrated that micro rectangular passages have a higher heat transfer coefficient in laminar regimes in comparison with turbulent flow through macroscale channels. Therefore, they are capable of dissipating significant high heat fluxes.

Since, the regime of the flow has a noticeable influence on studying determination of heat dissipation rate; several researches have been conducted in this field. Wu and Little [19] measured the heat transfer characteristics for gas flows in miniature channels with inner diameter ranging from 134 (μm) to 164 (μm). The tests involved both laminar and turbulent flow regimes. Their results showed that the turbulent convection occurs at Reynolds number of approximately 1,000.

They also found that the convective heat transfer characteristics depart from the predictions of the established empirical correlations for the macroscale tubes. They attributed these deviations to the large asymmetric relative roughness of the microchannel walls. Harms et al. [20] tested a 2.5 (cm) long, 2.5 (cm) wide silicon heat sink having 251(μm) wide and 1030 (μm) deep microchannels. A relatively low Reynolds number of 1,500 marked transition from laminar to turbulent flow which was attributed to a sharp inlet, relatively long entrance region, and channel surface roughness. They concluded the classical relation for Nusselt number was fairly accurate for modeling microchannel flows.

Fedrov and Viskanta [21] reported that the thermal resistance decreases with Reynolds number and reaches an asymptote at high Reynolds numbers.

Qu et al. [22] investigated heat transfer and flow characteristics in trapezoidal silicon microtubes. In comparison, the measured friction factors were found to be higher than the numerical predictions. The difference was attributed to the wall roughness. Based on a roughness-viscosity model, they explained that the numerically predicted Nusselt numbers are smaller than the experimentally determined ones.

Choi et al. [23] measured the convective heat transfer coefficients for flow of nitrogen gas in microtubes for both laminar and turbulent regimes. They found that the measured Nusselt number in laminar flow exhibits a Reynolds number dependence in contrast with the conventional prediction for the fully developed laminar flow, in which Nusselt number is constant.

Adam et al. [24] conducted single-phase flow studies in microtubes using water as the working fluid. Two diameters of the circular microtubes, namely 0.76 (mm) and 1.09 (mm), were used in the investigation. It was found that the Nusselt numbers are larger than those encountered in microtubes. Peng and Peterson [25] investigated water flows in rectangular microtubes with hydraulic diameters ranging from 0.133 to 0.336 (mm). In laminar flows, it was found that the heat transfer depends on the aspect ratio and the ratio of the hydraulic diameter to the center-to-center distance of the microchannel.

Mala et al. [26] considered the electrical body forces resulting from the double layer field in the equations of motion. These effects are negligible in the macroscale as the dimensions of the electric double layer, EDL, is very small with respect to tube dimensions. They solved the PoissonBolzmann equation for the steady state flow. It was found that without the double layer a higher heat transfer rate is obtained. They proposed to consider the effects of the EDL on liquid flows and heat transfer in microtubes to prevent the overestimation of the heat transfer capacity of the system. Xu et al. [27] investigated the effects of viscous dissipation on the micro scale dimensions. They used a 2D microtube and considered the viscous dissipation term in energy equation. The results show that this term plays a significant role in temperature, pressure and velocity distributions. Therefore, the relationships between the friction factor and the Reynolds number change when the hydraulic diameter of the micrrotube is very small. The viscous dissipation effects are brought about by rises in the velocity gradient as hydraulic diameter reduces for a constant Reynolds number, a numerical study is conducted based on the experimental results of Tuckerman [28]. A FV method is used to solve the conjugate heat transfer through the heat sinks. The flow and heat transfer development regions inside the tubes are considered. The numerical results are then compared with the available experimental data. The effects of liquid velocity through channels and their effects on heat transfer and pressure drop along microchannels are investigated. Finally, the effects of aspect ratio on heat dissipation and pressure drop in microtubes are predicted.

The characteristics of the model are depicted in (Fig. 12). The width and thickness of microchannels are W_1, andH_1respectively. The thickness of the silicon substrate is H_2-H_1; and the total length of the microchannels is L. The heat supplies by a 1 × 1 cm heat source located at the entrance of the channels and were centered across the whole channel heat sink. A uniform heat flux of q is provided to heat the microchannels. The heat is removed by flowing water through channels. The inlet temperature of the cooling water is 20°C. The analysis is performed for five different cases.

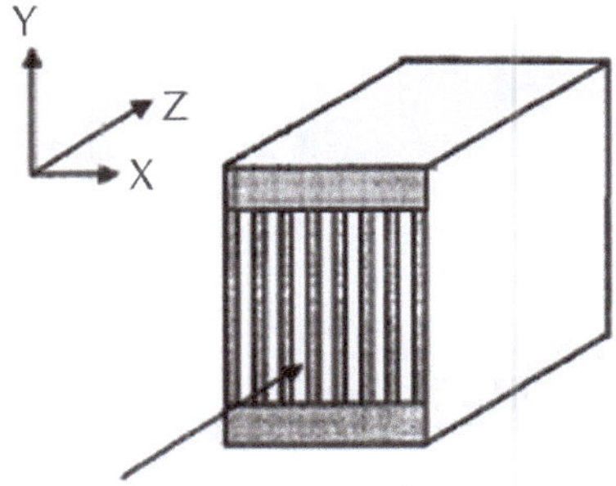

FIGURE 12 A sample microtube.

The dimensions related to each case are given in Table 15. By these dimensions; there will be 150 microchannels for cases 0 and 1 and 200 microchannels for cases 2, 3, and 4.

For the second section, case 2 was considered as the base geometry. The water velocity changed from 50 (cm/s) to 400 (cm/s).

In the third section, the Tuckerman's geometries were solved by the unique Reynolds of 150.As performing a numerical method for the whole microtubes heat sink is hard; a certain computational domain is considered (Fig. 13).

To prevent the various boundary condition effects, the computational domain is taken at the center of the heat sink to have the quoted uniform heat flux in Table 15. This is because there is very little spreading of heat towards the heat sinks. There is also some geometrical symmetry, which simplifies the computation. So only a semichannel and semisilicon substrate will be considered and the results will be the same for the other half. The whole substrate is made of silicon with thermal conductivity (k) of 148 (W/m. K). At the top of the channel $y = H_2$, there is a Pyrex plate to make an adiabatic condition (its thermal conductivity is two orders lower than silicon). There are two different boundary conditions at the bottom.

TABLE 15 Four different cases of micro channels

	Case				
	0	**1**	**2**	**3**	**4**
L(cm)	2	2	1.4	1.4	1.4
W_1 (μm)	64	64	56	55	50
W_2(μm)	36	36	44	45	50
H_1(μm)	280	280	320	287	302
H_2(μm)	489	489	533	430	458
$\dot{Q}(cm^3 / s)$	1.277	1.86	4.7	6.5	8.6
q(W/cm²)	34.6	34.6	181	277	790
Number of channels	150	150	200	200	200

For $z < L_h$ a uniform heat flux of q is imposed over the heat sink and the rest is assumed to be adiabatic. Water flows through the channel from the entrance in z direction.

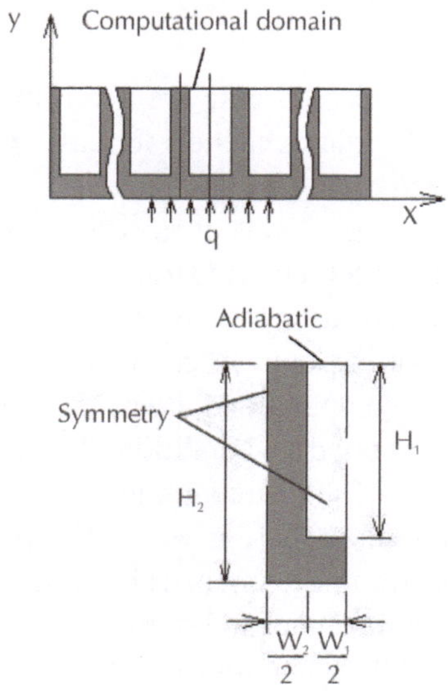

FIGURE 13 Dimensions and computational domain.

The transverse velocities of the inlet are assumed to be zero. The axial velocity is considered to be evenly distributed through the whole channel. The velocities at the top and bottom of tubes are zero.

Several simplifying assumptions are incorporated before establishing the governing equations for the fluid flow and heat transfer in a unit cell:

1. Steady fluid flow and heat transfer;
2. Incompressible fluid;
3. Laminar flow;
4. Negligible radiative heat transfer; and
5. Constant solid and fluid properties;

In the last assumption the solid and liquid properties are assumed to be constant because of the small variations within the temperature range tested.

Based on the above assumptions, the governing differential equations and Nomenclatures used to describe the fluid flow and heat transfer in a unit cell are expressed as follows:

A = flow area, T = Temperature, cp = Specific heat, Tin = inlet temperature, Dh = hydraulic diameter, $Tmax$ = maximum temperature, EDL = Electric Double Layer, w = velocity (z direction), H_1= channel height, W_1= channel width, H_2= total height, W_2= substrate width, k = thermal conductivity, L = heat sink length, Φ = viscous dissipation terms in energy equation, Lh = heated length, p = pressure, μ= viscosity, P = wetted perimeter, ρ = density, q = heat flux Subscripts, Q = total average volumetric flow rate, R = thermal resistance, in = inlet, Re = Reynolds number, max = maximum.

Continuity Equation

$$\frac{\partial w}{\partial z}=0 \tag{37}$$

Momentum Equation

$$\rho\frac{\partial(w_i w_j)}{\partial z_j}=-\frac{\partial p}{\partial z_i}+\mu\frac{\partial}{\partial z_j}\left(\frac{\partial w_i}{\partial z_j}+\frac{\partial w_j}{\partial z_i}\right)$$

$$-\frac{2}{3}\mu\frac{\partial}{\partial z_i}\left(\frac{\partial w_k}{\partial z_k}\right) \tag{38}$$

Energy Equation

$$\rho c_p\frac{\partial w_j T}{\partial z_j}=k\frac{\partial^2 T}{\partial x_j^{\,2}}+\mu\Phi \tag{39}$$

where

$$\Phi = 2\left[\left(\frac{\partial u}{\partial x}\right)^2 + \left(\frac{\partial v}{\partial y}\right)^2 + \left(\frac{\partial w}{\partial z}\right)^2\right] + \left(\frac{\partial u}{\partial y} + \frac{\partial v}{\partial x}\right)^2 + \left(\frac{\partial u}{\partial z} + \frac{\partial w}{\partial x}\right)^2 + \left(\frac{\partial v}{\partial z} + \frac{\partial w}{\partial y}\right)^2 \quad (40)$$

The finite volume method (FVM) is used to solve the continuity, momentum, and energy equations. A very brief description of the method used is given here.

In this method the domain is divided into a number of control volumes such that there is one control volume surrounding each grid point. The grid point is located in the center of a control volume. The governing equations are integrated over the individual control volumes to construct algebraic equations for the discrete dependent variables such as velocities, pressure, and temperature. The discretization equation then expresses the conservation principle for a finite control volume just as the partial differential equation expresses it for an infinitesimal control volume. In the present study, a solution is deemed converged when the mass imbalance in the continuity equation is less than 10^{-6}.

As the thermal specifications and flow characteristics are of the great importance in design of microchannel heat sinks, the results are concentrated in these fields.

For solving the equations several grid structures were used. The grid density of 120 × 40 × 20 in z, y, and$x$ directions is considered to be appropriate.

The thermal resistance is calculated as follows:

$$R(z) = \frac{T_{\text{max}}(z) - T_{\text{in}}}{q} \quad (41)$$

In Eq. (5), $R(z)$, $T_{\text{max}}(z)$, T_{in} and q the thermal resistance at z(cm) from the entrance, the inlet water temperature and the heat flux at the heating area.

In addition to thermal resistance, the Reynolds number is calculated as:

$$\text{Re} \equiv \frac{\rho w_{ave} D_h}{\mu} \quad (42)$$

where

$$D_h \equiv \frac{4A}{P} = \frac{2H_1W_1}{H_1 + W_1} \tag{43}$$

First, the temperature distribution, thermal resistances, and pressure drops based on the numerical results are demonstrated. Then the numerical results are compared with Tuckerman's experiments data.

6.3.1 TEMPERATURE DISTRIBUTION AND THERMAL RESISTANCE

The temperature distributions at four *x-y* cross sections along the channel are shown in Fig. 14 for Case 0. The four sections, $z_{=}1$, 3, 6, 9(mm) are all in the heated area. As all heat exchangers, isotherms are closest at the entrance of the channels. This means that heat transfer rate is the highest at the entrance and it decreases along the channel.

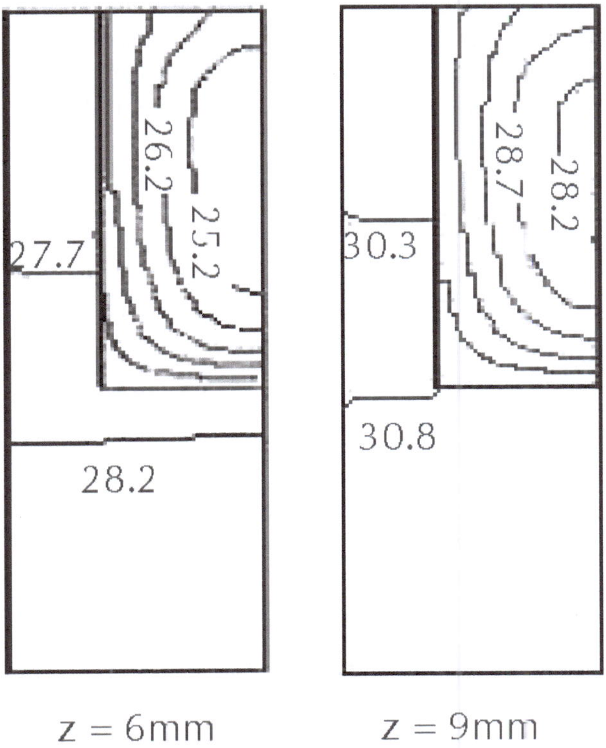

FIGURE 14 Temperature distribution along the tube for case 0.

The thermal resistance along the tube is shown in Fig. 14 for cases 0 and 1. It can be seen that the thermal resistance has increased by increasing the z value and attained the maximum value at $z = 9$(mm). It is consistent with Tuckerman's experiments. It may be seen that the thermal resistance increased linearly through the channel except the entry region and near its peak. The sharper slopes in the experimental data of Fig. 15(a) and the numerical prediction in Fig. 15(b) are evidently due to the entrance region effects. So the flow may be considered thermally fully developed with a proper precision. The maximum thermal resistance is occurred in $z = 9$(mm)which is consistent with Tuckerman's experiments. For the other Cases the results are given in Table 16. The numerical values of resistances are predicted well [155–164].

6.3.2 PRESSURE DROP

Pressure drop is linear along the channel. Fig. 16(a) shows the pressure drop for Case 4. The pressure drop increases by increasing the inlet velocity. The slope of the pressure line in the entrance of the channel is maximum. This is due to the entry region effects. The velocity field will be fully developed after a small distance from the entrance. So the assumption of fully developed flow is acceptable. Figure16(b) shows the pressure drop for all five cases. These amounts are tabulated in Table 3.

TABLE 16 Thermal resistance comparison.

Case	$q(\frac{W}{cm^2})$	$R(cm^2 K/W)$		*Error (%)*
		Experimental	**Numerical**	
0	34.6	0.277	0.253	5.8
1	34.6	0.280	0.246	12.1
2	181	0.110	0.116	5.0
3	277	0.113	0.101	8.1
4	790	0.090	0.086	3.94

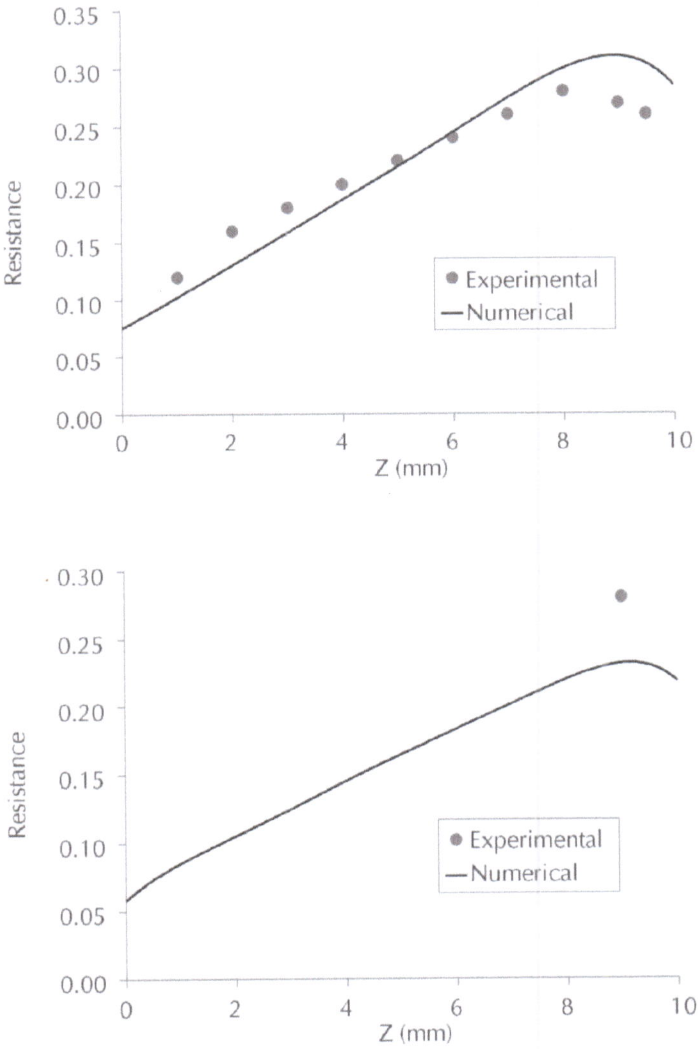

FIGURE 15 Numerical and experimental thermal resistances for: (a) Case 0. and(b) Case 1.

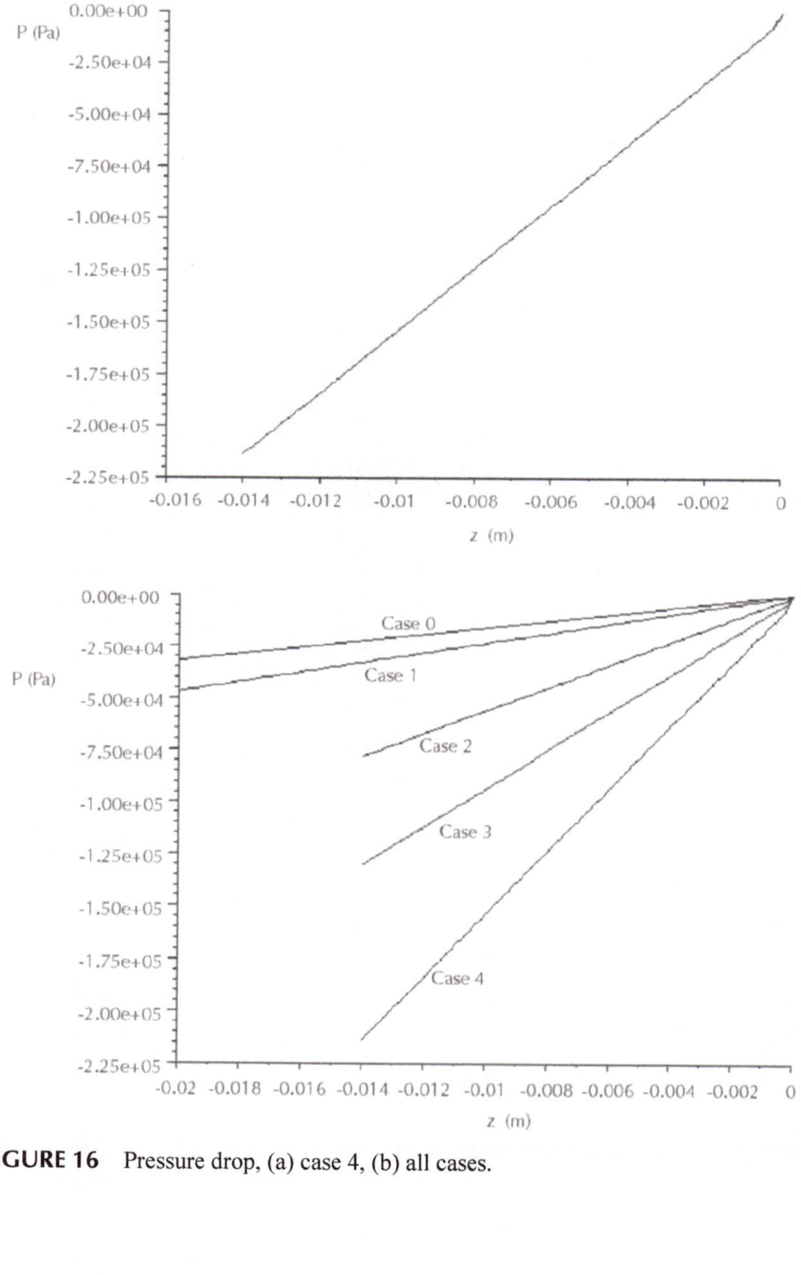

FIGURE 16 Pressure drop, (a) case 4, (b) all cases.

TABLE 17 Pressure drop in fivecases.

Case	*Pressure Drop (bar)*
0	0.322
1	0.469
2	0.784
3	1.302
4	2.137

The effects of velocity on the temperature rise and pressure drop in microtubes are displayed in Figs. 17 and 18. The amount of heat dissipation increases by increasing the velocity. But it can be seen that the amount of decrease in temperature falls drastically by increase in velocity.

The following function can predict the temperature rise:

$$(T_{\max} - T_{in}) = 265.67w^{-0.4997} \tag{44}$$

The pressure drop is a linear function of velocity. The amounts of temperature rise and pressure drop are given in Table 18.

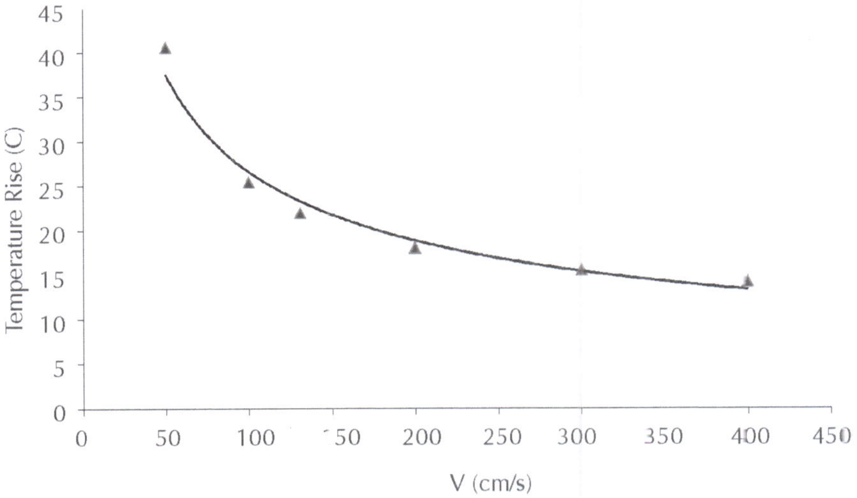

FIGURE 17 Temperature rise for different velocities.

TABLE 18 Temperature rise and pressure drop for section 2.

Velocity (cm/s)	*Re*	*Temp. Rise (°C)*	*Pressure Drop (bar)*
50	47	40.6	0.298
100	95	25.4	0.598
131	124	21.9	0.784
200	190	18.0	1.206
300	285	15.5	1.817
400	380	14.2	2.439

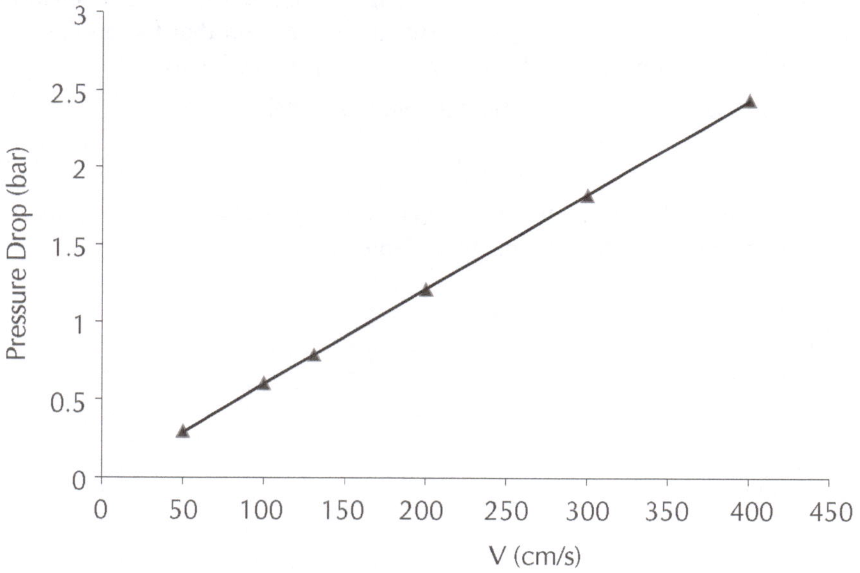

FIGURE 18 Pressure drop for different velocities.

Aspect ratio is an important factor in microtube design. In this section the inlet heat flux of 181 (W/cm^2)imposed over the heat sinks and the hydraulic diameter changes between 85.8 and 104.2. (Fig. 19) shows the maximum temperature of each case. It may be seen that with an identical heat flux the heat dissipation of the largest aspect ratio is the lowest. But this case has the minimum pressure drop too (Fig. 20). By increasing the amount of aspect ratio the ability of heat dissipation increases.

TABLE 19 Temperature rise and pressure drop for section 3.

Case	*Aspect Ratio*	*Hydraulic Diameter (μm)*	*Velocity (m/s)*	*Temp. Rise (°C)*	*Pressure Drop (bar)*
1	4.375	104.19	1.447	27.84	0.98
2	5.714	95.32	1.58	19.97	0.95
3	5.218	92.31	1.633	20.76	1.03
4	6.040	85.80	1.76	20.35	1.31

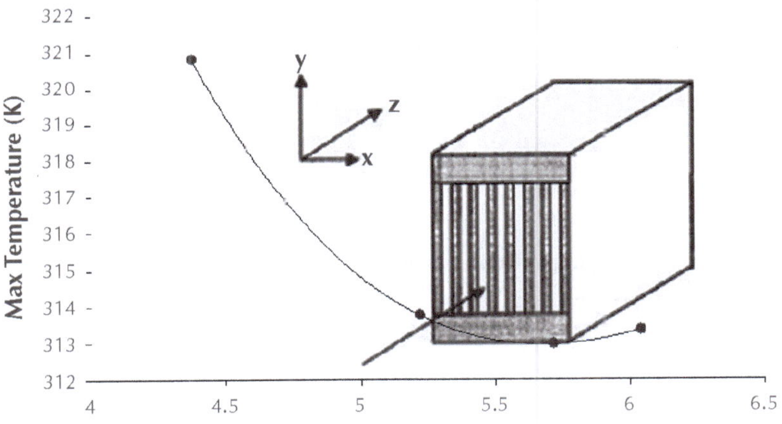

FIGURE 19 Temperature with respect to aspect ratio.

The third geometry has the minimum temperature rise. The temperature rise of second and forth geometries are approximately alike. But the pressure drop of these two cases has a noticeable difference. The pressure drop increases with increase in aspect ratio [165–179].The amounts of temperature rise and temperature rise of all geometries are given in Table 19.

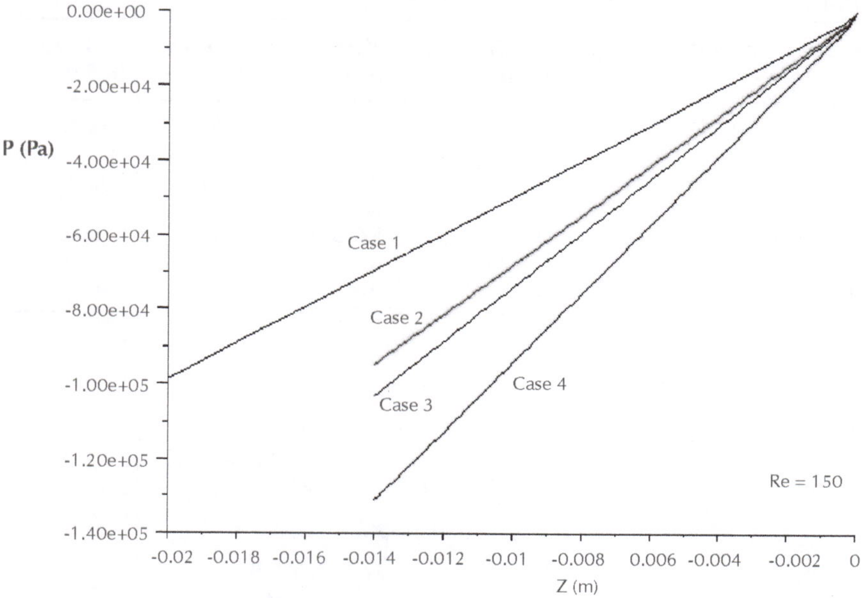

FIGURE 20 Pressure drop relating geometry in constant Reynolds of 150.

6.4 CONCLUSION

Generally, water hammer is manifested as a hydro-machines phenomenon which can leading to the destruction of pipelines. The cycles of increased and decreased in pressure iterates at intervals equal to time for dual-path shock wave length of the pipeline from the valve prior to the pipeline. Thus, the hydraulic impact of the liquid in the pipeline performed oscillatory motion. The cause of oscillatory motion was the hydraulic resistance and viscosity. It absorbed the initial energy of the liquid for overcoming the friction.

The effects of the penetrated air on the surge wave velocity in water pipeline. It showed that Eulerian based computational model is more accurate than the regression model. Hence in order to presentation for importance of penetrated air on water hammer phenomenon, it was compared the models for laboratory; computational and field tests experiments. As long as these procedure, it was showed that the Eulerian based model for water transmission line in comparison with the regression model. On the other hand, this idea were included the proper analysis to provide a dynamic response to the shortcomings of the system. It also performed the design protection equipments to manage the transition energy and determine the operational procedures to avoid transients. Consequently, the results will help to reduce the risk of system damage or failure at the water pipeline.

KEYWORDS

- **Eulerian based model**
- **method of characteristics**
- **Navier-Stokes equations**
- **surge wave**
- **water transmission**

CHAPTER 7

LAGRANGIAN AND EULERIAN TRANSITIONAL FLOW

CONTENTS

7.1 INTRODUCTION

The starting point of any numerical method is the mathematical model, the set of partial differential equations and boundary conditions. After selecting the mathematical model, one has to choose a suitable discretization method. The most important are: finite differences (FD), finite volume (FV) and finite element (FE) methods. The discrete locations at which the variables are to be calculated are defined by the numerical grid, which is essentially a discrete representation of the geometric domain on which the problem is to be solved. It divides the solution domain into a finite number of subdomains. Different types of grids are:

- Structured grid: A structured mesh is defined as a mesh where all the nodes have the same number of elements around it. This makes that the matrix of algebraic equation system has a regular structure. There is large number of efficient solvers applicable only to structured grids. Disadvantages: only for geometrically simple domains.
- Unstructured grid: For very complex geometries, can fit arbitrary boundaries. Grids made of triangles or quadrilaterals in 2D and tetrahedral or hexahedra in 3D are the most often used. Such grids can be generated automatically by existing algorithms. Disadvantage: irregularity of the data structure. The solvers for the algebraic equation system are usually slower than those for structured grids.
- Block-structured grid: Structured grid inside each block, but the order of blocks is irregular. Consistency The discretization should become exact, as the grid spacing tends to zero. The difference between the discretized equation and the exact one is called *truncation error*. For a method to be consistent, the truncation error must become zero when the mesh spacing. If the most important term of the truncation error is proportional to n the method is of n-th order. $n > 0$ is required for consistency. Even if the approximations are consistent, it does not necessarily mean that the approximated solution will become the exact solution in the limit of small step size. For this to happen, the method has to be stable also. Stability A numerical method is stable if does not magnify the errors that appear during the process. For temporal problems, stability guarantees that the method produces a bounded solution whenever the exact solution is also bounded. A numerical method is convergent if the discrete solution tends to the exact one as the grid spacing tends to zero. For linear initial value problems, the Lax equivalence theorem states:

Consistency, stability, and convergence for nonlinear problems the stability and convergence of a method are difficult to demonstrate. Therefore, convergence is usually checked using numerical experiments, that is, repeating the calculation on a series of successively refined grids. If the method is stable and if all approximations used in the discretization process are consistent, we usually find that the solution does converge to a grid-independent solution. Since the equations to be solved are conservation laws, the numerical scheme should also respect these laws. This means that, at steady state and in the absence of sources, the amount of a conserved quantity leaving a closed volume is equal to the amount entering that volume. If the strong conservation form of equations and a finite volume method are used, this is guaranteed for each individual control volume and for the domain as a whole. Accuracy numerical solutions of fluid flow are only approximate solutions. In addition to the errors that might be introduced in the development of the solution algorithm, in programming or setting up the boundary conditions, numerical solutions always include three kinds of systematic errors:

- Modeling errors: Difference between the actual flow and the exact solution of the mathematical model.
- Discretization errors: Difference between the exact solution of the conservation equations and the exact solution of the algebraic system of equations obtained by discretizing these equations.
- Iteration errors: Difference between the iterative and exact solutions of the algebraic equation systems.
- The governing equations of fluid flow represent mathematical statements of the conservation laws of physics.
- The mass of fluid is conserved.
- The rate of change of momentum equals the sum of the forces applied on a fluid particle (Newton's second law).
- The rate of change of energy is equal to the sum of the rate of heat addition and to the rate of work done on a fluid particle (first law of thermodynamics).

This book discusses on Lagrangian and Eulerian based model for water hammer. This model was defined by the method of characteristics "MOC", finite difference form. The method was encoded into an existing hydraulic simulation model. The surge wave was assumed as a failure factor in an elastic case of water pipeline with free water bubble. The results were compared by regression analysis. It indicated that the accuracy of the Eulerian based model for the process of interpenetration of two fluids into water transmission line.

These ranges of methods are included by approximate equations to numerical solutions of the nonlinear Navier–Stokes equations.

Various methods have been developed to solve transient flow in pipes. These ranges have been formed from approximate equations to numerical solutions of the nonlinear Navier–Stokes equations. Elastic theory describes the unsteady flow of a compressible liquid in an elastic system. Transient theory stem from the two governing equations. The continuity equation and the momentum equation are needed to determine velocity and surge pressure in a one-dimensional flow system. Solving these two equations produces a theoretical result that usually corresponds quite closely to actual system measurements, if the data and assumptions used to build the numerical model are valid. Among the approaches proposed to solve.

The single-phase (pure liquid) transient equations are the MOC FD, wave characteristic method (WCM), FE, and FV. One difficulty that commonly arises relates to the selection of an appropriate level of time step to use for the analysis. The obvious trade off is between computational speed and accuracy. In general, for the smaller the time step, there is the longer the run time but the greater the numerical accuracy [13–44].

7.2 MATERIALS AND METHODS

This chapter is started with the solving of approximate equations by numerical solutions of the nonlinear Navier-Stokes equations based on the MOC. Then it derived the Joukowski formula and velocity of surge or pressure wave in an elastic case with the high value of free water bubble. So the numerical modeling and simulation, which was defined by "MOC" provided a set of results. Basically the "MOC" approach transforms the water hammer partial differential equations into the ordinary differential equations along the characteristic lines.

7.2.1 FINITE ELEMENT METHOD (FEM)

This is a short introduction to the FEM, which is, besides others like the finite differences approximation, a technique to solve partial differential equations (PDE's) numerically.

There is a systematic account of changes in the mass, momentum and energy of the fluid element due to flow across the boundaries and the sources inside the element. Infinitesimal fluid element has six faces: North, South, East, West, Top, and Bottom.

Types of time integration methods for unsteady flows, initial value problem:

1. Explicit, values at time $n+1$ computed from values at time n.

Advantages:
- Direct computation without solving system of equation.
- Few numbers of operations per time step.

Disadvantage:
- Strong conditions on time step for stability.

2. Implicit, values at time $n+1$ computed from the unknown values at time $n+1$.

Advantage:
- Larger time steps possible, always stable.

Disadvantages:
- Every time step requires solution of system.
- More number of operations.

7.2.2 FINITE VOLUME METHOD (FVM)

- FVM uses integral form of conservation (transport) equation.
- Domain subdivided in control volumes (CV).
- Surface and volume integrals approximated by numerical quadrature.
- Interpolation used to express variable values at CV faces in terms of nodal values.
- It results in an algebraic equation per CV.
- Suitable for any type of grid.
- Conservative by construction.
- Commercial codes: CFX, Fluent, Phoenics, Flow3D.

7.2.3 TURBULENT FLOWS

- Most flows in practice are turbulent.
- With increasing Re, smaller eddies.
- Very fine grid necessary to describe all length scales.
- Even the largest supercomputer does not have (yet) enough speed and memory to simulate turbulent flows of high Re.

Computational methods for turbulent flows:

- Direct Numerical Simulation (DNS)
- Large Eddy Simulation (LES)
- Reynolds-Averaged Navier-Stokes (RANS)

The FEM was mainly developed for equations of elasticity and structural mechanics. In these fields problems have to be solved in complicated and irregular geometries. So one of the main advantages of the FEM, in comparison to the finite differences approximation, lies in the flexibility concerning the geometry of the domain where the PDE is to be solved.

Moreover, the FEM is perfectly suitable as adaptive method, because it makes local refinements of the solution easily possible. The method does not operate on the PDE itself, instead the PDE is transformed to an equivalent variational or weak form. This will be the topic of the second part: the variational principle.

For transient analysis by FE, first we choose a geometric shape and divide the domain into a finite number of regions. In one dimension, the domain is split into intervals. In two dimensions, the elements are usually of triangular or quadrilateral shape. And in three dimensions, tetrahedral or hexahedral forms are most common. Most elements used in practice have fairly simple geometries, because this result in very easy computation, since integrating over these shapes is quite basic.

The basic functions are usually not defined directly. Instead a function types, the so-called ansatz function, (e.g., linear or quadratic polynomial) is selected which our approximation should adopt on each of these elements. Most commonly a linear function is chosen, which means that it will be a linear function on each element and continuous over (but not continuously differentiable).

Each element possesses a set of distinguishing points called nodal points or nodes. Nodes define the element geometry, and are the degrees of freedom of the ansatz function. So the number of nodes in an element depends on the ansatz function as well as the geometry. They are usually located at the corners or end points of elements. For higher-order (higher than linear) ansatz functions, nodes are also placed on sides or faces, as well as perhaps the interior of the element.

The combination of the geometric shape of the finite element and their associated ansatz function on this region is referred as finite element type.

The basis arises from the choice of the finite element type for water hammer analysis. Water hammer is the result of sharp changes of fluid pressure by the instantaneous changes in the rate of flow in the pipeline [39–41] This phenomena occurs during transient are explained on the basis of compressibility of liquid drops. After closing the valves on the horizontal pipe of constant diameter, which moves the liquid with an average speed V_0, a liquid layer, located directly at the gate, immediately stops [42–44]. Then successively terminate movement of the liquid layers (turbulence, counter flows) to increase with time away from the gate. It is compacted before stopping the mass of liquid. As a result of increasing pressure somewhat expanded pipe. In the tube includes an additional volume of liquid. Since the fluid is compressible, the whole of its mass in the pipeline does not stop immediately. It moves from the gate along the pipeline with some velocity C, called the speed of propagation of the pressure wave.

A first approach to solve the variational or weak form was made by Ritz (1908). A discussion of this method is the subject in the third section: the Ritz method. Considering the disadvantages of the Ritz method will lead to the finite element method and to the fourth and last part: the finite element method PDE are separated into different types which behave very differently and demand an entirely own treatment. In the field of second-order linear differential equations three types are of fundamental interest, these are the hyperbolic, parabolic and elliptic equations. Depending on the type of the PDE boundary or initial conditions have to be given. The main focus of the finite element method is elliptic PDE's so we will concentrate on this type. The correct side conditions for elliptic PDE's are boundary conditions. In this book, an experimental and computational method was used. It applied for prediction of surge tank effects on pressure and flow variation [45–180] in the following case:

- Direct Numerical Simulation (DNS) applies for analysis of pressure and flow variation at surge tank location due to transient flow.
- Fluid transient flow in practice was turbulent.
- Fluid transient flow with smaller eddies was lead to increasing Re.
- Very fine grid described by all length scales.
- The main problem was the supercomputer, which has not enough speed and memory to simulate turbulent flows of high Re.

Navier-Stokes equations:

$$(\partial\rho/\partial t)+div(\rho\vec{u})=0, \text{ Mass (Continuity equation)}, \tag{1}$$

$$(\partial\rho u/\partial t)+div(\rho u\vec{u})=-(\partial p/\partial x)+div(\mu\, grad\, u)+q_x, \text{X (momentum equation)}, \tag{2}$$

$(\partial \rho v / \partial t) + div(\rho v \vec{u}) = -(\partial p / \partial y) + div(\mu grad v) + q_y$, Y (momentum equation), (3)

$(\partial \rho w / \partial t) + div(\rho w \vec{u}) = -(\partial p / \partial z) + div(\mu grad w) + q_z$, Z (momentum equation), (4)

$(\partial \rho i / \partial t) + div(\rho i \vec{u}) = -p div \vec{u} + div(k grad T) + \phi$, Z (internal energy equation), (5)

7.2.3.1 DIRECT NUMERICAL SIMULATION

- Discretize Navier-Stokes equation on a sufficiently fine grid for resolving all motions occurring in turbulent flow.
- No uses any models.
- Equivalent to laboratory experiment.
- Relationship between length η of smallest eddies and the length L of largest eddies,

$$\frac{L}{\eta} \approx (\mathrm{Re}_L)^{\frac{3}{4}}, \tag{6}$$

Number of elements necessary to discretize the flow field in industrial applications,

$\mathrm{Re} > 10^6$ then $n_{elem} > 10^{13}$

7.2.3.2 LARGE EDDY SIMULATION

- Only large eddies are computed.
- Small eddies are modelled, subgrid-scale (SGS) models.

7.2.3.3 REYNOLDS-AVERAGED NAVIER-STOKES

- $u = \bar{u} + u'$ Variables decomposed in a mean part and a fluctuating part.
- Navier-Stokes equations averaged over time.
- Turbulence models are necessary.

7.3 RESULTS

In this book, numerical analysis and simulation processes are applied for water hammer nonlinear heterogeneous model. Table 1 (located in the Appendix at the back of this book) shows the extreme heads of pressure variation due to water transmission modeling.

7.4 CONCLUSION

Generally, water hammer is manifested as a hydro-machines phenomenon, which can leading to the destruction of pipelines. The cycles of increased and decreased in pressure iterates at intervals equal to time for dual-path shock wave length of the pipeline from the valve prior to the pipeline. Thus, the hydraulic impact of the liquid in the pipeline performed oscillatory motion. The cause of oscillatory motion was the hydraulic resistance and viscosity. It absorbed the initial energy of the liquid for overcoming the friction.

The effects of the penetrated air on the surge wave velocity in water pipeline. It showed that Eulerian based computational model is more accurate than the regression model. Hence in order to presentation for importance of penetrated air on water hammer phenomenon, it was compared the models for laboratory; computational and field tests experiments. As long as these procedures, it was showed that the Eulerian based model for water transmission line in comparison with the regression model. On the other hand, this idea were included the proper analysis to provide a dynamic response to the shortcomings of the system. It also performed the design protection equipments to manage the transition energy and determine the operational procedures to avoid transients. Consequently, the results will help to reduce the risk of system damage or failure at the water pipeline.

KEYWORDS

- **Eulerian based model**
- **method of characteristics**
- **Navier-Stokes equations**
- **surge wave**
- **water transmission**

CHAPTER 8

DYNAMIC MODELING FOR WATER FLOW

CONTENTS

8.1 INTRODUCTION

The ranges of various methods for analysis of water hammer in pipes are included by approximate equations to numerical solutions of the nonlinear Navier-Stokes equations. In this chapter a case study with experimental and computational approach on hydrodynamics instability for a water pipeline have been presented. This book shows the water hammer effect on flow and pressure variations in surge tank as a surge protection device. Therefore, computational performances of a numerical method have been showed by a dynamic model for water transmission failure condition water hammer and surge-protection needs must be considered in the context of a water utility's risk management and environmental protection plan. Surge or water hammer, as it is commonly known is the result of a sudden change in liquid velocity. Water hammer usually occurs when a transfer system is quickly started, stopped or is forced to make a rapid change in direction. During a transient analysis, the fluid and system boundaries can be either elastic or inelastic: 1) Elastic theory, describes unsteady flow of a compressible liquid in an elastic system (e.g., where pipes can expand and contract). 2) Rigid-column theory, describes unsteady flow of an incompressible liquid in a rigid system. It is only applicable to slower transient phenomena. Both branches of transient theory stem from the same governing equations. Among the approaches proposed to solve the single-phase (pure liquid) water hammer equations are the Method of Characteristics (MOC), Finite Differences (FD), Wave Characteristic Method (WCM), Finite Elements (FE), and Finite Volume (FV). One difficulty that commonly arises relates to the selection of an appropriate level of time step to use for the analysis. The obvious trade-off is between computational speed and accuracy. In general the smaller the time step, the longer the run time but the greater the numerical accuracy. The challenge of selecting a time step is made difficult in pipeline systems by two conflicting constraints. First, to calculate many boundary conditions, such as obtaining the head and discharge at the junction of two or more pipes, it is necessary that the time step be common to all pipes. The second constraint arises from the nature of the MOC. If the adjective terms in the governing equations are neglected (as is almost always justified), the MOC requires that ratio of the distance Δx to the time step Δt be equal to the wave speed in each pipe. In other words, the Courant number should ideally be equal to one and must not exceed one by stability reasons. For most pipeline systems, having as they do a variety of different pipes with a range of wave speeds and lengths, it is impossible to satisfy exactly the Courant requirement in all pipes with a reasonable (and common) value of Δt. Faced with this challenge, researchers have sought for ways of relaxing the numerical constraints. Two contrasting strategies present themselves. The method of wave-speed adjustment changes one of the pipeline properties (usually the wave speed, though more rarely the pipe length is altered) so as to satisfy exactly the courant condition [1–55]. In this work an experimental and computational method was used. It applied for prediction of surge tank effects on pressure and flow variation at leakage condition in two cases: 1) pressure and flow variation at surge tank location, 2) pressure and flow variation at leakage location. Therefore, it mentioned to transient flow at water transmission pipeline failure condition.

8.2 MATERIALS AND METHODS

8.2.1 EXPERIMENTAL (FIELD TESTS MODEL CRITERIA)

In this book, water hammer numerical modeling and simulation processes are handled by nonlinear heterogeneous model.

8.2.2 APPROACHES TO TRANSIENT FLOW (METHOD OF CHARACTERISTICS "MOC" MODEL)

Model was defined by method of characteristics "MOC". In this work water hammer software was applied for numerical modeling. Specification of system was defined to water hammer software, version 07.00.049.00. In this case, water pipeline assumed in water leakage condition [56–67] and equipped with surge tank (real condition or existent condition). The method of characteristics "MOC" is based on a finite difference technique where pressures are computed along the pipe for each time step (Joukowski, 1904). The combined elasticity of both the water and the pipe walls is characterized by the pressure wave speed (Arithmetic method [3] combination of the Joukowski formula and Allievi formula, Wylie and Streeter, 1982). The method of characteristics "MOC" approach transforms the water hammer partial differential equations into the ordinary differential equations along the characteristic lines [Eq. (1)–(18)].

Balancing the energy across two points in the system yields the energy or Bernoulli equation for steady-state flow. The components of the energy equation can be combined to express two useful quantities, the hydraulic grade and the energy grade:

$$\left(P_1 / \gamma\right) + Z_1 + \left(V_1^2 / 2g\right) + h_p = \left(P_2 / \gamma\right) + Z_2 + \left(V_2^2 / 2g\right) + h_L, \tag{1}$$

$$(g / a)\left(dH / dt\right) + dv / dt + \left(f\, v|v| 2d\right) = 0 \Rightarrow (ds / dt) = c^+, \tag{2}$$

$$-(g / a)\left(dH / dt\right) + dv / dt + \left(f\, v|v| 2d\right) = 0 \Rightarrow (ds / dt) = c^-, \tag{3}$$

The method of characteristics is a finite difference technique where pressures were computed along the pipe for each time step. Calculation automatically subdivided the pipe into sections (intervals) and selected a time interval for computations.

$$\left(dp / dt\right) = \left(\partial p / \partial t\right) + \left(\partial p / \partial s\right)\left(ds / dt\right), \tag{4}$$

$$\left(dv / dt\right) = \left(\partial v / \partial t\right) + \left(\partial v / \partial s\right)\left(ds / dt\right), \tag{5}$$

P and *V* changes due to time are high and due to coordination are low then it can be neglected for coordination differentiation:

$$\left(\partial v / \partial t\right) + \left(1 / \rho\right)\left(\partial p / \partial s\right) + g\left(dz / ds\right) + \left(f / 2D\right) v|v| = 0, \text{ (Euler equation)}, \tag{6}$$

$$C^2\left(\partial v / \partial s\right) + \left(1 / P\right)\left(\partial P / \partial t\right) = 0, \text{ (Continuity equation)}, \tag{7}$$

By linear combination of Euler and continuity equations in characteristic solution method:

$$\lambda\left[\left(\partial v / \partial t\right) + \left(1 / \rho\right)\left(\partial p / \partial s\right) + g\left(dz / ds\right) + \left(f / 2D\right) v|v|\right] + C^2\left(\partial v / \partial s\right) + \left(1 / p\right)\left(\partial p / \partial t\right) = 0, \quad \lambda = {}^+c \;\&\; \lambda = {}^-c \tag{8}$$

$$\left(dv / dt\right) + \left(1 / cp\right)\left(dp / ds\right) + g\left(dz / ds\right) + \left(f / 2D\right) v|v| = 0, \tag{9}$$

$$\left(dv / dt\right) - \left(1 / cp\right)\left(\partial p / \partial s\right) + g\left(dz / ds\right) + \left(f / 2D\right) v|v| = 0, \tag{10}$$

Method of characteristics drawing in $(s\text{–}t)$ coordination:

$$\left(dv / dt\right) - \left(g / c\right)\left(dH / dt\right) = 0, \tag{11}$$

$$dH = \left(c / g\right) dv, \text{ (Joukowski Formula)}, \tag{12}$$

By Finite Difference method:

$$c+:((vp-v_{Le})(Tp-0))+((g/c)(Hp-H_{Le})/(Tp-0))+((fv_{Le}|v_{Le}|)/2D)=0|, \quad (13)$$

$$c-:((vp-vRi)(Tp-0))+((g/c)(Hp-HRi)/(Tp-0))+((fvRi|vRi|)/2D)=0|, \quad (14)$$

$$c+:(vp-v_{Le})+(g/c)(Hp-H_{Le})+(f\Delta t)(fv_{Le}|v_{Le}|)/2D)=0, \quad (15)$$

$$c-:(vp-vRi)+(g/c)(Hp-HRi)+(f\Delta t)(v_{Ri}|v_{Ri}|)/2D)=0, \quad (16)$$

$$V_P=1/2\begin{pmatrix}(V_{Le}+V_{ri})+(g/c)(H_{Le}-H_{ri})\\ -(f\ \Delta t/2D)(V_{Le}\ |V_{Le}|+V_{ri}|V_{ri}|)\end{pmatrix}, \quad (17)$$

$$H_P=1/2\begin{pmatrix}C/g(V_{Le}-V_{ri})+(H_{Le}+H_{ri})\\ -C/g(f\ \Delta t/2D)(V_{Le}\ |V_{Le}|-V_{ri}|V_{ri}|)\end{pmatrix}, \quad (18)$$

8.2.3 REGRESSION

The curve estimation procedure allows quick estimating regression statistics, and producing related plots for different models. Curve estimation is the most appropriate when the relationship between the dependent variable(s) and the independent variable is not necessarily linear. Linear regression is used to model the value of a dependent scale variable based on its linear relationship to one or more predictors. Non-linear regression is appropriate when the relationship between the dependent and independent variables is not intrinsically linear. Binary logistic regression is most useful in modeling of the event probability for a categorical response variable with two outcomes. The auto-regression procedure is an extension of ordinary least-squares regression analysis specifically designed for time series. One of the assumptions underlying ordinary least-squares regression is the absence of auto-correlation in the model residuals. Time series, however, often exhibit first-order auto-correlation of the residuals. In the presence of auto-correlated residuals, the linear regression procedure gives inaccurate estimates of how much of the series variability is accounted by the chosen predictors. This can adversely affect the choice of predictors, and hence the validity of the model. The auto-regression procedure accounts for first-order auto-correlated residuals.

8.3 RESULTS AND DISCUSSION

8.3.1 FIELD TESTS MODEL CRITERIA

At this work fast transients (down to 1 second), surge pressure and velocity of surge wave were recorded. Also flow and pressure reading were collected by laboratory model instrument. Those data have been compared by flow and pressure data, which have been detected from existent system [68–78].This chapter results showed sub atmospheric or even full-vacuum pressures. Air also sucked into the piping. Collapse of the vapor pocket caused a dramatic high-pressure transient when the water column rejoins vary rapidly. Analysis showed sub atmospheric or even full-vacuum pressures was generated in the near of reservoir. It provides reliable estimates of both goodness of fit measures and significant levels of chosen predictor variables. The auto-regression procedure by regression software "SPSS 10.0.5" has been selected for the curve estimation procedure in this chapter. The regression model has been built based on field test data and in the final procedure it has been compared with the method of characteristics "MOC" numerical modeling and simulation results (Tables 1–55).

For initial head and minimum head there were not any changes. The experimental and computational method was used for prediction of surge tank effects on pressure and flow variation. Therefore, defined model of this chapter shows condition due to transient flow at water transmission failure. This chapter was conformed to the results of the Kodura, Weinerowska and Leon's work (Figs. 1–9).

TABLE 1 Variables Entered/Removed.[a]

Model	Variables Entered	Variables Removed	Method
1	Flow[b]	Pressure[a]	Enter

[a]Dependent Variable: pressure.
[b]All requested variables entered.

TABLE 2 ANOVA.[a]

Model		Sum of Squares	df	Mean Square	F	Sig.
1	Regression	0.044	1	0.044 0.3	55	0.557[b]
	Residual	2.735	22	0.124		
	Total	2.780	23			

[a]Dependent Variable: pressure.
[b]Predictors: (Constant), flow.

TABLE 3 Model Summary.

Model	R	R Square	Adjusted R Square	Std. Error of the Estimate
1	0.126[a]	0.016	–0.029	0.35262

[a]Predictors: (Constant), flow.

Curve Fit

TABLE 4 Model Description.

Model Name		MOD_2
Dependent Variable	1	Pressure
Equation	1	Linear
Independent Variable		Flow
Constant		Included
Variable Whose Values Label Observations in Plots		Unspecified

TABLE 5 Case Processing Summary.

	N
Total Cases	24
Excluded Cases[a]	0
Forecasted Cases	0
Newly Created Cases	0

[a]Cases with a missing value in any variable are excluded from the analysis.

TABLE 6 Variable Processing Summary.

		Variables	
			Independent
		Dependent Pressure	Flow
Number of Positive Values		24	23
Number of Zeros		0	1
Number of Negative Values		0	0
Number of Missing Values	User-Missing	0	0
	System-Missing	0	0

TABLE 7 Model Summary and Parameter Estimates.

Dependent Variable: Pressure							
	Model Summary					Parameter Estimates	
Equation	R Square	F	df1	df2	Sig.	Constant	b1
Linear	0.016	0.355	1	22	0.557	7.223	–0.037

The independent variable is flow.

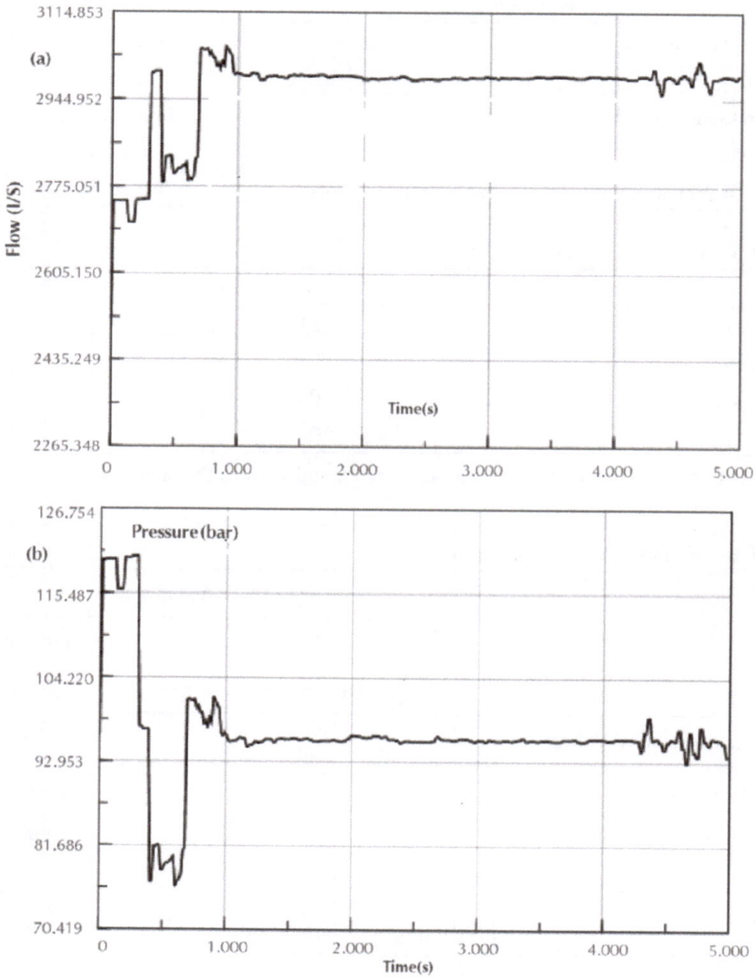

FIGURE 1 Pipeline with surge tank: (a) flow variation, (b) pressure variation.

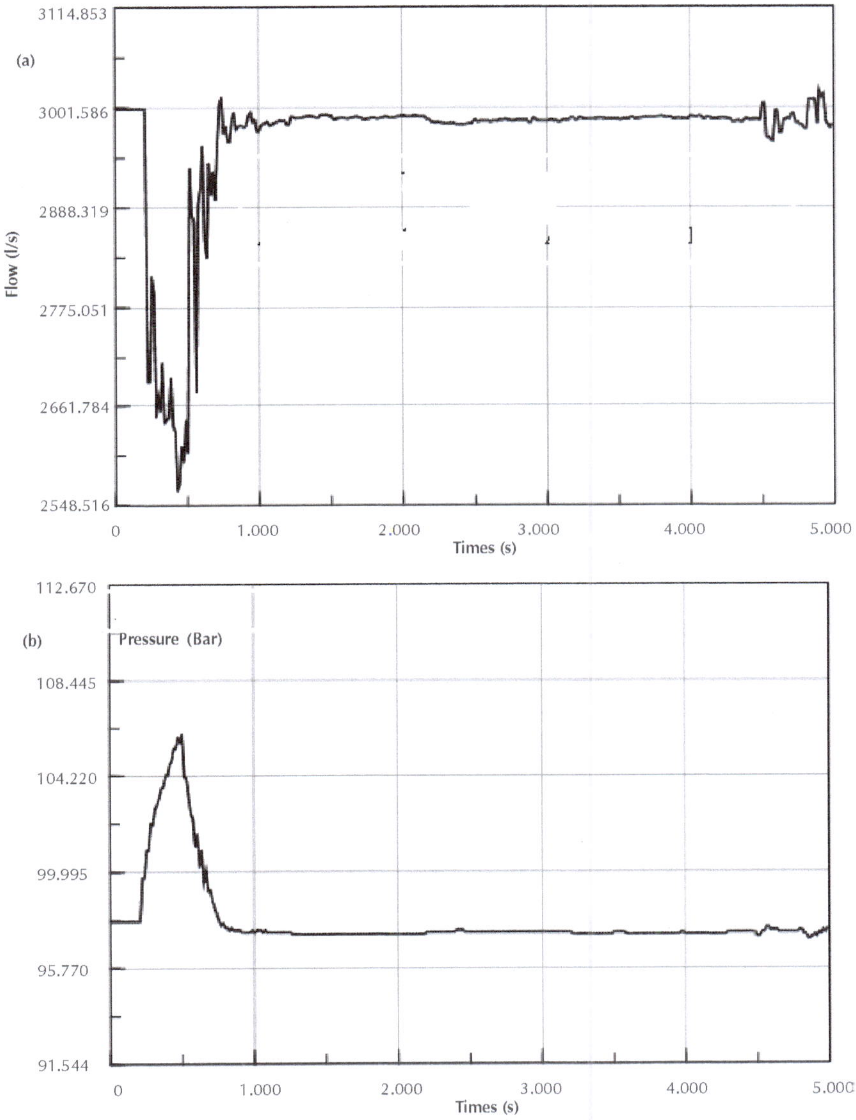

FIGURE 2 Surge tank on pipeline: (a) flow variation, (b) pressure variation

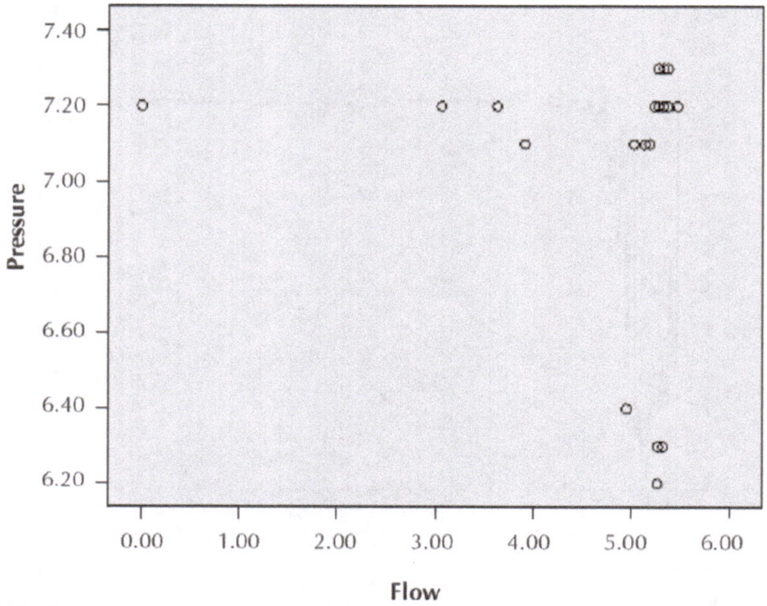

FIGURE 3 Scatter diagram of water pipeline transient flow analysis.

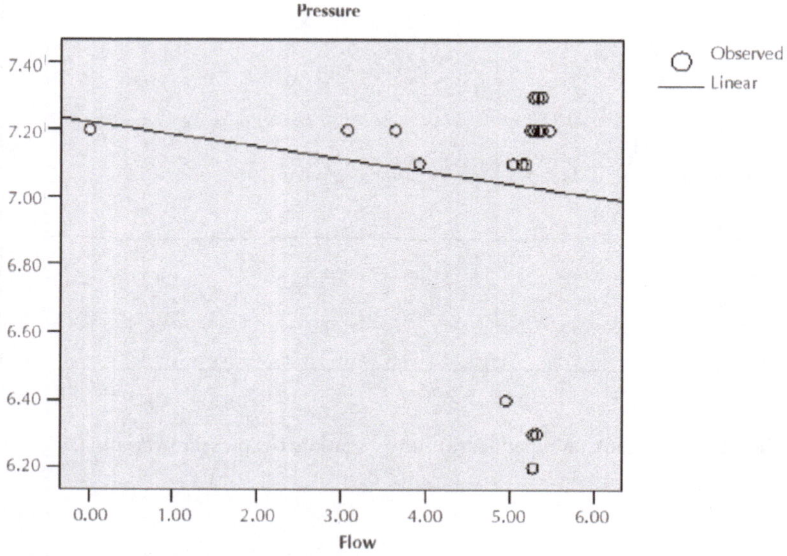

FIGURE 4 Flow and pressure variation.

Weighted Least Squares

MODEL: MOD_3.

C

Source variable... pressure

Log-likelihood Function =	–6.002132	POWER value = –2.000
Log-likelihood Function =	–6.519016	POWER value = –1.500
Log-likelihood Function =	–7.030830	POWER value = –1.000
Log-likelihood Function =	–7.537144	POWER value = –.500
Log-likelihood Function =	–8.037531	POWER value = .000
Log-likelihood Function =	–8.531560	POWER value = .500
Log-likelihood Function =	–9.018804	POWER value = 1.000
Log-likelihood Function =	–9.498839	POWER value = 1.500
Log-likelihood Function =	–9.971241	POWER value = 2.000

The Value of POWER Maximizing Log-likelihood Function = -2.000

C

Source variable...	Pressure	POWER value = -2.000
Dependent variable...	Pressure	

List wise Deletion of Missing Data

Multiple R	.10581
R Square	.01120
Adjusted R Square	-.03375
Standard Error	2.27964

Analysis of Variance:

DF Sum of Squares	Mean Square		
Regression	1	1.29446	1.2944641
Residuals	22	114.32864	5.1967566
F = .24909	Sign if F = .6227		

------------------------------ Variables in the Equation ------------------------------

Variable	B	SE B	Beta	T Sig T
Flow	–.027661	.055424	–.105809	–.499 .6227
(Constant)	7.210458	.274847	26.234	.0000

Log-likelihood Function = -6.002132

Two-stage Least Squares

MODEL: MOD_4.

C

Equation number: 1

Dependent variable... Pressure

List wise Deletion of Missing Data

Multiple R	.12602		
R Square	.01588		
Adjusted R Square	–.02885		
Standard Error	.35262		
Analysis of Variance:			
DF Sum of Squares	Mean Square		
Regression	1	.0441416	.04414158
Residuals	22	2.7354418	.12433826

F = .35501 Sign if F = .5574

-------------------------------- Variables in the Equation ---------------------------

Variable	B	SE B	Beta	T Sig	T
Flow	-.036685	.061570	-.126018	–.596	.5574
(Constant)	7.222932	.305821	23.618	.0000	

Correlation Matrix of Parameter Estimates

Flow

Flow 1.0000000

Scatter Diagram Water Pipeline Transient Flow Analysis and Regression

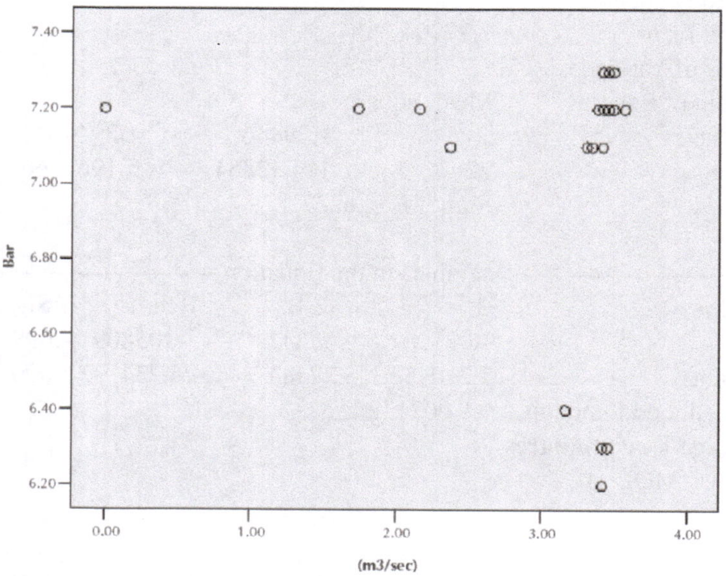

FIGURE 5 Pipeline flow variation.

Regression

TABLE 8 Variables Entered/Removed.[a]

Model	Variables Entered	Variables Removed	Method
1	(m³/sec)[b]	.	Enter

[a]Dependent Variable: (bar).
[b]All requested variables entered.

TABLE 9 Model Summary.

Model	R	R Square	Adjusted R Square	Std. Error of the Estimate
1	.125 (a)	.016	–.029	.35264

[a]Predictors: (Constant), (m3/sec)

TABLE 10 ANOVA[a].

Model		Sum of Squares	df	Mean Square	F	Sig.
1	Regression	.044	1	.044	.352	.559[b]
	Residual	2.736	22	.124		
	Total	2.780	23			

[a]Dependent Variable: (bar).
[b]Predictors: (Constant), (m³/sec).

Curve Fit

TABLE 11 Model Description.

Model Name		MOD_1
Dependent Variable	1	(bar)
Equation	1	Linear
	2	Quadratic
Independent Variable		(m³/sec)
Constant		Included
Variable Whose Values Label Observations in Plots		Unspecified
Tolerance for Entering Terms in Equations		.0001

TABLE 12 Case Processing Summary.

	N
Total Cases	24
Excluded Cases[a]	0
Forecasted Cases	0
Newly Created Cases	0

[a]Cases with a missing value in any variable are excluded from the analysis.

TABLE 13 Variable Processing Summary.

		Variables	
		Dependent (bar)	Independent (m^3/sec)
Number of Positive Values		24	23
Number of Zeros		0	1
Number of Negative Values		0	0
Number of Missing Values	User-Missing	0	0
	System-Missing	0	0

TABLE 14 Model Summary and Parameter Estimates.

Dependent Variable: (bar)								
	Model Summary					Parameter Estimates		
Equation	R Square	F	df1	df2	Sig.	Constant	b1	b2
Linear	.016	.352	1	22	.559	7.213	–.054	
Quadratic	.017	.180	2	21	.837	7.244	–.109	.014

The independent variable is (m^3/sec).

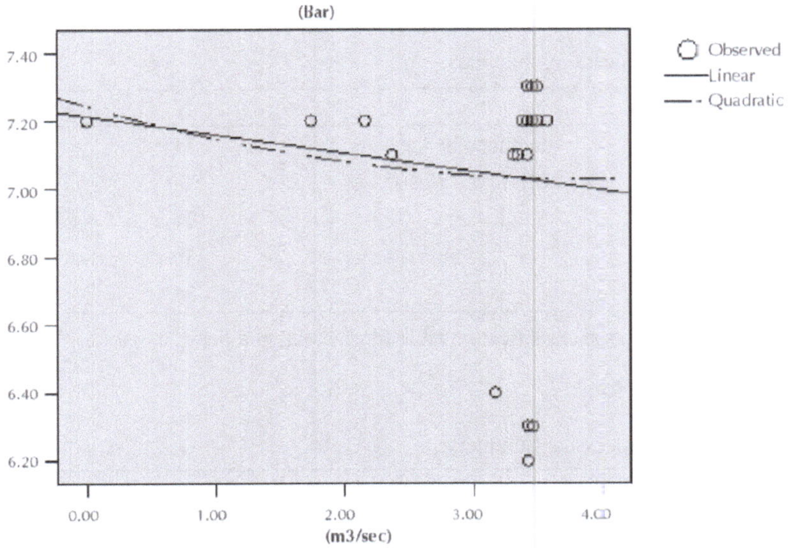

FIGURE 6 Pipeline flow variation and pressure variation.

Curve Fit

TABLE 15 Model Description.

Model Name		MOD_2
Dependent Variable	1	(bar)
Equation	1	Logarithmic
Independent Variable		(m^3/sec)
Constant		Included
Variable Whose Values Label Observations in Plots		Unspecified

TABLE 16 Variable Processing Summary.

		Variables	
		Dependent (bar)	Independent (m^3/sec)
Number of Positive Values		24	23
Number of Zeros		0	1[a]
Number of Negative Values		0	0
Number of Missing Values	User-Missing	0	0
	System-Missing	0	0

[a]The Logarithmic or Power model cannot be calculated.

Logistic Regression

TABLE 17 Case Processing Summary.

Unweighted Cases[a]		N	Percent
Selected Cases	Included in Analysis	24	100.0
	Missing Cases	0	0
	Total	24	100.0
Unselected Cases		0	0
Total		24	100.0

[a]If weight is in effect, see classification table for the total number of cases.

Nominal Regression

TABLE 18 Case Processing Summary.

		N	Marginal Percentage
(bar)	6.20	1	4.2%
	6.30	2	8.3%
	6.40	1	4.2%
	7.10	5	20.8%
	7.20	11	45.8%
	7.30	4	16.7%
Valid		24	100.0%
Missing		0	
Total		24	
Subpopulation		12[a]	

[a]The dependent variable has only one value observed in 9 (75.0%) subpopulations.

TABLE 19 Model Fitting Information.

	Model Fitting Criteria	Likelihood Ratio Tests		
Model	–2 Log Likelihood	Chi-Square	df	Sig.
Intercept Only	50.686			
Final	45.272	5.414	5	.368

TABLE 20 Pseudo R-Square.

Cox and Snell	.202
Nagelkerke	.214
McFadden	.078

TABLE 21 Likelihood Ratio Tests.

	Model Fitting Criteria	Likelihood Ratio Tests		
Effect	**-2 Log Likelihood of Reduced Model**	**Chi-Square**	**df**	**Sig.**
Intercept	53.647	8.375	5	.137
Flow	50.686	5.414	5	.368

The chi-square statistic is the difference in −2 log-likelihoods between the final model and a reduced model. The reduced model is formed by omitting an effect from the final model. The null hypothesis is that all parameters of that effect are 0.

TABLE 22 Case Processing Summary.

		N	Marginal Percentage
(bar)	6.20	1	4.2%
	6.30	2	8.3%
	6.40	1	4.2%
	7.10	5	20.8%
	7.20	11	45.8%
	7.30	4	16.7%
Valid		24	100.0%
Missing		0	
Total		24	

TABLE 23 Model Fitting Information.

Model	–2 Log Likelihood	Chi-Square	df	Sig.
Intercept Only	50.686			
Final	50.666	.020	1	.888

Link function: Logit.

TABLE 24 Goodness-of-Fit.

	Chi-Square	df	Sig.
Pearson	55.344	54	.424
Deviance	38.673	54	.943

Link function: Logit.

TABLE 25 Pseudo R-Square.

Cox and Snell	.001
Nagelkerke	.001
McFadden	.000

Link function: Logit.

TABLE 26 Parameter Estimates.

		Estimate	Std. Error	Wald	df	Sig.	95% Confidence Interval	
							Lower Bound	Upper Bound
Threshold	[pressure = 6.20]	–3.325	1.806	3.391	1	.066	–6.864	.214
	[pressure = 6.30]	–2.134	1.610	1.756	1	.185	–5.290	1.022
	[pressure = 6.40]	–1.796	1.584	1.286	1	.257	–4.901	1.308
	[pressure = 7.10]	–.695	1.541	.203	1	.652	–3.714	2.325
	[pressure = 7.20]	1.426	1.568	.827	1	.363	–1.648	4.499
Location	Flow	–.061	.474	.016	1	.898	–.991	.869

Link function: Logit.

Nonlinear Regression Analysis

TABLE 27 Iteration History.[b]

		Parameter
Iteration Number[b]	**Residual Sum of Squares**	**Pressure Flow**
1.0	.000	1.000

Derivatives are calculated numerically.
[a]Run stopped after 1 model evaluations and 1 derivative evaluations because the relative reduction between successive parameter estimates is at most PCON = 1.00E-008.
[b]Major iteration number is displayed to the left of the decimal, and minor iteration number is to the right of the decimal.

TABLE 28 Parameter Estimates.

			95% Confidence Interval	
Parameter	**Estimate**	**Std. Error**	**Lower Bound**	**Upper Bound**
Pressure Flow	1.000	.000	1.000	1.000

TABLE 29 ANOVA.[a]

Source	Sum of Squares	df	Mean Squares
Regression	1194.230	1	1194.230
Residual	.000	23	.000
Uncorrected Total	1194.230	24	
Corrected Total	2.780	23	

Dependent variable: (bar)
[a]R squared = 1 – (Residual Sum of Squares) / (Corrected Sum of Squares) = 1.000.

Weighted Least Squares

MODEL: MOD_3.
C
Source variable.. pressure
Log-likelihood Function = –6.003869 POWER value = –2.000
Log-likelihood Function = –6.520760 POWER value = –1.500
Log-likelihood Function = –7.032574 POWER value = –1.000
Log-likelihood Function = –7.538882 POWER value = –.500
Log-likelihood Function = –8.039253 POWER value = .000
Log-likelihood Function = –8.533258 POWER value = .500
Log-likelihood Function = –9.020470 POWER value = 1.000
Log-likelihood Function = –9.500462 POWER value = 1.500

Log-likelihood Function = –9.972812 POWER value = 2.000
The Value of POWER Maximizing Log-likelihood Function = -2.000

C
Source variable.. pressure POWER value = -2.000
Dependent variable.. pressure
List wise Deletion of Missing Data
Multiple R .10513
R Square .01105
Adjusted R Square -.03390
Standard Error 2.27980
Analysis of Variance:

DF	Sum of Squares	Mean Square	
Regression	1	1.27792	1.2779177
Residuals	22	114.34519	5.1975087
F = .24587	Signif F = .6249		

------------------------------ Variables in the Equation ------------------------------

Variable	B	SE B	Beta	T Sig T	
flow	–.040499	.081675	–.105131	–.496	.6249
(Constant)	7.203079	.262111	27.481	.0000	

Log-likelihood Function = -6.003869

Two-stage Least Squares

TABLE 30 Coefficients.

	Standardized Coefficients				
	Beta	**Std. Error**	**df**	**F**	**Sig.**
(m^3/sec)	–.188	.214	2	.766	.477

Dependent Variable: (bar)

TABLE 31 Correlations and Tolerance.

	Correlations				Tolerance	
	Zero-Order	**Partial**	**Part**	**Importance**	**After Transformation**	**Before Transformation**
(m^3/sec)	–.188	–.188	–.188	1.000	1.000	1.000

Dependent Variable: (bar)

Two-stage Least Squares

MODEL: MOD_5.
Equation number: 1
Dependent variable.. pressure
List wise Deletion of Missing Data

Multiple R	.12546
R Square	.01574
Adjusted R Square	-.02900
Standard Error	.35264

Analysis of Variance:

DF	Sum of Squares	Mean Square	
Regression	1	.0437490	.04374897
Residuals	22	2.7358344	.12435611
F =	.35180	Signif F = .5591	

------------------------------ Variables in the Equation ------------------------------

Variable	B	SE B	Beta	T Sig T	
flow	-.053802	.090708	-.125457	-.593	.5591
(Constant)	7.213423	.291576	24.739	.0000	

Correlation Matrix of Parameter Estimates

	flow
flow	1.0000000

Scatter Diagram Water Pipeline Transient Flow Analysis and Regression

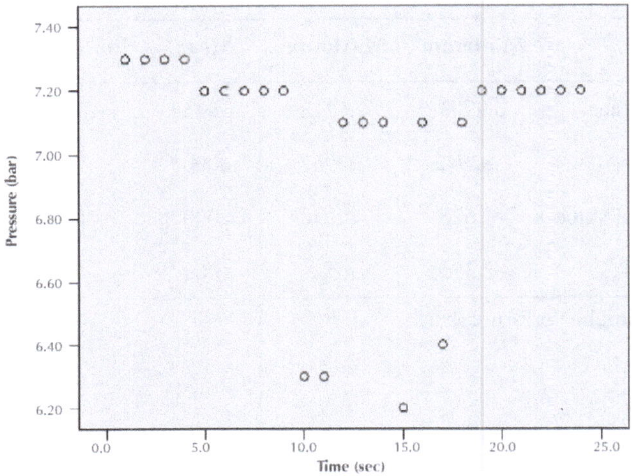

FIGURE 7 Pipeline pressure variation.

Regression

TABLE 32 Variables Entered/Removed.[a]

Model	Variables Entered	Variables Removed	Method
1	Time (sec)[b]	.	Enter

[a]Dependent Variable: pressure (bar).
[b]All requested variables entered.

TABLE 33 Model Summary.[a]

Model	R	R Square	Adjusted R Square	Std. Error of the Estimate
1	.134[b]	.018	–.027	.35227

[a]Dependent Variable: pressure (bar).
[b]Predictors: (Constant), time (sec).

TABLE 34 ANOVA.[a]

Model		Sum of Squares	df	Mean Square	F	Sig.
1	Regression	.050	1	.050	.399	.534[b]
	Residual	2.730	22	.124		
	Total	2.780	23			

[a]Dependent Variable: pressure (bar).
[b]Predictors: (Constant), time (sec).

TABLE 35 Residuals Statistics.[a]

	Minimum	Maximum	Mean	Std. Deviation	N
Predicted Value	6.9703	7.1213	7.0458	.04642	24
Residual	–.82942	.22967	.00000	.34452	24
Std. Predicted Value	–1.626	1.626	.000	1.000	24
Std. Residual	–2.355	.652	.000	.978	24

[a]Dependent Variable: pressure (bar).

Curve Fit

TABLE 36 Model Description.

Model Name		**MOD_4**
Dependent Variable	1	pressure (bar)
Equation	1	Linear
	2	Quadratic
Independent Variable		time (sec)
Constant		Included
Variable Whose Values Label Observations in Plots		Unspecified
Tolerance for Entering Terms in Equations		.0001

TABLE 37 Case Processing Summary.

	N
Total Cases	35
Excluded Cases[a]	11
Forecasted Cases	0
Newly Created Cases	0

[a]Cases with a missing value in any variable are excluded from the analysis.

TABLE 38 Variable Processing Summary.

		Variables	
		Dependent pressure (bar)	**Independent time (sec)**
Number of Positive Values		24	24
Number of Zeros		0	0
Number of Negative Values		0	0
Number of Missing Values	User-Missing	0	0
	System-Missing	11	11

TABLE 39 Model Summary and Parameter Estimates.
Dependent Variable: pressure (bar)

	Model Summary					**Parameter Estimates**		
Equation	**R Square**	**F**	**df1**	**df2**	**Sig.**	**Constant**	**b1**	**b2**
Linear	.018	.399	1	22	.534	7.128	−.007	
Quadratic	.293	4.353	2	21	.026	7.580	−.111	.004

The independent variable is time (sec).

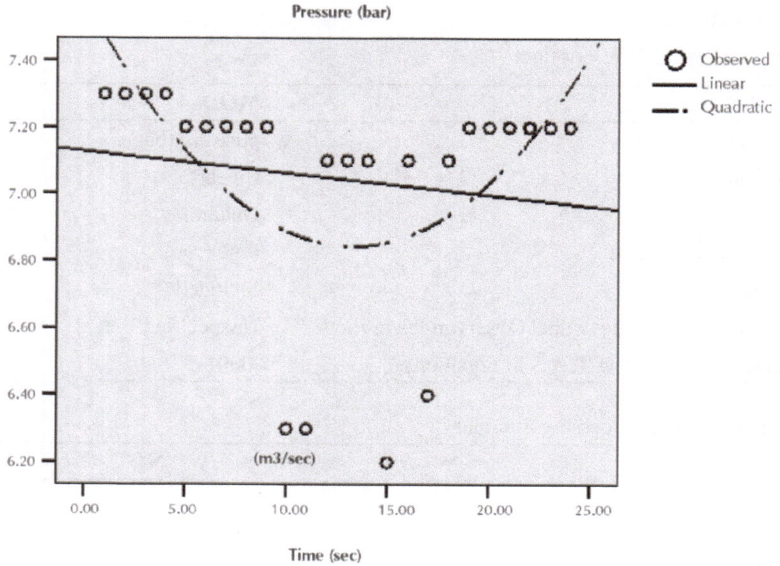

FIGURE 8 Surge tank effect on pressure variation.

TABLE 40 Logistic Regression.

Unweighted Cases[a]		N	Percent
Selected Cases	Included in Analysis	24	68.6
	Missing Cases	11	31.4
	Total	35	100.0
Unselected Cases		0	.0
Total		35	100.0

[a]If weight is in effect, see classification table for the total number of cases.

Nominal Regression

TABLE 41 Model Fitting Information.

	Model Fitting Criteria	Likelihood Ratio Tests		
Model	–2 Log Likelihood	Chi-Square	df	Sig.
Intercept Only	69.836			
Final	46.921	22.915	5	.000

TABLE 42 Pseudo R-Square.

Cox and Snell	.615
Nagelkerke	.651
McFadden	.328

TABLE 43 Likelihood Ratio Tests.

	Model Fitting Criteria	Likelihood Ratio Tests		
Effect	**–2 Log Likelihood of Reduced Model**	**Chi-Square**	**df**	**Sig.**
Intercept	66.999	20.073	5	.001
Time	69.836	22.915	5	.000

The chi-square statistic is the difference in −2 log-likelihoods between the final model and a reduced model. The reduced model is formed by omitting an effect from the final model. The null hypothesis is that all parameters of that effect are 0.

PLUM – Ordinal Regression

TABLE 44 Case Processing Summary.

		N	Marginal Percentage
pressure (bar)	6.20	1	4.2%
	6.30	2	8.3%
	6.40	1	4.2%
	7.10	5	20.8%
	7.20	11	45.8%
	7.30	4	16.7%
time (sec)	1.00	1	4.2%
	2.00	1	4.2%
	3.00	1	4.2%
	4.00	1	4.2%
	5.00	1	4.2%
	6.00	1	4.2%
	7.00	1	4.2%
	8.00	1	4.2%
	9.00	1	4.2%
	10.00	1	4.2%

TABLE 44 *(Continued)*

		N	Marginal Percentage
	11.00	1	4.2%
	12.00	1	4.2%
	13.00	1	4.2%
	14.00	1	4.2%
	15.00	1	4.2%
	16.00	1	4.2%
	17.00	1	4.2%
	18.00	1	4.2%
	19.00	1	4.2%
	20.00	1	4.2%
	21.00	1	4.2%
	22.00	1	4.2%
	23.00	1	4.2%
	24.00	1	4.2%
Valid		24	100.0%
Missing		11	
Total		35	

TABLE 45 Model Fitting Information.

Model	–2 Log Likelihood	Chi-Square	df	Sig.
Intercept Only	69.836			
Final	.000	69.836	23	.000

Link function: Logit.

TABLE 46 Goodness-of-Fit.

	Chi-Square	df	Sig.
Pearson	.290	92	1.000

Link function: Logit.

TABLE 47 Credit.

	Chi-Square	df	Sig.
Deviance	.565	92	1.000

TABLE 48 Pseudo R-Square.

Cox and Snell	.946
Nagelkerke	1.000
McFadden	1.000

Link function: Logit.

TABLE 49 Parameter Estimates.

		Estimate	Std. Error	Wald	df	Sig.	95% Confidence Interval	
							Lower Bound	Upper Bound
Threshold	[pressure = 6.20]	–30.734	21.091	2.124	1	.145	–72.071	10.603
	[pressure = 6.30]	–21.976	19.057	1.330	1	.249	–59.326	15.375
	[pressure = 6.40]	–16.203	18.048	.806	1	.369	–51.576	19.170
	[pressure = 7.10]	–6.142	16.238	.143	1	.705	–37.967	25.683
	[pressure = 7.20]	6.222	16.345	.145	1	.703	–25.813	38.257
Location	[time=1.00]	12.755	30.928	.170	1	.680	–47.864	73.373
	[time=2.00]	12.755	30.928	.170	1	.680	–47.864	73.373
	[time=3.00]	12.755	30.928	.170	1	.680	–47.864	73.373
	[time=4.00]	12.755	30.928	.170	1	.680	–47.864	73.373
	[time=5.00]	1.58E-014	22.059	.000	1	1.000	–43.235	43.235
	[time=6.00]	1.93E-014	22.059	.000	1	1.000	–43.235	43.235
	[time=7.00]	2.01E-014	22.059	.000	1	1.000	–43.235	43.235
	[time=8.00]	1.82E-014	22.059	.000	1	1.000	–43.235	43.235
	[time=9.00]	2.17E-014	22.059	.000	1	1.000	–43.235	43.235
	[time=10.00]	–26.200	20.464	1.639	1	.200	–66.309	13.910
	[time=11.00]	–26.200	20.464	1.639	1	.200	–66.309	13.910
	[time=12.00]	–11.299	18.992	.354	1	.552	–48.523	25.925
	[time=13.00]	–11.299	18.992	.354	1	.552	–48.523	25.925
	[time=14.00]	–11.299	18.992	.354	1	.552	–48.523	25.925
	[time=15.00]	–36.574	28.117	1.692	1	.193	–91.681	18.534
	[time=16.00]	–11.299	18.992	.354	1	.552	–48.523	25.925

TABLE 49 *(Continued)*

	Estimate	Std. Error	Wald	df	Sig.	95% Confidence Interval	
						Lower Bound	Upper Bound
[time=17.00]	–19.103	18.598	1.055	1	.304	–55.554	17.348
[time=18.00]	–11.299	18.992	.354	1	.552	–48.523	25.925
[time=19.00]	1.35E-014	22.059	.000	1	1.000	–43.235	43.235
[time=20.00]	1.69E-014	22.059	.000	1	1.000	–43.235	43.235
[time=21.00]	2.05E-014	22.059	.000	1	1.000	–43.235	43.235
[time=22.00]	1.35E-014	22.059	.000	1	1.000	–43.235	43.235
[time=23.00]	2.33E-014	22.059	.000	1	1.000	–43.235	43.235
[time=24.00]	0[a]	.	.	0	.	.	.

Link function: Logit.
[a]This parameter is set to zero because it is redundant.

Weighted Least Squares

MODEL: MOD_5.
C
Source variable.. pressure
Log-likelihood Function = –5.883800 POWER value = -2.000
Log-likelihood Function = –6.423830 POWER value = -1.500
Log-likelihood Function = –6.959075 POWER value = -1.000
Log-likelihood Function = –7.489160 POWER value = -.500
Log-likelihood Function = –8.013705 POWER value = .000
Log-likelihood Function = –8.532326 POWER value = .500
Log-likelihood Function = –9.044641 POWER value = 1.000
Log-likelihood Function = –9.550262 POWER value = 1.500
Log-likelihood Function = –10.048805 POWER value = 2.000
The Value of POWER Maximizing Log-likelihood Function = –2.000
C
------------------------------ Variables in the Equation ------------------------------
Source variable.. pressure POWER value = -2.000
Dependent variable.. pressure
List wise Deletion of Missing Data
Multiple R .14456
R Square .02090
Adjusted R Square -.02361

Standard Error 2.26843
Analysis of Variance:

DF	Sum of Squares	Mean Square
Regression 1	2.41632	2.4163153
Residuals 22	113.20679	5.1457633
F = .46957	Signif F =	.5003

-------------------------------- Variables in the Equation ----------------------------

Variable	B	SE B	Beta	T	Sig T
time	-.006361	.009282	-.144562	-.685	.5003
(Constant)	7.156235	.132588	53.973	.0000	

Log-likelihood Function = -5.883800

Two-stage Least Squares

MODEL: MOD_6.

C

Equation number: 1

Dependent variable.. pressure

List wise Deletion of Missing Data

Multiple R .13354

R Square .01783

Adjusted R Square -.02681

Standard Error .35227

Analysis of Variance:

DF	Sum of Squares	Mean Square	
Regression	1	.0495674	.04956739
Residuals	22	2.7300159	.12409163
F =	.39944	Signif F =	.5339

TABLE 50 Variables in the Equation.

Variable	B	SE B	Beta	T	Sig T
Time	-.006565	.010388	-.133539	-.632	.5339
(Constant)	7.127899	.148428	48.023	.0000	
Correlation Matrix of Parameter Estimates					
Time	1.0000000				

TABLE 51 Case Processing Summary.

Unweighted Cases[a]		N	Percent
Selected Cases	Included in Analysis	24	68.6
	Missing Cases	11	31.4
	Total	35	100.0
Unselected Cases		0	0
Total		35	100.0

[a]If weight is in effect, see classification table for the total number of cases.

TABLE 52 Model Summary.

Multiple R	R Square	Adjusted R Square
.915	.836	.821

Dependent Variable: pressure (bar).
Predictors: time (sec).

TABLE 53 ANOVA.

	Sum of Squares	df	Mean Square	F	Sig.
Regression	20.072	2	10.036	53.659	.000
Residual	3.928	21	.187		
Total	24.000	23			

Dependent Variable: pressure (bar).
Predictors: time (sec).

TABLE 54 Coefficients.

	Standardized Coefficients				
	Beta	Std. Error	df	F	Sig.
Time (sec)	–.915	.088	2	107.318	.000

Dependent Variable: pressure (bar).

TABLE 55 Correlations and Tolerance.

	Correlations				Tolerance	
	Zero-Order	Partial	Part	Importance	After Transformation	Before Transformation
Time (sec)	–.915	–.915	–.915	1.000	1.000	1.000

Dependent Variable: pressure (bar).

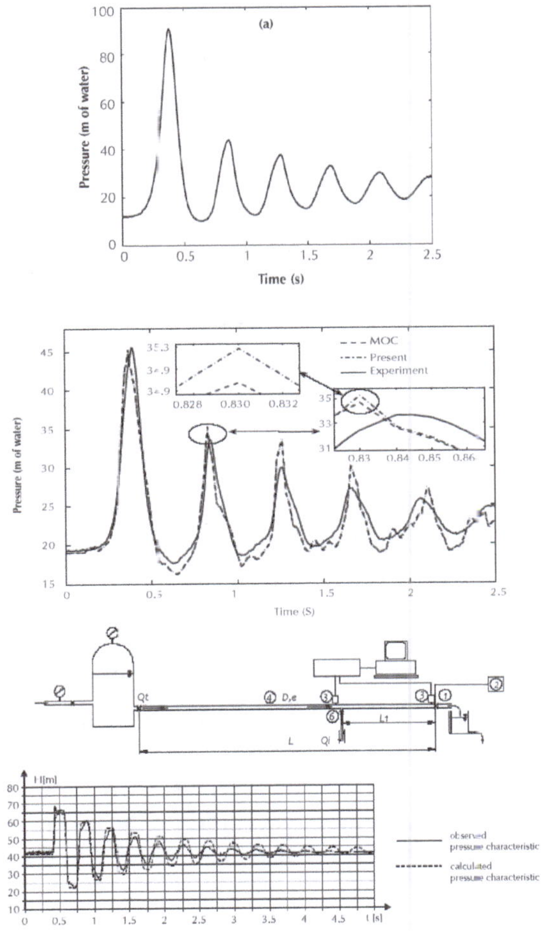

FIGURE 9 (a, b) Experimental absolute pressure trace at downstream end (Arturo S. Leon); (c, d) pipeline with local leak (Apoloniusz Kodura and Katarzyna Weinerowska).

8.4 COMPARISON OF PRESENT WORK WITH OTHER EXPERTS' RESEARCH

Comparison of present work results with other experts' research results shows significant points which are as (Fig. 3):

In this chapter [1, 5], water hammer has been run in pressurized pipeline and the results of experiments and computational numerical analysis were compared for pressure and flow variations at two cases: 1) Pressure and flow variation at surge tank location, 2) pressure and flow variation at pipeline. About sucking air in negative phase, the works of Kodura, Weinerowska, 2005, and Arturo S. Leon, 2007, were studied [79–88]. Detailed conclusions were drawn on the basis of experiments and calculations for the pipeline. Hence the most important effects that have been observed are as flowing: The influence of the ratio of discharge in relation with surge tank effects has been analyzed (Figs. 1 and 2).

The pressure wave speed was a fundamental parameter for hydraulic transient modeling at present work. It was determined how quickly disturbances propagate throughout the system. This affected whether or not different pulses may superpose or cancel each other as they meet at different times and locations. Wave speed was affected by pipe material and bedding, as well as by the presence of fine air bubbles in the fluid. On the other hand, the effects of total transmission flow on the periods of wave oscillations have been investigated (Fig. 3).

The outflow to the overpressure reservoir was effected the value of wave celerity. The water hammer pressure characteristics were measured by extensometers, and were recorded in computer's memory. The supply of the water to the system was realized with use of reservoir, which enabled inlet pressure stabilization [79–99].

8.5 A FLOW OF THE MIXTURE OF AIR AND VAPOR

For heat flow analysis of wet porous materials, the liquid is water and the gas is air. Evaporation or condensation occurs at the interface between the water and air so that the air is mixed with water vapor. A flow of the mixture of air and vapor may be caused by external forces, for instance, by an imposed pressure difference. The vapor will also move relative to the gas by diffusion from regions where the partial pressure of the vapor is higher to those where it is lower.

8.6 COMPUTER MODELS FOR HEAT FLOW IN WET POROUS MATERIALS

Heat flow in porous media is the study of energy movement in the form of heat, which occurs in many types of processes. The transfer of heat in porous media occurs from the high to the low temperature regions. Therefore, a temperature gradient has to exist between the two regions for heat transfer to happen. It can be done by conduction (within one porous solid or between two porous solids in contact), by convection (between two fluids or a fluid and a porous solid in direct contact with the fluid), by radiation (transmission by electromagnetic waves through space) or by combination of the above three methods.

The general equation for heat transfer in porous media is (Eqs. 19–38):

$$\text{(rate of heat in)} + \text{(rate of generation of heat)} =$$

$$\text{(rate of heat out)} + \text{(rate of generation of heat)}, \quad (19)$$

When a wet porous material is subjected to thermal drying two processes occur simultaneously, namely:

a) Transfer of heat to raise the wet porous media temperature and to evaporate the moisture content.
b) Transfer of mass in the form of internal moisture to the surface of the porous material and its subsequent evaporation.

The rate at which drying is accomplished is governed by the rate at which these two processes proceed. Heat is a form of energy that can across the boundary of a system. Heat can, therefore, be defined as "the form of energy that is transferred between a system and its surroundings as a result of a temperature difference." There can only be a transfer of energy across the boundary in the form of heat if there is a temperature difference between the system and its surroundings. Conversely, if the system and surroundings are at the same temperature there is no heat transfer across the boundary.

Strictly speaking, the term *"heat"* is a name given to the particular form of energy crossing the boundary. However, heat is more usually referred to in thermodynamics through the term "heat transfer", which is consistent with the ability of heat to raise or lower the energy within a system.

There are three modes of heat flow in porous media:

– convection

– conduction

– radiation

All three are different. Convection relies on movement of a fluid in porous material. Conduction relies on transfer of energy between molecules within a porous solid or fluid. Radiation is a form of electromagnetic energy transmission and is independent of any substance between the emitter and receiver of such energy. However, all three modes of heat flow rely on a temperature difference for the transfer of energy to take place.

The greater the temperature difference the more rapidly will the heat be transferred. Conversely, the lower the temperature difference, the slower will be the rate at which heat is transferred. When discussing the modes of heat transfer it is the rate of heat transfer Q that defines the characteristics rather than the quantity of heat.

As it was mentioned earlier, there are three modes of heat flow in porous structures, convection, conduction and radiation. Although two, or even all three, modes of heat flow may be combined in any particular thermodynamic situation, the three are quite different and will be introduced separately.

The coupled heat and liquid moisture transport of porous material has wide industrial applications. Heat transfer mechanisms in porous textiles include conduction by the solid material of fibers, conduction by intervening air, radiation, and convection. Meanwhile, liquid and moisture transfer mechanisms include vapor diffusion in the void space and moisture sorption by the fiber, evaporation, and capillary effects. Water vapor moves through porous textiles as a result of water vapor concentration differences. Fibers absorb water vapor due to their internal chemical compositions and structures. The flow of liquid moisture through the textiles is caused by fiber-liquid molecular attraction at the surface of fiber materials, which is determined mainly by surface tension and effective capillary pore distribution and pathways. Evaporation and/or condensation take place, depending on the temperature and moisture distributions. The heat transfer process is coupled with the moisture transfer processes with phase changes such as moisture sorption/desorption and evaporation/condensation.

All three of the mechanisms by which heat is transferred- conduction, radiation and convection, may enter into drying. The relative importance of the mechanisms varies from one drying process to another and very often one mode of heat transfer predominates to such extent that it governs the overall process.

As an example, in air-drying the rate of heat transfer is given by:

$$q = h_s A(T_a - T_s), \tag{20}$$

where q is the heat transfer rate in Js^{-1}, h_s is the surface heat-transfer coefficient in $Jm^{-2} s^{-1} °C^{-1}$, A is the area through which heat flow is taking place, m^{-2}, T_a is the air temperature and T_s is the temperature of the surface which is drying, °C.

To take another example, in a cylindrical dryer where moist material is spread over the surface of a heated cylinder, heat transfer occurs by conduction from the cylinder to the porous media, so that the equation is

$$q = UA(T_i - T_s), \tag{21}$$

where U is the overall heat-transfer coefficient, T_i is the cylinder temperature (usually very close to that of the steam), T_s is the surface temperature of textile and A is the area of the drying surface on the cylinder. The value of U can be estimated from the conductivity of the cylinder material and of the layer of porous solid.

Mass transfer in the drying of a wet porous material will depend on two mechanisms: movement of moisture within the porous material which will be a function of the internal physical nature of the solid and its moisture content; and the movement of water vapor from the material surface as a result of water vapor from the material surface as a result of external conditions of temperature, air humidity and flow, area of exposed surface and supernatant pressure.

Some porous materials such as textiles exposed to a hot air stream may be cooled evaporatively by bleeding water through its surface. Water vapor may condense out of damp air onto cool surfaces. Heat will flow through an air-water mixture in these situations, but water vapor will diffuse or convect through air as well. This sort of transport of one substance relative to another called mass transfer. The moisture content, X, is described as the ratio of the amount of water in the materials, m_{H2O} to the dry weight of material, $m_{material}$:

$$X = \frac{m_{H2O}}{m_{material}}, \tag{22}$$

There are large differences in quality between different porous materials depending on structure and type of material. A porous material such as textiles can be hydrophilic or hydrophobic. The hydrophilic fibers can absorb water, while hydrophobic fibers do not. A textile that transports water through its porous structures without absorbing moisture is preferable to use as a first layer. Mass transfer during drying depends on the transport within the fiber and from the textile surface, as well as on how the textile absorbs water, all of which will affect the drying process.

As the critical moisture content or the falling drying rate period is reached, the drying rate is less affected by external factors such as air velocity. Instead, the internal factors due to moisture transport in the material will have a larger impact. Moisture is transported in porous media during drying through:

- capillary flow of unbound water
- movement of bound water and
- vapor transfer

Unbound water in a porous media will be transported primarily by capillary flow. As water is transported out of the porous material, air will be replacing the water in the pores. This will leave isolated areas of moisture where the capillary flow continues.

Moisture in a porous structure can be transferred in liquid and gaseous phases. Several modes of moisture transport can be distinguished (Bejan et al., 2004):

- transport by liquid diffusion;
- transport by vapor diffusion;
- transport by effusion (Knudsen-type diffusion);
- transport by thermo-diffusion;
- transport by capillary forces;
- transport by osmotic pressure; and
- transport due to pressure gradient.

A very common method of removing water from porous structures is convective drying. Concevtion is a mode of heat transfer that takes place as a result of motion within a fluid. If the fluid, starts at a constant temperature and the surface is suddenly increased in temperature to above that of the fluid, there will be convective heat transfer from the surface to the fluid as a result of the temperature difference. Under these conditions the temperature difference causing the heat transfer can be defined as: ΔT = surface temperature mean fluid temperature.

Using this definition of the temperature difference, the rate of heat transfer due to convection can be evaluated using Newton's law of cooling:

$$Q = h_c A \Delta T, \tag{23}$$

where A is the heat transfer surface area and h_c is the coefficient of heat transfer from the surface to the fluid, referred to as the "convective heat transfer coefficient."

The units of the convective heat transfer coefficient can be determined from the units of other variables:

$$\begin{aligned} Q &= h_c A \Delta T \\ W &= (h_c) m^2 K \end{aligned}, \tag{24}$$

so the units of h_c are W/m^2K.

The relationships given in equations (0.4 and 0.5) are also true for the situation where a surface is being heated due to the fluid having higher temperature than the surface. However, in this case the direction of heat transfer is from the fluid to the surface and the temperature difference will now be T = mean fluid temperature surface temperature.

The relative temperatures of the surface and fluid determine the direction of heat transfer and the rate at which heat transfer take place.

As given in previous equations, the rate of heat transfer is not only determined by the temperature difference but also by the convective heat transfer coefficient h_c. This is not a constant but varies quite widely depending on the properties of the fluid and the behavior of the flow. The value of h_c must depend on the thermal capacity of the fluid particle considered, that is, mC_p for the particle. So the higher the density and C_p of the fluid the better the convective heat transfer.

Two common heat transfer fluids are air and water, due to their widespread availability. Water is approximately 800 times denser than air and also has a higher value of C_p. If the argument given above is valid then water has a higher thermal capacity than air and should have a better convective heat transfer performance. This is borne out in practice because typical values of convective heat transfer coefficients are as follows:

TABLE 56 Heat transfer coefficients

Fluid	$h_c\left(W/m^2K\right)$
Water	500–10,000
Air	5–100

The variation in the values reflects the variation in the behavior of the flow, particularly the flow velocity, with the higher values of h_c resulting from higher flow velocities over the surface.

When a fluid is in forced or natural convective motion along a surface, the rate of heat transfer between the solid and the fluid is expressed by the following equation:

$$q = h.A\left(T_W - T_f\right), \tag{25}$$

The coefficient h is dependent on the system geometry, the fluid properties and velocity and the temperature gradient. Most of the resistance to heat transfer happens in the stationary layer of fluid present at the surface of the solid, therefore the coefficient h is often called film coefficient.

Correlations for predicting film coefficient h are semi empirical and use dimensionless numbers, which describe the physical properties of the fluid, the type of flow, the temperature difference and the geometry of the system.

The Reynolds Number characterizes the flow properties (laminar or turbulent). L is the characteristic length: length for a plate, diameter for cylinder or sphere.

$$N_{\mathrm{Re}} = \frac{\rho L v}{\mu}, \tag{26}$$

The Prandtl Number characterizes the physical properties of the fluid for the viscous layer near the wall.

$$N_{\mathrm{Pr}} = \frac{\mu c_p}{k}$$

The Nusselt Number relates the heat transfer coefficient h to the thermal conductivity k of the fluid.

$$N_{Nu} = \frac{hL}{k}, \tag{27}$$

The Grashof Number characterizes the physical properties of the fluid for natural convection.

$$N_{Gr} = \frac{L^3 \Delta\rho g}{\rho\gamma^2} = \frac{L^3 \rho^2 g \beta \Delta T}{\mu^2}, \tag{28}$$

In capillary porous materials, moisture migrates through the body as a result of capillary forces and gradients of moisture content, temperature and pressure. This movement contributes to other heat transfer mechanisms while eventual phase change occurring within the material act as heat sources or sinks.

Drying is fundamentally a problem of simultaneous heat and mass transfer under transient conditions resulting in a system of coupled nonlinear partial differential equations.

Scientists defined a coupled system of partial differential equations for heat and mass transfer in porous bodies. Although they used different approaches to obtain equations, their formulations don't differ substantially from each other. Many numerical works have been executed in this field, on basis of these two theories.

Researchers have developed a one-dimensional model for simultaneous heat and moisture transfer in porous materials. Also many scientists have used one-dimensional in studying heat and mass transfer during convective drying of porous media.

Researchers studied drying problem of timber, with a two-dimensional model. Many scientists used finite element method for solution of two-dimensional heat and mass transfer in porous media.

All of above listed studies, have estimated heat and mass transfer between porous materials and drying fluid, by coefficients obtained from standard correlations based on boundary layer equations, and more of them assumed analogy between heat and mass transfer coefficients. However, since the actual process of drying is a conjugate problem, the heat and mass transfer, to and from the porous solid have to be studied along with the flow field.

Scientists in a conjugate study of paper drying have shown that results of solution by conjugate view differ considerably from those of decoupled system. Also the analogy between heat and mass transfer coefficient may not exist in reality, even in drying of one-dimensional objects.

Researchers found that the mentioned analogy holds good only for initial period of unsaturated sand drying, and for the later part of drying, the heat and mass transfer coefficients at the interface may doesn't satisfy the analogy, due to the non-uniformity of moisture and temperature distribution at the interface resulting from conjugate nature of transfers.

Scientists studied drying of wood as a conjugate problem. They have used boundary layer equations for flow field, and presented temperature and moisture contours during the process.

Researchers applied two-dimensional model for brick drying. They used Navier-Stokes equations for flow field including buoyancy terms in their conjugate analysis. They concluded that restricting heat and mass transfer to top surface of two-dimensional porous body will cause considerable errors into solution. They also have shown that neglecting buoyant forces in flow analysis, leads to considerable differences in heat and mass transfer values and lower drying rate, in Reynolds number of 200.

In the majority of the previous conjugate studies, the buoyancy effects have been simply ignored in flow field analysis except the study performed by some researchers. The solution method used is finite volume approach and is related to an unsteady problem. In this chapter we have tried to use a much conservative method for calculation of energy and momentum fluxes. Regarding the weak capability of the finite element methods (specially in their flux averaging steps), the motivation of the current work concerns on solving the same two dimensional conjugate problem with a finite volume approach which fundamentally guaranties the energy and the mass fluxes to be conserved during the solution and the discretization procedures. However, using the same mesh and consequently the same cells, entire the solution domain highlights the ability of finite volume approach on using unstructured grids. On the other hand, the solution is extended to a higher Reynolds numbers to give a wider range study of the buoyancy effects not only on temperature and the moisture content, but also on flow patterns during the drying process.

8.7 MODELING

The problem model, including corresponding boundary conditions is shown in (Figs. 10–25). The problem considers a sample of rectangular brick exposed to convective airflow. The brick is assumed to be saturated with water initially. The governing moisture removal from brick to air exceeds in cause of concentration gradient between air in vicinity of body and free stream air. Flow is incompressible and thermophysical properties are taken to be constant. The initial moisture content of the brick is 0.13 kg/kg of dry solid, and the solid temperature is set to 293K and the drying air has a 50% of relative humidity. At the Reynolds 200 the air velocity is 0.02 m/s. The thermal conductivity of the brick is 1.8 W/mK, and the thermal capacity is set to 1200 J/kgK. The brick density is set to 1800 kg/m^3.

Non-isothermal diffusion coefficient of porous body in vapor phase, $D_{tv} = 1(10)^{-12}$

Non-isothermal diffusion coefficient of porous body in liquid phase, $D_{tl} = 1(10)^{-12}$

Iso-thermal diffusion Coefficient of porous body in vapor phase, $D_{mv} = 1(10)^{-12}$

Iso-thermal diffusion Coefficient of porous body in liquid phase, $D_{ml} = 1(10)^{-8}$

Enthalpy of Evaporation (Initial value), $H_{fg} = 2454KJ / KgK$

Mass diffusion coefficient of vapor in air, $Diff = 0.256(10)^{-4}$

Fluid thermal conductivity, $K_f = 0.02568$

Porous body thermal conductivity, $K_s = 1.8$

Buoyancy coefficient of temperature, $\beta = 3.4129(10)^{-3}\ 1/K$

Buoyancy coefficient of Concentration, $\beta' = 0.0173$

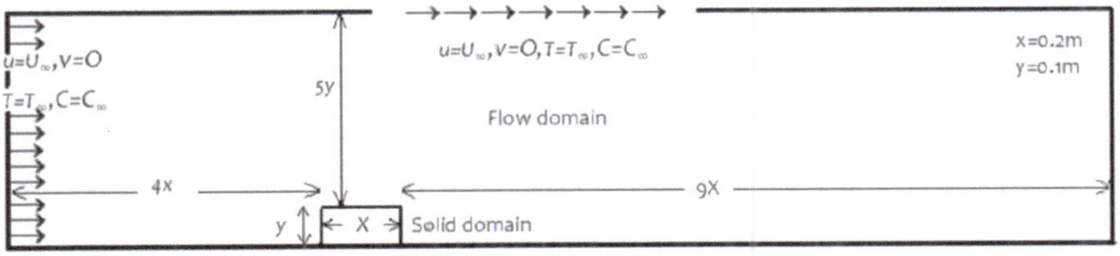

FIGURE 10 Geometry of computational field.

8.7.1 GOVERNING EQUATIONS FOR SOLID (BRICK)

The equations for porous solid phase as obtained by researchers on the basis of continuum approach, were applied for numerical solution

- Energy equation

$$c^* \frac{\partial T}{\partial t} = \left(\frac{k}{\rho_0} + h_{fg} D_{tv} \right) \left(\frac{\partial^2 T}{\partial x^2} + \frac{\partial^2 T}{\partial y^2} \right) + h_{fg} D_{mv} \left(\frac{\partial^2 M}{\partial x^2} + \frac{\partial^2 M}{\partial y^2} \right). \qquad (29)$$

where

$c^* = c_0 + m_l c_l + m_v c_v$

- Moisture conservation equation

$$\frac{\partial M}{\partial t}=\left(D_{tl}+D_{tv}\right)\left(\frac{\partial^2 T}{\partial x^2}+\frac{\partial^2 T}{\partial y^2}\right)+\left(D_{ml}+D_{mv}\right)\left(\frac{\partial^2 M}{\partial x^2}+\frac{\partial^2 M}{\partial y^2}\right), \tag{30}$$

8.7.1.1 GOVERNING EQUATIONS FOR FLOW FIELD

- Continuity

$$\frac{\partial u}{\partial x}+\frac{\partial v}{\partial y}=0, \tag{31}$$

- Momentum equation (2D Navier-Stokes)

$$\frac{\partial u}{\partial t}+u\frac{\partial u}{\partial x}+v\frac{\partial u}{\partial y}=-\frac{1}{\rho}\frac{\partial P}{\partial x}+\upsilon\left(\frac{\partial^2 u}{\partial x^2}+\frac{\partial^2 u}{\partial y^2}\right), \tag{32}$$

$$\frac{\partial v}{\partial t}+u\frac{\partial v}{\partial x}+v\frac{\partial v}{\partial y}=-\frac{1}{\rho}\frac{\partial P}{\partial y}+\upsilon\left(\frac{\partial^2 v}{\partial x^2}+\frac{\partial^2 v}{\partial y^2}\right)+g\beta\left(T-T_\infty\right)+g\beta'\left(C-C_\infty\right), \tag{33}$$

- Energy equation

$$\frac{\partial T}{\partial t}+u\frac{\partial T}{\partial x}+v\frac{\partial T}{\partial y}=\alpha\left(\frac{\partial^2 T}{\partial x^2}+\frac{\partial^2 T}{\partial y^2}\right), \tag{34}$$

- Vapor concentration equation

$$\frac{\partial C}{\partial t}+u\frac{\partial C}{\partial x}+v\frac{\partial C}{\partial y}=D\left(\frac{\partial^2 C}{\partial x^2}+\frac{\partial^2 C}{\partial y^2}\right), \tag{35}$$

8.7.1.2 BOUNDARY AND INITIAL CONDITIONS

Initially the porous material is assumed to be at uniform moisture content (saturation value), and temperature (equal to air temperature).

$$T\left(x,y,0\right)=T_0;\ M\left(x,y,0\right)=M_0, \tag{36}$$

The boundary condition at interface of solid and fluid are:

- No slip condition

$$u = 0 \; ; v = 0$$

- Continuity of temperature

$$T_f = T_s$$

- Continuity of concentration

$$C = C(T, M)_s$$

- Heat balance

$$\left(k + \rho_0 h_{fg} D_{tv}\right)\frac{\partial T}{\partial n} + \rho_0 h_{fg} D_{mv}\frac{\partial M}{\partial n} = k_f \frac{\partial T_f}{\partial n} + h_{fg} D \frac{\partial C}{\partial n}. \tag{37}$$

- Species flux balance

$$\rho_0\left(D_{tv}\frac{\partial T}{\partial n} + D_{mv}\frac{\partial M}{\partial n}\right) = D\frac{\partial C}{\partial n}, \tag{38}$$

8.7.1.3 BOUNDARY CONDITIONS FOR FLOW FIELD

- Inlet boundary condition

  $u = U_\infty,\; v = 0, T = T_\infty, C = C_\infty$

- Far stream boundary condition (upper boundary)

  $u = U_\infty,\; T = T_\infty, C = C_\infty$

- Outflow boundary condition

$$\frac{\partial u}{\partial x} = 0$$

  Also bottom surface of solid is taken adiabatic.

8.8 GRID DEPENDENCY

A structured mesh was used for computational work. Mesh clustering around body is illustrated in Fig. 12.

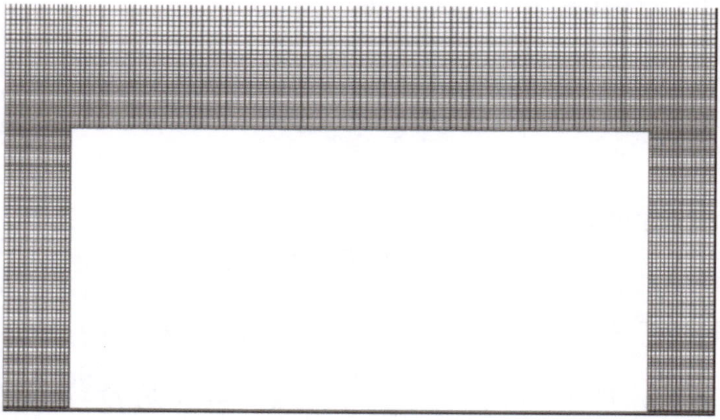

FIGURE 11 Mesh structure over the porous material.

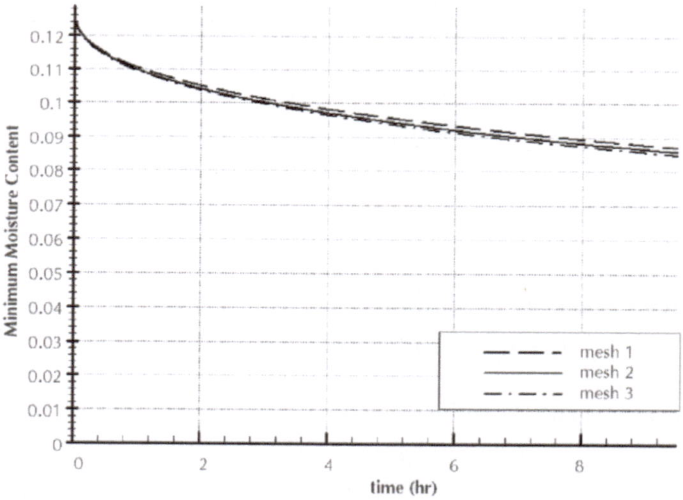

FIGURE 12 Value of moisture content in leading edge obtained with three different meshes.

To clarify effect of mesh refinement on numerical solution, three meshes with different precision were used in numerical analysis:

Mesh 1, with 69,120 total nodes.

Mesh 2 with 108,000 total nodes.

Mesh 3 with 155,520 total nodes.

The value of moisture content in leading edge (minimum value of moisture content) obtained by numerical solution with mesh 1 and mesh 2 in 9 h have maximum difference of 1.28 %, while for mesh 2 and mesh 3 the maximum difference is 0.66 %. So mesh 2 seems to be optimum in accuracy and run-time, and therefore was decided to continue the computational work.

8.9 NUMERICAL SOLUTION

In each time-step, following items should be carried out:

1. Solving two-dimensional flow field equations (continuity + NS) by SIMPLE algorithm, with finite volume scheme.
2. Solution of energy equation for fluid by ADI technique with finite difference scheme. Neumann boundary condition was used for interface, which obtained from last time-step derivative value of solid temperature.
3. Then energy equation for porous field is solved by ADI technique with finite difference scheme. Boundary condition at interface is the known value of fluid temperature.
4. Determining concentration distribution for the flow field. The solution is similar with step 2, but boundary condition is obtained by explicit moisture content derivative at interface.
5. Calculation of moisture content distribution for porous body. The solution is similar with step 3, but boundary value of moisture is obtained by known fluid concentration (from previous step) value at interface.

Because of using explicit values in the solution procedure, we have to do internal repetition in each time-step, until internal conversion is reached for all four variables. After that, solution for next time step starts.

8.10 VALIDATION AND GENERAL RESULTS

Drying curve of modeled brick in Re=200 is verified here. As illustrated, this comparison reveals that samples possessed almost the nearly same trend of moisture content reduction in the overall drying time of 15 h. Nevertheless, the accuracy of above mentioned curves are within 0.6%.

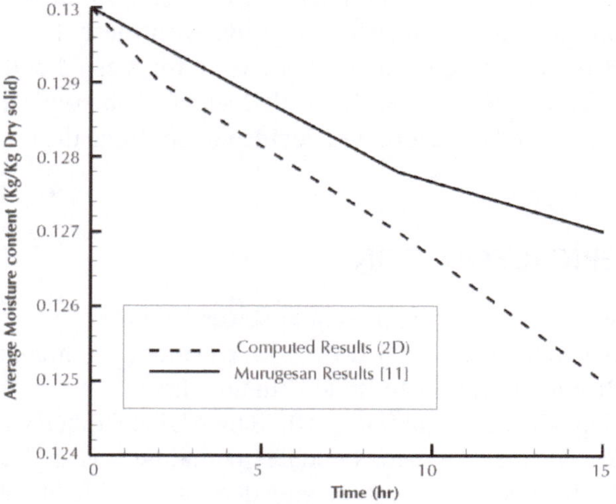

FIGURE 13 Drying curve.

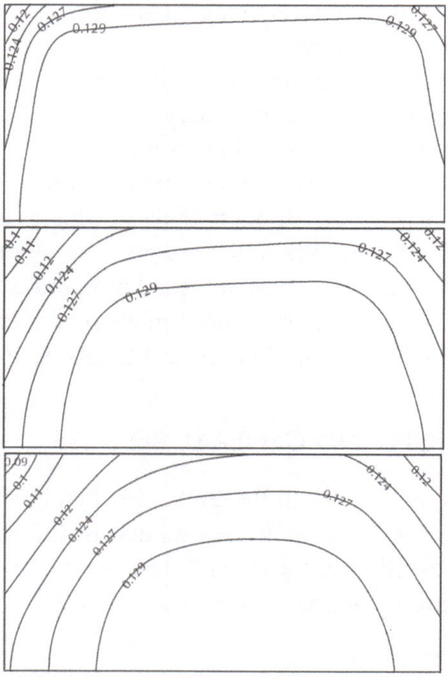

FIGURE 14 Moisture content distributions (kg moisture/kg dry) during drying. (a) t = 2 1/2 hr (b) t = 9 hr (c) t = 15 1/2 hr.

The computational results of moisture content profile for Re=200 illustrates that drying rate in region near leading edge (which corresponds to maximum concentration gradient in adjacent air) is more than other regions in porous body. Gradually, drying spreads from that region to centric regions of body.

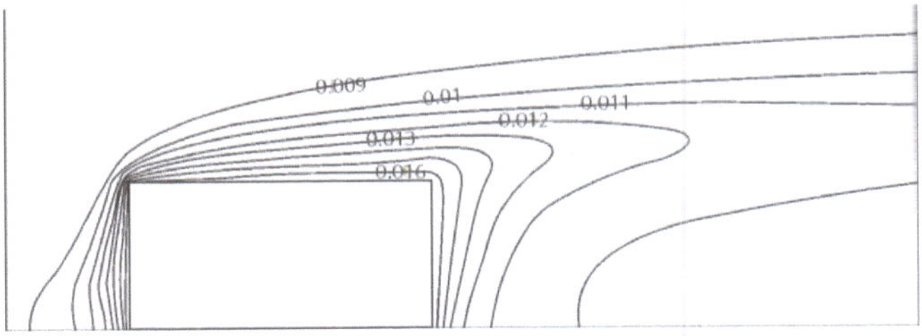

FIGURE 15 Concentration contours in air around porous body.

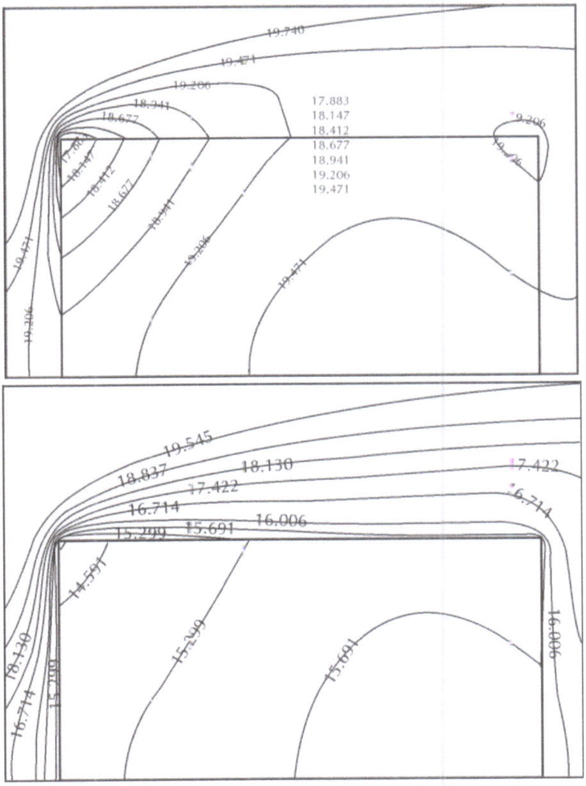

FIGURE 16 Temperature distributions in porous body and around. (a) t = $2^1/_2$ hr (b) t = $15^1/_2$ hr.

Concentration values in air around porous body concludes that gradient of concentration in the body surface has shown strong effect on moisture content distribution in body (as seen for leading edge).

Temperature distributes in porous body and around air in the course of drying. It's evident that, temperature value of porous body near leading edge decreases quickly as a result of higher moisture vaporization from surface there, and this temperature drop transfer to centric regions of porous body.

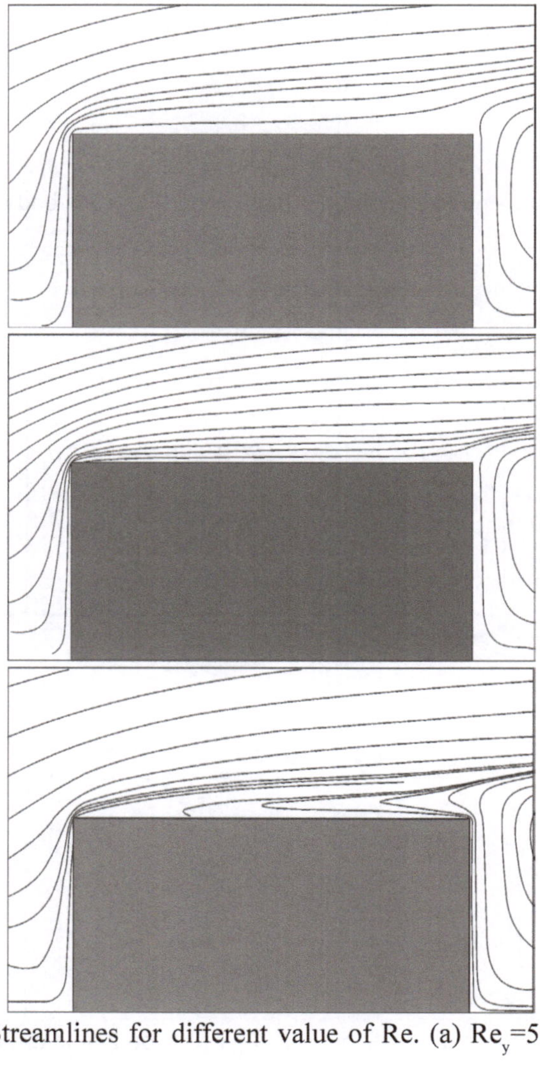

FIGURE 17 Streamlines for different value of Re. (a) Re_y=50 (b) Re_y=100 (c) Re_y=200.

8.11 DIFFERENT AIR VELOCITIES

Numerical solution was executed for different velocities of drying fluid (different Reynolds number) to clarify the effect of this parameter. It shows streamlines around body for Reynolds number of 50, 100 and 200. As shown in Fig. 18, velocity increment in Re=50 to Re=100, results in more compactness of streamlines above body, and then in Re=200 separation occurs, while a vortex forms on upper surface.

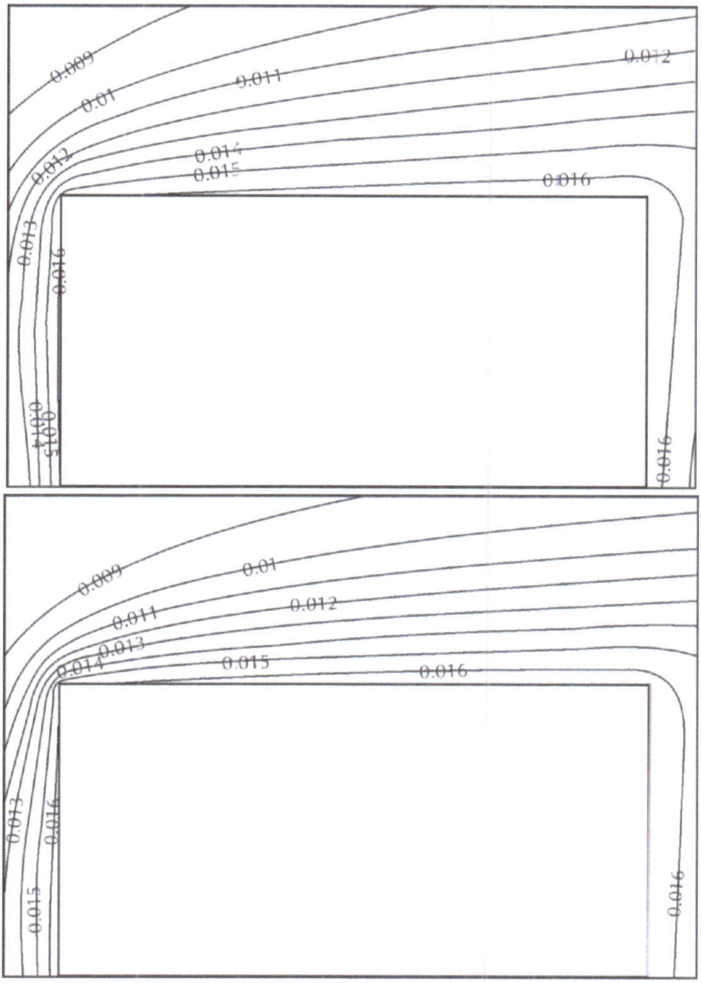

FIGURE 18 *(Continued)*

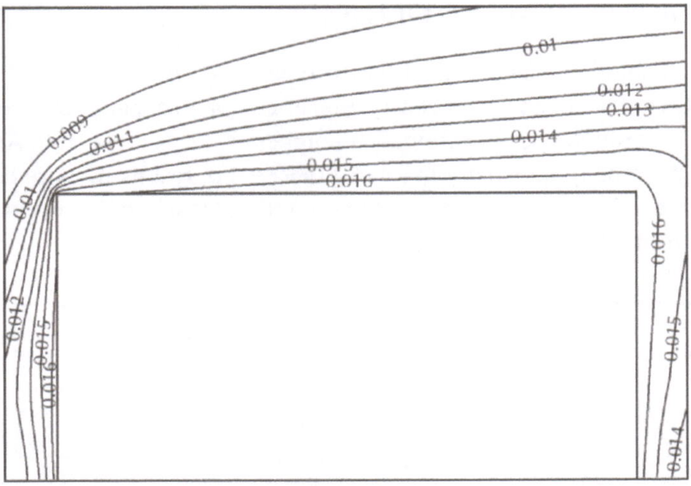

FIGURE 18 Concentration contours for different value of Re. (a) Re_y=50 (b) Re_y=100 (c) Re_y=200.

In contours of concentration around body for denoted numbers of Reynolds are illustrated. It is obvious that Reynolds increment results a significant decrement in thickness of concentration boundary layer on the upper side (especially for leading edge) due to changes in streamlines and velocity boundary layer.

For Re=100, moisture contents are less than those obtained for Re=50, especially for regions nearby leading edge. This is mainly due to thinner concentration boundary layer aforementioned. In transition to Re=200 from Re=100, this matter satisfies just for leading edge and left side of body (as a result of vortex formation above body in Re=200).

In curves are shown for various Reynolds number ranging from 50 to 1000. Effect of drying fluid velocity on process speed could be clearly analyzed. For example, removed moisture after 5 h of drying process for Re=100 is 15.4% more than corresponding value of Re=50. This difference is 17.5% for two Reynolds numbers of 500 and 200.

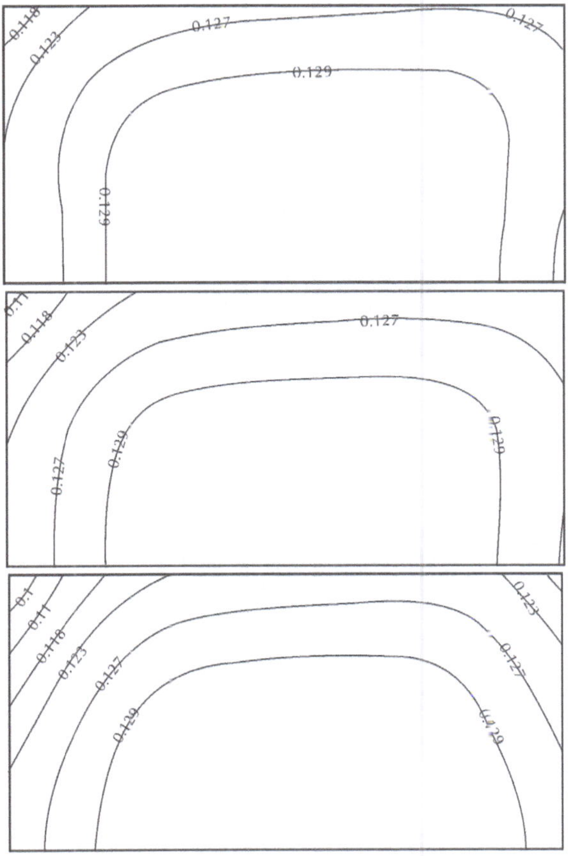

FIGURE 19 Moisture content contours (kg moisture /kg dry) for different value of Re. (a) Re_y=50 (b) Re_y=100 (c) Re_y=200.

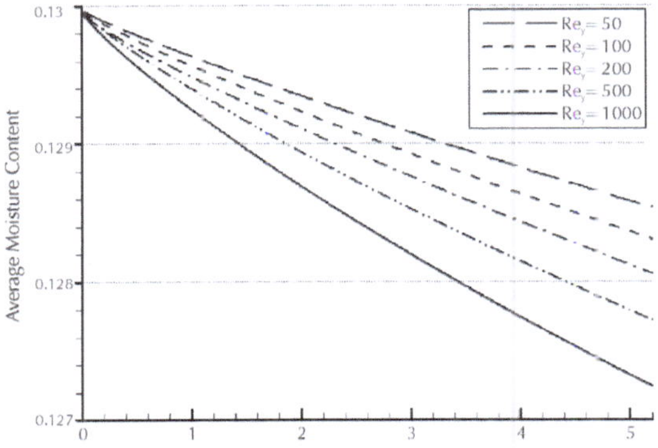

FIGURE 20 Drying curves for Re=50 to 100.

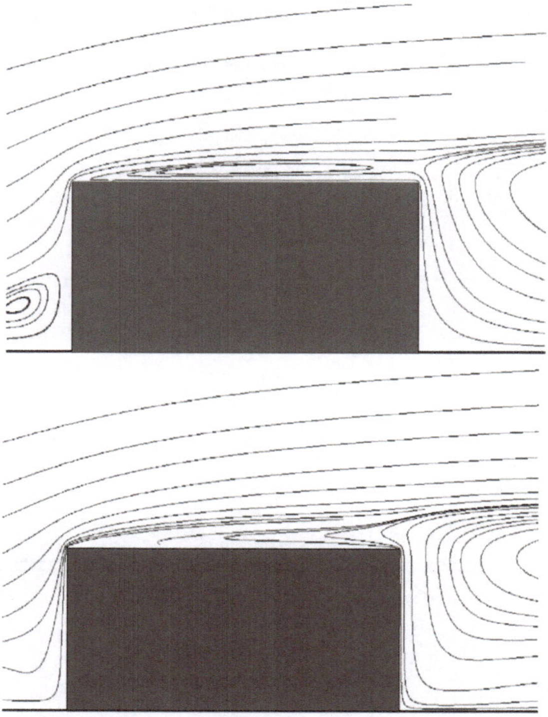

FIGURE 21 Effect of buoyancy forces on streamlines in Re=200. (a) forced convection (b) mixed convection.

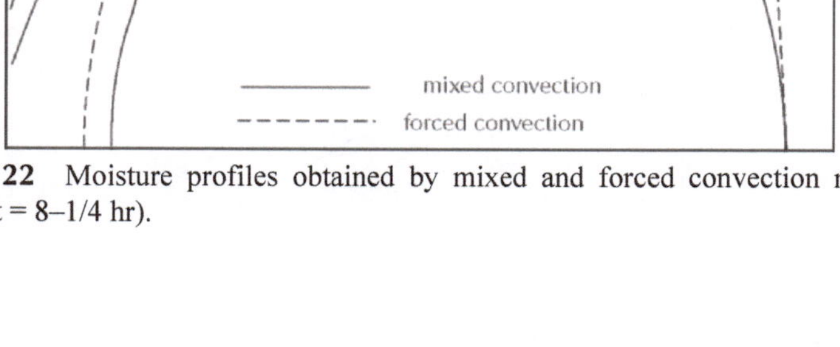

FIGURE 22 Moisture profiles obtained by mixed and forced convection models in Re=200 (t = 8–1/4 hr).

8.12 THE EFFECT OF BUOYANCY

To study the contribution of buoyancy on flow patterns and consequently on drying process, the computations were performed with and without buoyancy terms in flow equations (i.e., mixed and forced convection, respectively). It shows streamlines around porous body in Re=200 for both of forced and mixed convection assumptions. As shown, buoyancy forces clustered the streamlines near vertical walls. The moisture content (MC) distributions were shown for 8 1/4 duration for both cases (Re=200). Obviously, the mixed convection results show a higher drying performance.

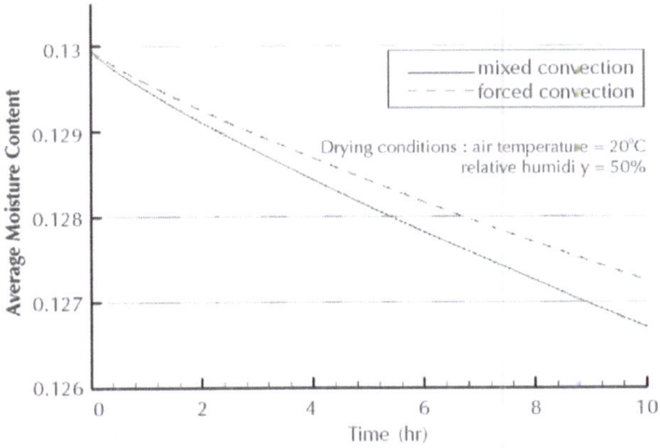

FIGURE 23 Drying curve for mixed and forced convection models in Re=200.

The comparison between drying curves of mixed and forced assumptions. The decrease of (MC) has a lower rate in forced convection case for a 10 h period, (e.g., 19.7% after 10 h of drying). Consequently, the forced convection model underestimates drying rate noticeably in Re=200.

As stated above, the buoyancy forces has a great contribution in drying prediction in Re=200. To investigate effect intensity in various drying fluid velocities (different Reynolds numbers), drying curves resulted by mixed and forced convection models in Reynolds numbers 50 and 1000 (a practical range in drying) are illustrated. It shows average moisture fluxes (during initial 5 h of process) obtained by each of two models, and so percentage increase in average moisture flux by taking buoyancy into account, are listed for different Reynolds numbers. These figures and table implies that despite relative decreasing in drying rate with increasing Reynolds, the effect of buoyancy on drying process in whole of Reynolds range of 50 to 1,000, are considerable.

So, in the denoted range of Re, which include most practical velocities in porous bodies drying (especially clay products drying), ignoring buoyancy effects in flow analysis, will impose noticeable error into computations. In other words, forced convection model hasn't enough accuracy for drying process analysis in governing range.

TABLE 57 Effect of buoyancy on average moisture flux for different Reynolds numbers.

Re	Percentage increase in average moisture flux
50	26%
100	20%
200	19%
500	16%
1000	15%

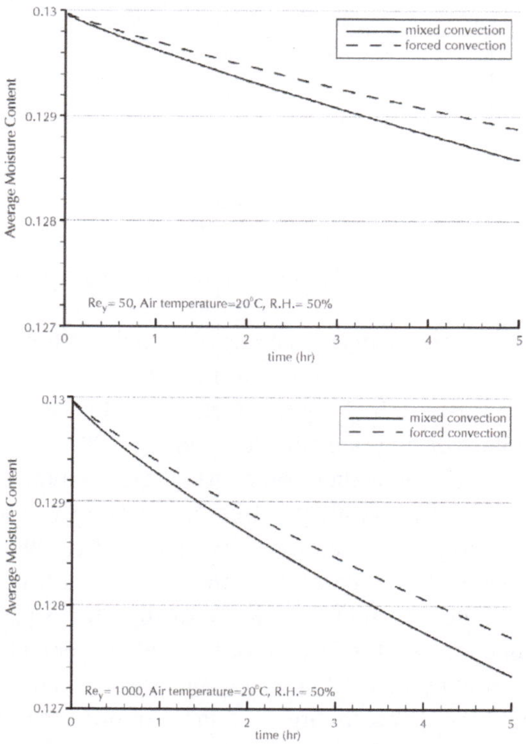

FIGURE 24 Drying curve for mixed and forced convection models. (a) Re =50 (b) Re=1,000.

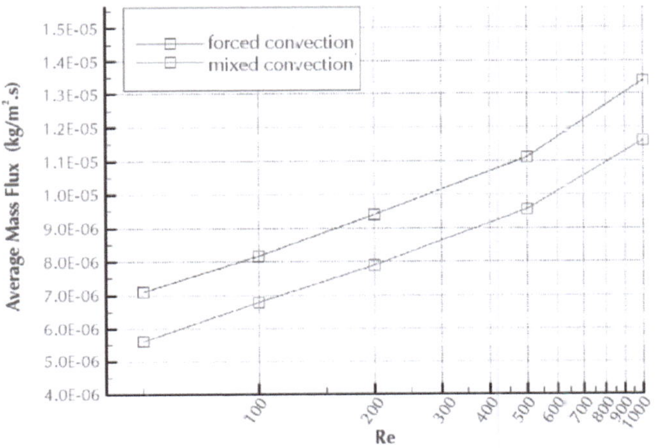

FIGURE 25 Average mass flux during initial 5 h of drying for mixed and forced convection models.

Using a FV method, which employed in this article the pressure based conservative algorithms, performed a two dimensional conjugated solution that guaranties the conservation laws despite of the finite elements methods.

By studying drying process in a variety of flow velocity (various Reynolds number), It's observed that airflow velocity increment has proportional effect on drying rate, with a factor between 1/4 and 1/5.

Moisture profiles and drying rate are considerably affected by buoyancy forces. Moisture removal from porous body surface for mixed convection model is more than forced convection in whole of Reynolds range of 50 to 1,000. So, it's suggested to taking into consideration buoyancy effects in analysis of flow around porous bodies in drying or other similar processes.

Present work although showed the surge tank effective role as a water hammer protection devise. Water flows have been interred and exited to the surge tank (for transmission line with surge tank and in leakage condition case). The surge pressure was 11(bar) nearby the pump station (i.e., at the start of transmission line). Results showed that minimum pressure line curve was under the transmission line profile, in the near of water reservoir. Hence, this showed that there was minus pressure in that zone of transmission line. Maximum transient pressure line was completely over the steady flow pressure line. Maximum pressure in the system was 156.131(m). Comparison showed similarity in results between this work discussion and other expert works. Field tests results showed in leakage condition at negative phase, air was sucked into the water pipeline [100–128].

KEYWORDS

- **computational method**
- **hydraulic transient modeling**
- **hydraulic transition**
- **surge tank**
- **water hammer**

CHAPTER 9

MODELING FOR FLOW PROCESS

CONTENTS

9.1 INTRODUCTION

Many factors have to be taken into account in selecting the most suitable type of dryer to install and the problem requires to be analyzed from several standpoints. Even an initial analysis of the possibilities must be backed up by pilot-scale tests unless previous experience has indicated the type most likely to be suitable. The accent today, due to high labor costs is on continuously operating unit equipment, to what extent possible automatically controlled. In any event, the selection of a suitable dryer should be made in two stages, a preliminary selection based on the general nature of the problem and the textile material to be handled, followed by a final selection based on pilot-scale tests or previous experience combined with economic considerations.

A leather industry involves a crucial energy-intensive drying stage at the end of the process to remove moisture left from dye setting. Determining drying characteristics for leather, such as temperature levels, transition times, total drying times, and evaporation rates, is vitally important so as to optimize the drying stage. Meanwhile, a textile material undergoes some physical and chemical changes that can affect the final leather quality [1–11].

In considering a drying problem, it is important to establish at the earliest stage, the final or residual moisture content of the textile material, which can be accepted. This is important in many hygroscopic materials and if dried below certain moisture content they will absorb or "regain" moisture from the surrounding atmosphere depending upon its moisture and humidity. The material will establish a condition in equilibrium with this atmosphere and the moisture content of the material under this condition is termed the equilibrium moisture content. Equilibrium moisture content is not greatly affected at the lower end of the atmospheric scale but as this temperature increases the equilibrium moisture content figure decreases, which explains why materials can in fact be dried in the presence of superheated moisture vapor. Meanwhile, drying medium temperatures and humidities assume considerable importance in the operation of direct dryers [12–21].

It should be noted that two processes occur simultaneously during the thermal process of drying a wet leather material, namely, heat transfer in order to raise temperature of the wet leather and to evaporate its moisture content together with mass transfer of moisture to the surface of the textile material and its evaporation from the surface to the surrounding atmosphere which, in convection dryers, is the drying medium. The quantity of air required to remove the moisture as liberated, as distinct from the quantity of air which will release the required amount of heat through a drop in its temperature in the course of drying, however, has to be determined from the known capacity of air to pick up moisture at a given temperature in relation to its initial content of moisture. For most practical purposes, moisture is in the form of water vapor but the same principles apply, with different values and humidity charts, for other volatile components [22–31].

Thermal drying consumes from 9–25% of national industrial energy consumption in the developed countries. In order to reduce net energy consumption in the drying operation there are attractive alternatives for drying of heat sensitive materials. Leather industry involves a crucial energy-intensive drying stage to remove the moisture left. Synthetic leather drying is the removal of the organic solvent and water. Determining drying characteristics for leathers is vitally important so as to optimize the drying stage. This book describes a way to determine the drying characteristics of leather with analytical method developed for this purpose. The model presented, is based on fundamental heat and mass transfer equations. Altering air velocity varies drying conditions. The work indicates closest agreement with the theoretical model. The results from the parametric study provide a better understanding of the drying mechanisms and may lead to a series of recommendations for drying optimization. Among the many processes that are performed in the leather industry, drying has an essential role: by this means, leathers can acquire their final texture, consistency and flexibility. However, some of the unit operations involved in leather industry, especially the drying process, are still based on empiricism and tradition, with very little use of scientific principles. Widespread methods of leather drying all over the world are mostly convective methods requiring a lot of energy. Specific heat energy consumption increases, especially in the last period of the drying process, when the moisture content of the leather approaches the value at which the product is storable. However, optimizing the drying process using mathematical analysis of temperature and moisture distribution in the material can reduce energy consumption in a convective dryer. Thus, development of a suitable mathematical model to predict the accurate performance of the dryer is important for energy conservation in the drying process [22–40].

The manufacturing of new-generation synthetic leathers involves the extraction of the filling polymer from the polymer-matrix system with an organic solvent and the removal of the solvent from the highly porous material. In this paper, a mathematical model of synthetic leather drying for removing the organic solvent is proposed. The model proposed adequately describes the real processes. To improve the accuracy of calculated moisture distributions a velocity correction factor (VCF) introduced into the calculations. The VCF reflects the fact that some of the air flowing through the bed does not participate very effectively in drying, since it is channelled into low-density areas of the inhomogeneous bed. This chapter discusses the results of experiments to test the deductions that increased rates of drying and better agreement between predicted and experimental moisture distributions in the drying bed can be obtained by using higher air velocities. Thus, this chapter focuses on reviewing convective heat and mass transfer equations in the industrial leather drying process with particular reference to VCF [41–50].

9.2 MATERIALS AND METHODS

The theoretical model (1–42) proposed in this model is based on fundamental equations to describe the simultaneous heat and mass transfer in porous media. It is possible to assume the existence of a thermodynamic quasi equilibrium state, where the temperatures of gaseous, liquid and solid phases are equal (Figs. 1–11),

$$\text{That is, } T_S = T_L = T_G = T, \tag{1}$$

Liquid Mass Balance:

$$\frac{\partial(\varepsilon_L \rho_L)}{\partial t} + \nabla(\rho_L \vec{u}_L) + \dot{m} = 0, \tag{2}$$

Water Vapor Mass Balance:

$$\frac{\partial[(\varepsilon - \varepsilon_L) X_V \rho_G]}{\partial t} + \nabla(X_V \rho_G \vec{u}_G + \vec{J}_V) - \dot{m} = 0, \tag{3}$$

$$\vec{J}_V = -\rho_G(\varepsilon - \varepsilon_L) D_{EFF} \nabla X_V, \tag{4}$$

Air Mass Balance:

$$\frac{\partial((\varepsilon - \varepsilon_L) X_A \rho_G)}{\partial t} + \nabla(X_A \rho_G \vec{u}_G - \vec{J}_V) = 0, \tag{5}$$

Liquid Momentum Equation (Darcy's Law):

$$\vec{u}_L = -\left(\frac{\alpha_G}{\mu_G}\right)\nabla(P_G), \tag{6}$$

Thermal Balance:

The thermal balance is governed by Eq. (7).

$$\frac{\partial\left\{\left[\rho_S C_{p_S} + (\varepsilon-\varepsilon_L)\rho_G\left(X_V C_{p_V} + X_A C_{p_A}\right) + \varepsilon_L \rho_L C_{p_L}\right]T\right\}}{\partial t} - \nabla(k_E \nabla T) + \nabla\left[\left(\rho_L \vec{u}_L C_{p_L} + \rho_G \vec{u}_G\left(X_V C_{p_V} + X_A C_{p_A}\right)\right)T\right] + (\varepsilon-\varepsilon_L)\frac{\partial P_G}{\partial t} + \dot{m}\Delta H_V = 0, \tag{7}$$

Thermodynamic Equilibrium-Vapor mass Fraction:

In order to attain thermal equilibrium between the liquid and vapor phase, the vapor mass fraction should be such that the partial pressure of the vapor (P'_V) should be equal to its saturation pressure (P_{VS}) at temperature of the mixture. Therefore, thermodynamic relations can obtain the concentration of vapor in the air/vapor mixture inside the pores. According to Dalton's Law of Additive Pressure applied to the air/vapor mixture, one can show that:

$$\rho_G = \rho_V + \rho_A, \tag{8}$$

$$X_V = \frac{\rho_V}{\rho_G}, \tag{9}$$

$$\rho_V = \frac{P'}{R_V T}, \tag{10}$$

$$\rho_A = \frac{(P_G - P'_V)}{R_A T}, \tag{11}$$

Combining Eqs. (8)–(11), one can obtain:

$$X_V = \frac{1}{1+\left(\frac{P_G R_V}{P'_V R_A}\right)-\left(\frac{R_V}{R_A}\right)}, \tag{12}$$

Mass Rate of Evaporation:

The mass rate of evaporation was obtained in two different ways, as follows:

First of all, the mass rate of evaporation $\dot{m}$ was expressed explicitly by taking it from the water vapor mass balance (2), since vapor concentration is given by Eq. (12).

$$\dot{m} = \frac{\partial\left[(\varepsilon-\varepsilon_L)X_V\rho_G\right]}{\partial t} + \nabla\left(X_V\rho_G\vec{u}_G + \vec{J}_V\right), \tag{13}$$

Secondly, an equation to compute the mass rate of evaporation can be derived with a combination of the liquid mass balance (1) with a first-order-Arrhenius type equation. From the general kinetic equation:

$$\frac{\partial\alpha}{\partial t} = -kf(\alpha), \tag{14}$$

$$k = A\exp\left(-\frac{E}{RT_{SUR}}\right), \tag{15}$$

$$\alpha = 1 - \frac{\varepsilon_L(t)}{\varepsilon_0}, \tag{16}$$

Drying Kinetics Mechanism Coupling:

Using thermodynamic relations, according to Amagat's law of additive volumes, under the same absolute pressure,

$$m_V = \frac{V_V P_G}{R_V T} \tag{17}$$

$$m_A = \frac{V_A P_G}{R_A T}, \tag{18}$$

$$m_V = X_V m_T, \tag{19}$$

$$m_T = m_V + m_A, \tag{20}$$

$$V_G = V_V + V_A, \tag{21}$$

$$V_G = (\varepsilon - \varepsilon_L)V_S, \tag{22}$$

Solving the set of algebraic (17–22), one can obtain the vapor-air mixture density:

$$\rho_G = \frac{(m_V + m_A)}{V_G}, \tag{23}$$

$$\rho_V = \frac{m_V}{V_G}, \tag{24}$$

$$\rho_A = \frac{m_A}{V_G}, \tag{25}$$

Equivalent Thermal Conductivity:

It is necessary to determine the equivalent value of the thermal conductivity of the material as a whole, since no phase separation was considered in the overall energy equation. The equation we can propose now which may be used to achieve the equivalent thermal conductivity of materials K_E, composed of a continued medium with a uniform disperse phase. It is expressed as follows in Eq. (26).

$$K_E = \left[\frac{k_S + \varepsilon_L k_L \left(\frac{3k_S}{2k_S + k_L} \right) + k_G (\varepsilon - \varepsilon_L) \left(\frac{3k_S}{2k_S + k_G} \right)}{1 + \varepsilon_L \left(\frac{3k_S}{2k_S + k_L} \right) + (\varepsilon - \varepsilon_L) \left(\frac{3k_S}{2k_S + k_G} \right)} \right], \tag{26}$$

$$k_G = X_V k_V + X_A k_A, \tag{27}$$

Effective Diffusion Coefficient Equation:

The binary bulk diffusivity D_{AV} of air-water vapor mixture is given by:

$$D_{AV} = (2.20)(10^{-5}) \left(\frac{P_{ATM}}{P_G} \right) \left(\frac{T_{REF}}{273.15} \right)^{1.75}, \tag{28}$$

Factor α_F can be used to account for closed pores resulting from different nature of the solid, which would increase gas outflow resistance, so the equation of effective diffusion coefficient D_{EFF} for fiber drying is:

$$D_{EFF} = \alpha_F D_{AV}, \tag{29}$$

The convective heat transfer coefficient can be expressed as:

$$h = Nu_{\delta}\left(\frac{k}{\delta}\right), \tag{30}$$

The convective mass transfer coefficient is:

$$h_M = \left(\frac{h}{C_{PG}}\right)\left(\frac{\Pr}{Sc}\right)^{2/3}, \tag{31}$$

$$\Pr = \frac{C_{PG}\mu_G}{k_G}, \tag{32}$$

$$Sc = \frac{\mu_G}{\rho_G D_{AV}}, \tag{33}$$

The deriving force determining the rate of mass transfer inside the fiber is the difference between the relative humidities of the air in the pores and the fiber. The rate of moisture exchange is assumed to be proportional to the relative humidity difference in this study.

The heat transfer coefficient between external air and fibers surface can be obtained by: $h = Nu_{\delta}\left(\frac{k}{\delta}\right)$.

The mass transfer coefficient was calculated using the analogy between heat transfer and mass transfer as $h_M = \left(\frac{h}{C_{PG}}\right)\left(\frac{\Pr}{Sc}\right)^{2/3}$. The convective heat and mass transfer coefficients at the surface are important parameters in drying processes; they are functions of velocity and physical properties of the drying medium.

Describing kinetic model of the moisture transfer during drying as follows:

$$-\frac{dX}{dt} = k(X - X_e), \tag{34}$$

where, X is the material moisture content (dry basis) during drying (kg water/kg dry solids), X_e is the equilibrium moisture content of dehydrated material (kg water/kg dry solids), k is the drying rate (min^{-1}), and t is the time of drying (min). The drying rate is determined as the slope of the falling rate-drying curve. At zero time, the moisture content (dry basis) of the dry material X (kg water/kg dry solids) is equal to X_i, and Eq. (34) is integrated to give the following expression:

$$X = X_e - (X_e - X_i)e^{-kt}, \tag{35}$$

Using above equation Moisture Ratio can be defined as follows:

$$\frac{X - X_e}{X_i - X_e} = e^{-kt}, \tag{36}$$

This is the Lewis's formula introduced in 1921. But by using experimental data of leather drying it seemed that there was an error in curve fitting of e^{-at}.

The experimental moisture content data were nondimensionlized using the equation:

$$MR = \frac{X - X_e}{X_i - X_e}, \tag{37}$$

Where MR is the moisture ratio. For the analysis it was assumed that the equilibrium moisture content, X_e, was equal to zero.

Selected drying models, detailed in Tables 1–3, were fitted to the drying curves (MR versus time), and the equation parameters determined using non-linear least squares regression analysis, as shown in Table 2.

TABLE 1 Drying models fitted to experimental data.

Model	Mathematical Expression
Lewis (1921)	$MR = \exp(-at)$
Page (1949)	$MR = \exp(-at^b)$
Henderson and Pabis (1961)	$MR = a\exp(-bt)$
Quadratic function	$MR = a + bt + ct^2$
Logarithmic (Yaldiz and Eterkin, 2001)	$MR = a\exp(-bt) + c$
Third Degree Polynomial	$MR = a + bt + ct^2 + dt^3$
Rational function	$MR = \frac{a + bt}{1 + ct + dt^2}$
Gaussian model	$MR = a\exp(\frac{-(t-b)^2}{2c^2})$
Present model	$MR = a\exp(-bt^c) + dt^2 + et + f$

TABLE 2 Estimated values of coefficients and statistical analysis for the drying models.

Model	Constants	T = 50	T = 65	T = 80
Lewis	a	0.08498756	0.1842355	0.29379817
	S	0.0551863	0.0739872	0.0874382
	r	0.9828561	0.9717071	0.9587434
Page	a	0.033576322	0.076535988	0.14847589
	b	1.3586728	1.4803604	1.5155253
	S	0.0145866	0.0242914	0.0548030
	r	0.9988528	0.9972042	0.9856112
Henderson	a	1.1581161	1.2871764	1.4922171
	b	0.098218605	0.23327801	0.42348755
	S	0.0336756	0.0305064	0.0186881
	r	0.9938704	0.9955870	0.9983375
Logarithmic	a	1.246574	1.3051319	1.5060514
	b	0.069812589	0.1847816	0.43995186
	c	–0.15769402	–0.093918118	0.011449769
	S	0.0091395	0.0117237	0.0188223
	r	0.9995659	0.9993995	0.9985010
Quadratic function	a	1.0441166	1.1058544	1.1259588
	b	–0.068310663	–0.16107942	–0.25732004
	c	0.0011451149	0.0059365371	0.014678241
	S	0.0093261	0.0208566	0.0673518
	r	0.9995480	0.9980984	0.9806334
Third Degree Polynomial	a	1.065983	1.1670135	1.3629748
	b	–0.076140508	–0.20070291	–0.45309695
	c	0.0017663191	0.011932525	0.053746805
	d	–1.335923e–005	–0.0002498328	–0.0021704758
	S	0.0061268	0.0122273	0.0320439
	r	0.9998122	0.9994013	0.9961941
Rational function	a	1.0578859	1.192437	1.9302135
	b	–0.034944627	–0.083776453	–0.16891461
	c	0.03197939	0.11153663	0.72602847
	d	0.0020339684	0.01062793	0.040207428

TABLE 2 *(Continued)*

Model	Constants	T = 50	T = 65	T = 80
	S	0.0074582	0.0128250	0.0105552
	R	0.9997216	0.9993413	0.9995877
Gaussian model	A	1.6081505	2.3960741	268.23939
	B	–14.535231	–9.3358707	–27.36335
	C	15.612089	7.7188252	8.4574493
	S	0.0104355	0.0158495	0.0251066
	R	0.9994340	0.9989023	0.9973314
Present model	A	0.77015136	2.2899114	4.2572457
	B	0.073835826	0.58912095	1.4683178
	C	0.85093985	0.21252159	0.39672164
	D	0.00068710356	0.0035759092	0.0019698297
	E	–0.037543605	–0.094581302	–0.03351435
	F	0.3191907	–0.18402789	0.04912732
	S	0.0061386	0.0066831	0.0092957
	R	0.9998259	0.9998537	0.9997716

The experimental results for the drying of leather are given in Fig. 7. Fitting curves for two sample models (Lewis model and present model) and temperature of 80°C are given in Figs. 8 and 9. Two criteria were adopted to evaluate the goodness of fit of each model, the Correlation Coefficient (r) and the Standard Error (S).The standard error of the estimate is defined as follows:

$$S = \sqrt{\frac{\sum_{i=i}^{n_{points}} (MR_{\exp,i} - MR_{Pred,i})^2}{n_{points} - n_{param}}}, \tag{38}$$

where $MR_{\exp,i}$ is the measured value at point i, and $MR_{Pred,i}$ is the predicted value at that point, and n_{param} is the number of parameters in the particular model (so that the denominator is the number of degrees of freedom).

To explain the meaning of correlation coefficient, we must define some terms used as follow:

$$S_t = \sum_{i=1}^{n_{points}} (\bar{y} - MR_{\exp,i})^2, \tag{39}$$

where, the average of the data points ($\bar{y}$) is simply given by

$$\bar{y} = \frac{1}{n_{points}} \sum_{i=1}^{n_{points}} MR_{\exp,i}, \tag{40}$$

The quantity S_t considers the spread around a constant line (the mean) as opposed to the spread around the regression model. This is the uncertainty of the dependent variable prior to regression. We also define the deviation from the fitting curve as:

$$S_r = \sum_{i=1}^{n_{points}} (MR_{\exp,i} - MR_{pred,i})^2, \tag{41}$$

Note the similarity of this expression to the standard error of the estimate given above; this quantity likewise measures the spread of the points around the fitting function. In view of the above, the improvement (or error reduction) due to describing the data in terms of a regression model can be quantified by subtracting the two quantities [51–108]. Because the magnitude of the quantity is dependent on the scale of the data, this difference is normalized to yield

$$r = \sqrt{\frac{S_t - S_r}{S_t}}, \tag{42}$$

where, r is defined as the correlation coefficient. As the regression model better describes the data, the correlation coefficient will approach unity. For a perfect fit, the standard error of the estimate will approach S=0 and the correlation coefficient will approach r=1.

The standard error and correlation coefficient values of all models are given in Figs. 10 and 11.

9.3 RESULTS AND DISCUSSION

Synthetic leathers are materials with much varied physical properties. As a consequence, even though a lot of research of simulation of drying of porous media has been carried out, the complete validation of these models are very difficult. The drying mechanisms might strongly influence by parameters such as permeability and effective diffusion coefficients. The unknown effective diffusion coefficient of vapor for fibers under different temperatures may be determined by adjustment of the model's theoretical alpha correction factor and experimental data. The mathematical model can be used to predict the effects of many parameters on the temperature variation of the fibers. These parameters include the operation conditions of the dryer, such as the initial moisture content of the fibers, heat and mass transfer coefficients, drying air moisture content, and dryer air temperature. From Figs. 1–6, it can be observed that the shapes of the experimental and calculated curves are somewhat different.

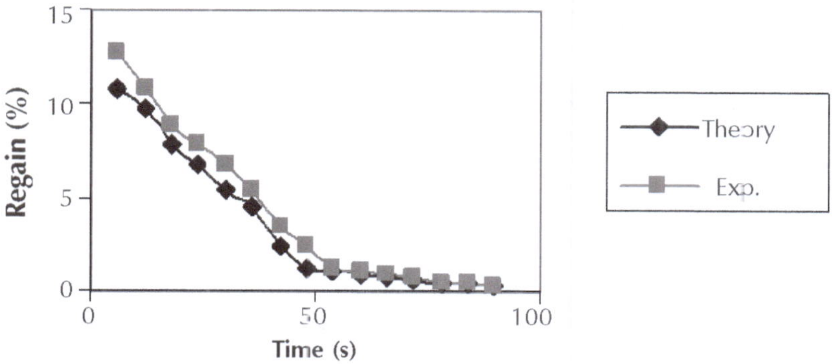

FIGURE 1 Comparison of the theoretical and experimental distribution at air velocity of 0.75 m/s and VCF = 0.39.

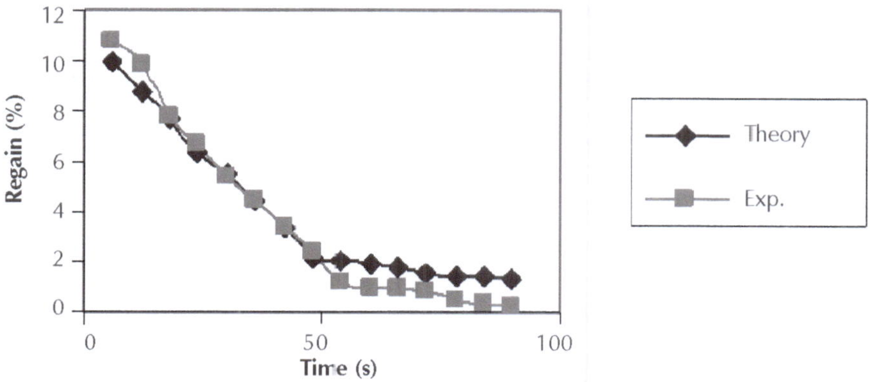

FIGURE 2 Comparison of the theoretical and experimental distribution at air velocity of 0.89 m/s and VCF = 0.44.

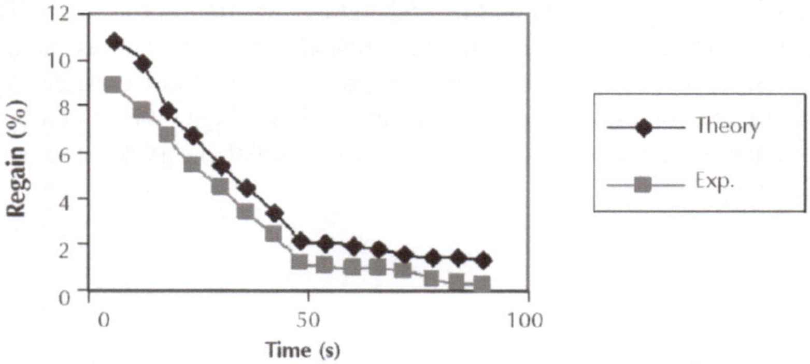

FIGURE 3 Comparison of the theoretical and experimental distribution at air velocity of 0.95 m/s and VCF = 0.47.

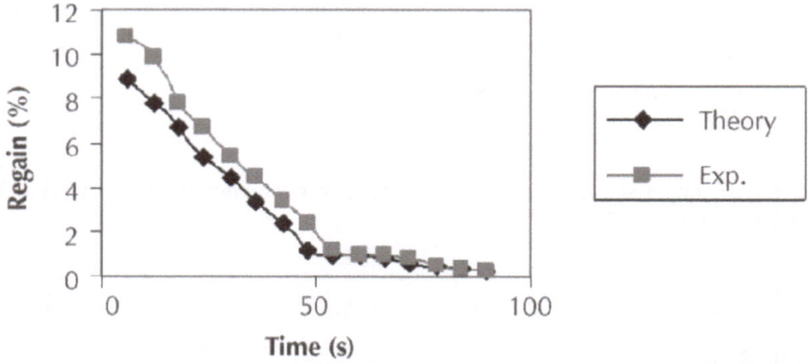

FIGURE 4 Comparison of the theoretical and experimental distribution at air velocity of 1.59 m/s and VCF = 0.62.

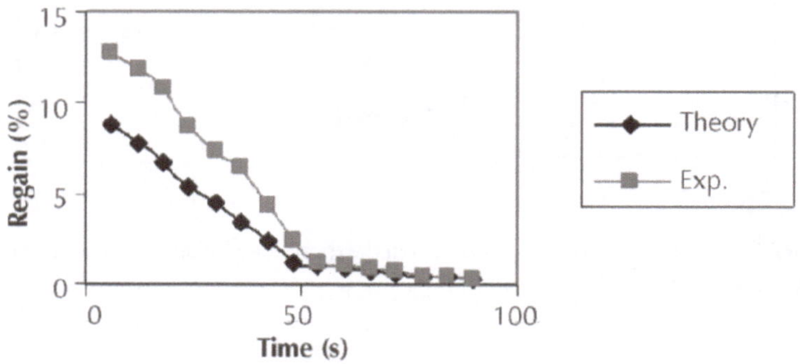

FIGURE 5 Comparison of the theoretical and experimental distribution at air velocity of 2.10 m/s and VCF = 0.62.

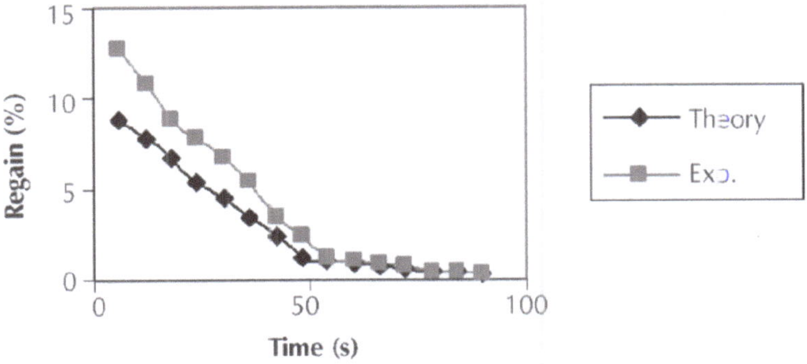

FIGURE 6 Comparison of the theoretical and experimental distribution at air velocity of 2.59 m/s and VCF = 0.61.

It can be seen that as the actual air velocity used in this experiment increases, the value of VCF necessary to achieve reasonable correspondence between calculation and experiment becomes closer to unity; that is, a smaller correction to air velocity is required in the calculations as the actual air velocity increases. This appears to confirm the fact that the loss in drying efficiency caused by bed in homogeneity tends to be reduced as air velocity increases. Figure 7 shows a typical heat distribution during convective drying. Table 3 relates the VCF to the values of air velocity actually used in the experiments It is evident from the table that the results show a steady improvement in agreement between experiment and calculation (as indicated by increase in VCF) for air velocities up to 1.59 m/s, above which to be no further improvement with increased flow.

TABLE 3 Variation of VCF with air velocity.

Air velocity, m/s	0.75	0.89	0.95	1.59	2.10	2.59
VCF used	0.39	0.44	0.47	0.62	0.62	0.61

In this chapter, the analytical model has been applied to several drying experiments. The results of the experiments and corresponding calculated distributions are shown in Figs. 7–11. It is apparent from the curves that the calculated distribution is in reasonable agreement with the corresponding experimental one. In view of the above, it can be clearly observed that the shapes of experimental and calculated curves are somewhat similar.

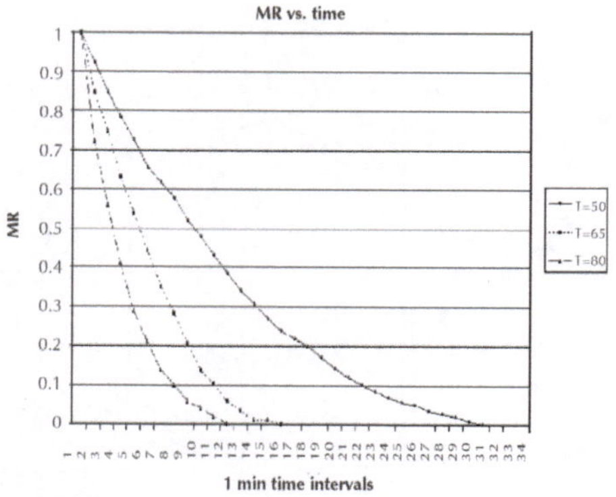

FIGURE 7 Moisture ratio vs. time.

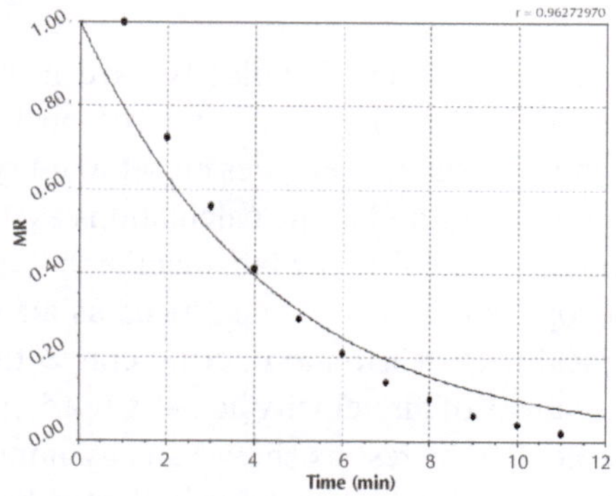

FIGURE 8 Lewis model.

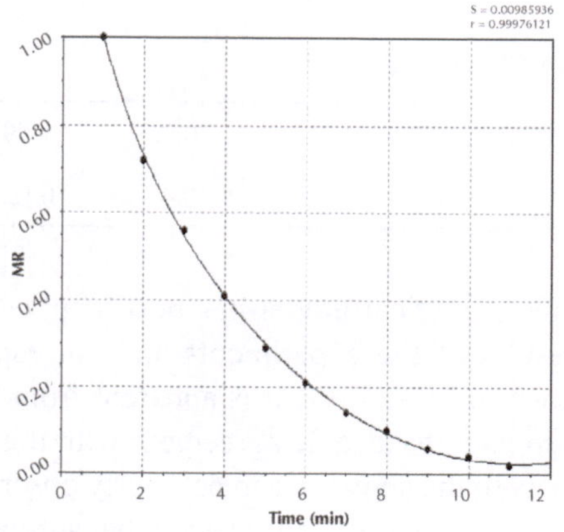

FIGURE 9 Present model.

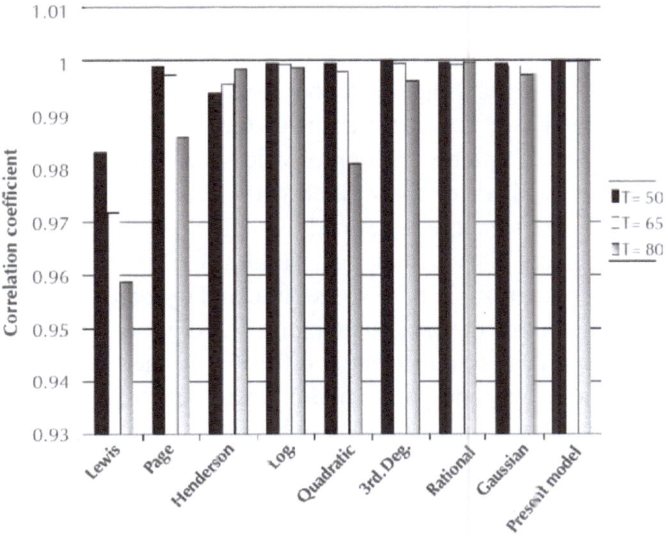

FIGURE 10 Correlation coefficient of all models.

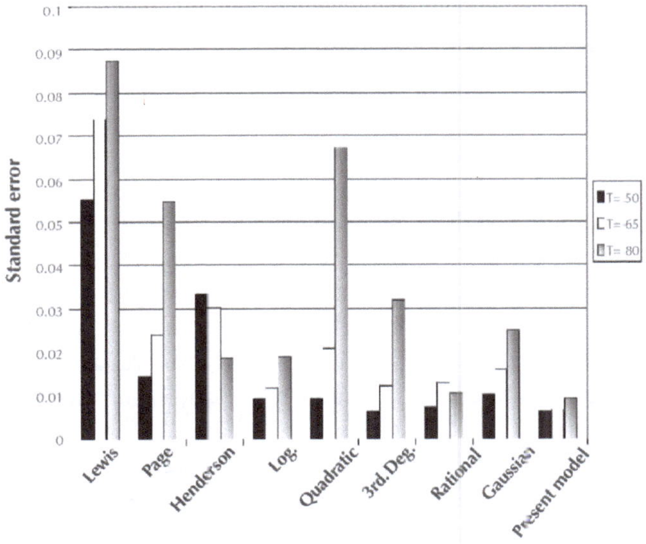

FIGURE 11 Standard error of all models.

9.4 NON-HOMOGENEOUS MATERIAL AND HEAT FLOW

Wood is a hygroscopic, porous, anisotropic and nonhomogeneous material (Figs. 12–16). After log sawing, the lumber contains liquid water in fiber cavities (capillary water) and bound water inside the fiber wall (hygroscopic water). Porosity refers to volume fraction of void space. This void space can be actual space filled with air or space filled with both water and air. Capillary-porous materials are sometimes defined as those having pore diameter less than 10^{-7} m. Capillary porous materials (Tables 4–6) were defined as those having a clearly recognizable pore space. In capillary porous material, transport of water is a more complex phenomena. In addition to molecular diffusion, water transport can be due to vapor diffusion, surface diffusion, Knudsen diffusion, capillary flow, and purely hydrodynamic flow. In hygroscopic materials, there is large amount of physically bound water and the material often shrinks during heating.

9.5 COMPUTER MODELS FOR HEAT FLOW IN NON-HOMOGENEOUS MATERIAL

In hygroscopic materials there is a level of moisture saturation below which the internal vapor pressure is a function of saturation and temperature. These relationships are called equilibrium moisture isotherms. Above this moisture saturation, the vapor pressure is a function of temperature only (as expressed by the Clapeyron equation) and is independent of the moisture level. Thus, above certain moisture level, all materials behave nonhygroscopic.

Green wood contains a lot of water. In the outer parts of the stem, in the sapwood, spruce and pine have average moisture content of about 130%, and in the inner parts, in the heartwood, the average moisture content is about 35%. Wood drying is the art of getting rid of that surplus water under controlled forms. It will dry to an equilibrium moisture content of 8–16% fluid content when left in air, which improves its stability, reduces its weight for transport, prepares it for chemical treatment or painting and improves its mechanical strength.

Water in wood is found in the cell cavities and cell walls. All void spaces in wood can be filled with liquid water called free water. Free water is held by adhesion and surface tension forces. Water in the cell walls is called bound water. Bound water is held by forces at the molecular level. Water molecules attach themselves to sites on the cellulose chain molecules. It is an intimate part of the cell wall but does not alter the chemical properties of wood.

Hydrogen bonding is the predominant fixing mechanism. If wood is allowed to dry, the first water to be removed is free water. No bound water is evaporated until all free water has been removed. During removal of water, molecular energy is expended. Energy requirement for vaporization of bound water is higher than free water. Moisture content at which only the cell walls are completely saturated (all bound water) but no free water exists in all lumens is called the fiber saturation point (F.S.P.). Typically the F.S.P. of wood is within the range of 20–40% moisture content depending on temperature and wood species. Water in wood normally moves from high to low zones of moisture content. The surface of the wood must be drier than the interior if moisture is to be removed. Drying can be divided into two phases movement of water from the interior to the surface of the wood, and removal of water from the surface. Water moves through the interior of the wood as a liquid or water vapor through various air passageways in the cellular structure of wood and through the cell walls.

Drying is a process of simultaneous heat and moisture transfer with a transient nature. The evolution process of the temperature and moisture with time must be predicted and actively controlled in order to ensure an effective and efficient drying operation. Lumber drying can be understood as the balance between heat transfer from airflow to wood surface and water transport from the wood surface to the airflow. Reduction in drying time and energy consumption offers the wood industries a great potential for economic benefit. In hygroscopic porous material like wood, mathematical models describing moisture and heat movements may be used to facilitate experimental testing and to explain the physical mechanisms underlying such mass transfer processes. The process of wood drying can be interpreted as simultaneous heat and moisture transfer with local thermodynamic equilibrium at each point within the timber. Drying of wood is in its nature an unsteady-state non-isothermal diffusion of heat and moisture, where temperature gradients may counteract with the moisture gradient.

9.6 HEAT FLOW DURING DRYING PROCESS

First stage: When both surface and core MC are greater than the F.S.P. Moisture movement is by capillary flow. Drying rate is evaporation controlled.

Second stage: When surface MC is less than the FSP and core MC is greater than the F.S.P. Drying is by capillary flow in the core and by bound water diffusion near the surface as fiber saturation line recedes into wood, resistance to drying increases. Drying rate is controlled by bound water diffusion.

Third stage: When both surface and core MC is less than the F.S.P. Drying is entirely by diffusion. As the MC gradient between surface and core becomes less, resistance to drying increases and drying rate decreases.

Capillary pressure is a driving force in convective wood drying at mild conditions. The temperature is higher outside than inside. The moisture profile during convective drying is in the opposite direction, namely, the drier part is toward the exposed surface of wood. This opposite pattern of moisture and temperature profiles lead to the concept of the wet front that separates the outer area, where the water is bound to the cell wall, from the inner area, where free water exists in liquid and vapor form. A wet front that moves slowly from the surface toward the center of a board during convective drying leads to subsequent enhancement of the capillary transportation. Capillary transportation can then be justified due to the moisture gradients developed around that area. When the drying conditions are mild, the drying period is longer so the relative portion of the total moisture removal, due to the capillary phenomena, is high, and it seems that this is the most important mass transfer mechanism.

Credible data on the bound water diffusion coefficient in wood and the boundary condition for the interface between moist air and wood surface are very important for accurate description of timber drying as well as for the proper design and use of products, structures and buildings made of wood already dried below the fiber saturation point. During the last century, two groups of methods for measuring the bound water diffusion coefficient in wood were developed. The first one, traditionally called the cup method, uses data from the steady-state experiments of bound water transfer and is based on Fick's first law of diffusion. Unfortunately, the method is not valid for the bound water diffusion coefficient determination in wood because it cannot satisfy the requirements of the boundary condition of the first kind and the constant value of the diffusion coefficient. The second group of methods is based on the unsteady-state experiments and Fick's second law of diffusion. The common name of this group is the sorption method and it was developed to overcome the disadvantages of the cup technique.

In solving the diffusion Eqs. (43)–(100) for moisture variations in wood, some authors have assumed that the diffusion coefficient depends strongly on moisture content, while others have taken the diffusion coefficient as constant. It has been reported that the diffusion coefficient is influenced by the drying temperature, density and moisture content of timber. The diffusion coefficient of water in cellophane and wood substance was shown to increase with temperature in proportion to the increase in vapor pressure of water. It is also observed that the diffusion coefficient decreased with increasing wood density. Other factors affecting the diffusion coefficient that are yet to be quantified are the species (specific gravity) and the growth ring orientation. Literature has suggested that the ratios of radial and tangential diffusion coefficients vary for different tree species. The radial diffusion coefficient of New Zealand Pinus radiate has been estimated to be approximately 1.4 times the tangential diffusion coefficient. It is observed that for northern red oak, the diffusion coefficient is a function of moisture content only. It increases dramatically at low moisture content and tends to level off as the fiber saturation point is approached.

In a one-dimensional formulation with moisture moving in the direction normal to a specimen of a slice of wood of thickness 2a, the diffusion equation can be written as:

$$\frac{\partial(MC)}{\partial t}=\frac{\partial}{\partial X}\left(D\frac{\partial(MC)}{\partial X}\right)(0<X<a,t>0) \tag{43}$$

where MC is moisture content, t is time, D is diffusion coefficient, and X is space coordinate measured from the center of the specimen.

The moisture content influences on the coefficient D only if the moisture content is below the fiber saturation point (F.S.P.)(typically 20%–30% for softwoods):

$$D(u)=\begin{cases} f_D(u) & ,u<u_{fsp} \\ f_D(u_{fsp}) & ,u\geq u_{fsp} \end{cases}, \tag{44}$$

where u_{fsp} denotes the F.S.P. and $f_D(u)$ is a function which expresses diffusion coefficient in moisture content, temperature and may be some other parameters of ambient air climate. The expression of $f_D(u)$ depends on variety of wood.

It was assumed that the diffusion coefficient bellow F.S.P. can be represented by:

$$f_D(u)=A.e^{-\frac{5280}{T}}.e^{\frac{B.u}{100}}, \tag{45}$$

where T is the temperature in Kelvin, u is percent moisture content, A and B can be experimentally determined.

The regression equation of diffusion coefficient of Pinus radiatatimber using the dry bulb temperature and the density is:

$$D\left(10^{-9}\right) = 1.89 + 0127 \times T_{DB} - 0.00213 \times \rho_S \ (R^2 = 0.499), \tag{46}$$

The regression equations of diffusion coefficients below of Masson's pine during high temperature drying are:

$$D = 0.0046MC^2 + 0.1753MC + 4.2850 \left(R^2 = 0.9391\right), \tag{47}$$

Tangential diffusion:

$$D = 0.0092MC^2 + 0.3065MC + 4.9243 \left(R^2 = 0.9284\right), \tag{48}$$

Radial diffusion:

The transverse diffusion coefficient D can be expressed by the porosity of wood v, the transverse bound water diffusion coefficient D_{bt} of wood and the vapor diffusion coefficient D_v in the lumens:

$$D = \frac{\sqrt{v} D_{bt} D_v}{(1-v)\left(\sqrt{v} D_{bt} + \left(1-\sqrt{v}\right) D_v\right)}, \tag{49}$$

The vapor diffusion coefficient D_v in the lumens can be expressed as:

$$D_v = \frac{M_w D_a P_s}{SG_d \rho_w R T_k} \cdot \frac{d\varphi}{du}, \tag{50}$$

where M_w (kg/kmol) is the molecular weight of water.

$$D_a = \frac{9.2.10^{-9} T_k^{2.5}}{\left(T_k + 245.18\right)}, \tag{51}$$

is the inter diffusion coefficient of vapor in air,

$$SG_d = \frac{1.54}{\left(1 + 1.54u\right)}, \tag{52}$$

is the nominal specific gravity of wood substance at the given bound water content. ρ_w = 103 kg/m^3 is the density of water, R = 8314.3 kmol, K is the gas constant, T_k is the Kelvin temperature, ψ is the relative humidity (%/100), and P_{sat} is saturated vapor pressure given by:

$$p_{sat} = 3390 \exp\left(-1.74 + 0.0759 T_C - 0.000424 T_C^2 + 2.44.10^{-6} T_C^3\right), \tag{53}$$

The derivative of air relative humidity ψ with respect to moisture content MC is given as:

$$MC = \frac{18}{w}\left(\frac{k_1 k_2 \psi}{1 + k_1 k_2 \psi} + \frac{k_2 \psi}{1 - k_2 \psi}\right), \tag{54}$$

where:

$$k_1 = 4.737 + 0.04773T_C - 0.00050012T_C^2, \tag{55}$$

$$k_2 = 0.7059 + 0.001695T_C + -0.000005638T_C^2, \tag{56}$$

$$W = 223.4 + .6942T_C + 0.01853T_C^2, \tag{57}$$

The diffusion coefficient D_{bt} of bound water in cell walls is defined according to the Arrhenius equation as:

$$D_{bt} = 7.10^{-6} \exp\left(-E_b / RT_k\right), \tag{58}$$

where:

$$E_b = \left(40.195 - 71.179Mc + 291Mc^2 - 669.92Mc^3\right).10^6, \tag{59}$$

is the activation energy.

The porosity of wood is expressed as:

$$v = 1 - SG\left(0.667 + Mc\right), \tag{60}$$

where specific gravity of wood SG at the given moisture content u is defined as:

$$SG = \frac{\rho_s}{\rho_W\left(1 + Mc\right)} = \frac{\rho_0}{\rho_W + 0.883\rho_0 Mc}, \tag{61}$$

where ρ_s is density of wood, ρ_0 is density of oven-dry wood (density of wood that has been dried in a ventilated oven at approximately 104°C until there is no additional loss in weight).

Wood thermal conductivity (K_{wood}) is the ratio of the heat flux to the temperature gradient through a wood sample. Wood has a relatively low thermal conductivity due to its porous structure, and cell wall properties. The density, moisture content, and temperature dependence of thermal conductivity of wood and wood-based composites were demonstrated by several researchers. The transverse thermal conductivity can be expressed as:

$$K_{wood} = \left[SG \times \left(4.8 + 0.09 \times MC\right) + 0.57\right] \times 10^{-4} \frac{cal}{cm * Cs} \tag{62}$$

When moisture content of wood is below 40%.

$$K_{wood} = \left[SG \times \left(4.8 + 0.125 \times MC\right) + 0.57\right] \times 10^{-4} \frac{cal}{cm * Cs} \tag{63}$$

When moisture content of wood is above 40%.

The specific gravity and moisture content dependence of the solid wood thermal conductivity in the transverse (radial and tangential) direction is given by:

$$K_T = SG\left(K_{cw} + K_w.Mc\right) + K_a v, \tag{64}$$

where:
SG= specific gravity of wood,

K_{cw} = Conductivity of cell wall substance (0.217 J/m/s/K),

K_w = conductivity of water (0.4 J/m/s/K),

K_a = conductivity of air (0.024 J/m/s/K),

Mc = moisture content of wood (fraction),

v = porosity of wood.

The thermal conductivity of wood is affected by a number of basic factors: density, moisture content, extractive content, grain direction, structural irregularities such as checks and knots, fibril angle, and temperature. Thermal conductivity increases as density, moisture content, temperature, or extractive content of the wood increases. Thermal conductivity is nearly the same in the radial and tangential directions with respect to the growth rings.

The longitudinal thermal conductivity of solid wood is approximately 2.5 times higher than the transverse conductivity:

$$K_L = 2.5 K_T \tag{65}$$

For moisture content levels below 25%, approximate thermal conductivity K across the grain can be calculated with a linear equation of the form:

$$K_{wood} = G\left(B + CM\right) + A \tag{66}$$

Where SG is specific gravity based on oven dry weight and volume at a given moisture content MC (%) and A, B, and C are constants.

For specific gravity >0.3, temperatures around 24°C, and moisture content values <25%, $A = 0.01864$, $B = 0.1941$, and $C = 0.004064$ (with k in W/(m·K)). It was derived from measurements made by several researchers on a variety of species.

During the early stages of drying the material consists of so much water that liquid surfaces exist and drying proceeds at a constant rate. Constant drying rates are achieved when surface free water is maintained and the only resistance to mass transfer is external to the wood. The liquid water moves by capillary forces to the surface in same proportion of moisture evaporation. Moisture movement across the lumber will depend on the wood permeability and the drying rate itself is controlled by external conditions in this period. Energy received by the surface increase temperature in this region, and the heat transfer to the inner part of lumber starts. Since the moisture source for the surface is internal moisture, constant drying rates can only be maintained if there is sufficient moisture transport to keep the surface moisture content above the F.S.P. If this level is not maintained then some of the resistance to mass transfer becomes internal and neither the drying rate nor the surface temperature remains constant and drying proceeds to the falling rate period. As the lumber dries, the liquid water or wet line recedes into wood and the internal moisture movement involves the liquid flow and diffusion of water vapor and hygroscopic water. The effect of internal resistance on the drying rate increases. In the last phase (second falling rate period) there is no more liquid water in the lumber, and the drying rate is controlled only by internal resistance (material characteristics) until an equilibrium moisture content is reached.

A typical drying curve showing three stages of drying characteristic is illustrated in Fig. 1.

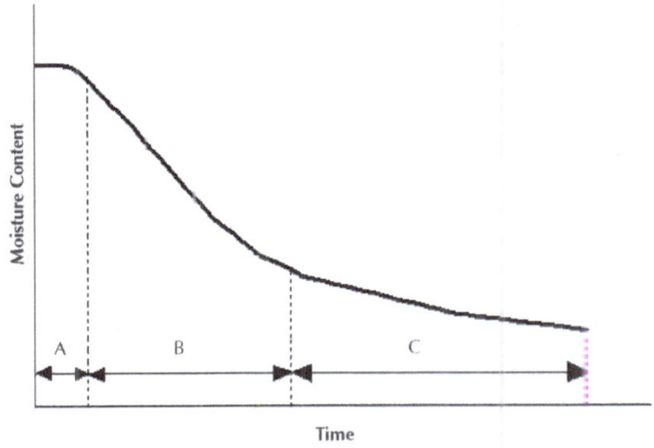

FIGURE 12 Drying characteristic of porous media: (A) constant rate region;(B) first falling rate region; (C) second falling rate region.

Pang et al. proposed that the three drying periods (constant rate, first falling rate and second falling rate) based on simulated drying of veneer be expressed by the following equations:

$$-\frac{d(MC)}{dt} = j_0 \text{ For MC} > M_{Cr1}, \tag{67}$$

$$-\frac{d(MC)}{dt} = A + B * MC \text{ For } M_{cr1} > MC > M_{cr2}, \tag{68}$$

$$-\frac{d(MC)}{dt} = \frac{A + B * M_{cr2}}{M_{cr2} - M_e} * (MC - M_e) \text{ For MC} < M_{cr2}, \tag{69}$$

where, j_0 is constant drying rate, M_{Cr1} is the first critical moisture content, M_{cr2} is the second critical moisture content, constants A and B also vary with wood thickness, wood density, and drying conditions.

Moisture content of wood is defined as the weight of water in wood expressed as a fraction, usually a percentage, of the weight of oven dry wood. Moisture exists in wood as bound water within the cell wall, capillary water in liquid form and water vapor in gas form in the voids of wood. Capillary water bulk flow refers to the flow of liquid through the interconnected voids and over the surface of a solid due to molecular attraction between the liquid and the solid. Moisture content varies widely between species and within species of wood. It varies particularly between heartwood and sapwood. The amount of moisture in the cell wall may decrease as a result of extractive deposition when a tree undergoes change from sapwood to heartwood. The butt logs of trees may contain more water than the top logs. Variability of moisture content exists even within individual boards cut from the same tree. Green wood is often defined as freshly sawn wood in which the cell walls are completely saturated with water. Usually green wood contains additional water in the lumens. Moisture content at which both the cell lumens and cell walls are completely saturated with water is the maximum moisture content. An average green moisture content value taken from the Wood Handbook (Forest products society, 1999) of southern yellow pine (loblolly) is 33 and 110% for heartwood and sapwood, respectively. Sweet gum is 79 and 137% while yellow poplar is 83 and 106% for heartwood and sapwood, respectively.

Permeability refers to the capability of a solid substance to allow the passage of gases or liquids under pressure. Permeability assumes the mass movement of molecules in which the pressure or driving force may be supplied by such sources as mechanically applied pressure, vacuum, thermal expansion, gravity, or surface tension. Under this condition, the permeability of wood is the dominant factor controlling moisture movement.

Fluid movement in wood is a very important process in wood products industries. An understanding of wood permeability is essential for determining lumber drying schedules for treating lumber and for producing high-quality wood products. The flow of gas inside the wood particle is limited due to the fact that wood consists of a large number of clustered small pores. The pore walls act as barriers largely preventing convective flow between adjacent pores. The wood annular rings also act as barriers for flow in the radial direction which makes flow in the axial direction more favorable and giving a lower permeability in the radial direction than in the axial direction where the axial flow is regarded as flow parallel to the wood fiber grains and the radial flow as flow perpendicular to the wood grains. The permeability in the wood cylinder is therefore an important parameter for the velocity field in the wood. The dry wood radial permeability is 10,000 times lower than the dry wood axial permeability. The chemical composition of the wood/char structure also affects the permeability, where the permeability in char is in order of 1,000 times larger than for wood.

Longitudinal flow becomes important, particularly in specimens having a low ratio of length to diameter, because of the high ratio of longitudinal to transverse permeability. Longitudinal permeability was found to be dependent upon specimen length in the flow direction, that is, the decrease of specimen length appears result in greater permeability in less permeable species.

The effect of drying conditions on gas permeability and preservative treatability was assessed on western hemlock lumber. Although there were no differences in gas permeability between lumber dried at conventional and high temperatures, there were differences in preservative penetration. High temperature drying significantly reduced drying time, but did not appear to affect permeability or shell-to-core MC differences compared with drying at conventional temperature. Pits have a major influence on softwood permeability. Across pits can be impeded by aspiration or occlusion by deposition extractives on the membrane. Drying conditions can significantly affect pit condition, sometimes inducing aspiration that blocks both air and fluid flow. Pressure treatment is presumed to enhance preservative uptake and flow across pits, but the exact impact of pit condition (i.e., open or aspirated) is unknown. Drying conditions may also alter the state of materials deposited on pits, thereby altering the effects of pressure and perhaps the nature of preservative wood interactions. The latter effect may be especially important, since changes in wood chemistry could affect the rates of preservative fixation, which could produce more rapid preservative deposition on pit membranes that would slow further fluid ingress. The longitudinal permeability of the outer heartwood of each species also was determined to evaluate the effect of growth rate on the decrease in longitudinal permeability following sapwood conversion to heartwood. Faster diameter growth produced higher longitudinal permeability in the sapwood of yellow poplar, but not in the sapwood of northern red oak or black walnut. Growth rate had no effect on either vessel lumen area percentage or decrease in longitudinal permeability in newly formed heartwood for all three species. Table 1 represents typical values for gas permeability. Values are given in orders of magnitude.

Darcy's law for liquid flow:

$$k = \frac{flux}{gradient} = \frac{V/(t \times A)}{\Delta P/L} = \frac{V \times L}{t \times A \times \Delta P}, \tag{70}$$

where:

k = Permeability [cm^3(liquid)/ (cm atm sec)]

V = Volume of liquid flowing through the specimen (cm^3)

t = Time of flow (sec)

A = Cross-sectional area of the specimen perpendicular to the direction of flow (cm^2)

ΔP = Pressure difference between ends of the specimen (atm)

L = Length of specimen parallel to the direction of flow (cm).

Darcy's law for gaseous flow:

$$K_g = \frac{V \times L \times P}{t \times A \times \Delta P \times \overline{P}}, \tag{71}$$

where:

K_g = Superficial gas permeability [cm^3(gas)/ (cm atm sec)]

V = Volume of gas flowing through the specimen (cm^3(gas))

P = Pressure at which V is measured (atm)

t = Time of flow (sec)

A = Cross-sectional area of the specimen perpendicular to the direction of flow (cm^2)

ΔP = Pressure difference between ends of the specimen (atm)

L = Length of specimen parallel to the direction of flow (cm)

$\overline{P}$ = Average pressure across the specimen (atm).

TABLE 4 Typical values for gas permeability.

Type of sample	Longitudinal gas permeability [cm^3(gas)/(cm at sec)]
Red oak (R = 150 micrometers)	10,000
Basswood (R = 20 micrometers)	1,000
Maple, Pine sapwood, Coast Douglas-fir sapwood	100
Yellow-poplar sapwood, Spruce sapwood, Cedar sapwood	10
Coast Douglas-fir heartwood	1
White oak heartwood, Beech heartwood	0.1
Yellow-poplar heartwood, Cedar heartwood, Inland Douglasfir heartwood	0.01
Transverse Permeability's	0.001–0.0001
(In approx. same species order as longitudinal)	

To simulate the heat and mass transport in drying, conservation equations for general nonhygroscopic porous media have been developed by Whitaker based on averaging procedures of all of the variables. These equations were further employed and modified for wood drying. Mass conservation equations for the three phases of moisture in local form are summarized in equations.

Water vapor:

$$\frac{\partial}{\partial t}\left(\varphi_g \rho_V\right) = -div\left(\rho_V V_V\right) + \dot{m}_{WV} + \dot{m}_{bV}, \tag{72}$$

Bound water:

$$\frac{\partial}{\partial t}\left(\varphi_s \rho_b\right) = -div\left(\rho_b V_b\right) + \dot{m}_{bV} + \dot{m}_{wb}, \tag{73}$$

Free water:

$$\frac{\partial}{\partial t}(\varphi_w \rho_w) = -div(\rho_w V_w) - \dot{m}_{wv} - \dot{m}_{wb}, \tag{74}$$

Where the velocity of the transported quantity is denoted by V_i, ρ_i is the density, and m_{ij} denotes the transition from phases i and j. From here on, the subscripts w, b, v, and$s$ refer, respectively, to free water, bound water, water vapor, and the solid skeleton of wood. Denoting the total volume by V and the volume of the phase i by V_i, the volumetric fraction of this phase is:

$$\phi_i = \frac{V_i}{V}, \tag{75}$$

with the geometrical constraint:

$$\phi_g + \phi_s + \phi_w = 1, \tag{76}$$

Darcy's law, by using relative permeabilities, provides expressions for the free liquid and gas phase velocities as follows:

$$V^l = -\frac{K_l K_r}{\mu_l} \nabla P_l, \tag{77}$$

and

$$V^v = -\frac{K_v K_r}{\mu_v} \nabla P_v, \tag{78}$$

where K is the intrinsic permeability (m^2), K_r is the relative permeability, P is the pressure (Pa), and μ is the viscosity (Pa.s).

The heat flux (q) and the moisture flux (N_v) are estimated by:

$$q = h(T_G - T_{surf}), \tag{79}$$

$$N_v = \psi K_0 (Y_{surf} - Y_G) = \beta (p_G^v - p_{ats}^v), \tag{80}$$

In which T_{surf}, Y_{surf} and p_s^v are respectively, the wood temperature, the air humidity and the vapor partial pressure at the wood surface and, T_G, Y_G and p_G^v are the corresponding parameters in the air stream. The heat-transfer coefficient is represented by h. The mass-transfer coefficient is β when vapor partial pressure difference is taken as driving force and is k_0 when humidity difference is taken as the driving force with ψ being the humidity factor. The mass-transfer coefficient related to humidity difference is a function of distance along the airflow direction from the inlet side. The heat-transfer coefficient is correlated to the mass-transfer coefficient, as shown by and can be calculated from it. The humidity coefficient φ has been found to vary from 0.70 to 0.76, depending on the drying schedules and board thickness.

For the moisture mass transfer and balance, the moisture loss from wood equals the moisture gain by the hot air, and the moisture transfer rate from the board is described by mass transfer coefficient multiplied by driving force (humidity difference, for example). These considerations yield:

$$-\frac{\partial}{\partial\tau}\left[MC.\rho_s.(1-\varepsilon)\right]=G.\frac{\partial \mathrm{Y}}{\partial X}=\begin{cases}-\psi K_0.a.\left(\mathrm{Y}_{surf}-\mathrm{Y}_G\right)(condensation)\\ \psi K_0.a.f.\left(\mathrm{Y}_{sat}-\mathrm{Y}_G\right)(evaporation)\end{cases}, \qquad (81)$$

Where MC is the wood moisture content, ρ_s is the wood basic density, ε is the void fraction in the lumber stack, a is the exposed area per unit volume of the stack and G is the dry air mass flow rate. In order to solve the above equations, the relative drying rate (f) needs to be defined which is a function of moisture content.

For the heat transfer and balance, the energy loss from the hot air equals the heat gain by the moist wood. The convective heat transfer is described by product of heat transfer coefficient and the temperature difference between the hot air and the wood surface. The resultant relationships are as follows:

$$\begin{aligned}\frac{\partial T_{wood}}{\partial\tau}&=\frac{\left(1+\alpha_R-\alpha_{LS}\right)}{\rho_s.(1-\varepsilon).C_{Pwood}}.\left[h.a.\left(T_G-T_{wood}\right)-G.\Delta H_{wv}.\frac{\partial \mathrm{Y}_G}{\partial X}\right]\\ \frac{\partial T_G}{\partial X}&=\frac{\left(h.a+G.C_{Pv}\frac{\partial \mathrm{Y}_G}{\partial Z}\right).\left(T_G-T_{wood}\right)}{G.\left(C_{Pv}+\mathrm{Y}_G.C_{Pv}\right)}\end{aligned}, \qquad (82)$$

In the above equations, T_{wood} is the wood temperature, α_R and α_{LS} are coefficients to reflect effects of heat radiation and heat loss, C_{Pwood} is the specific heat of wood, and $\Delta \mathrm{H}_{wv}$ is the water evaporation. These equations have been solved to determine the changes of air temperature and wood temperature along the airflow direction and with time.

The energy rate balance (kW) of a drying air adjacent to the wood throughout the wood board can be represented as follows:

$$\frac{1}{2}V_a\rho_{a,mt}cp_{a,mt}\frac{dT_a}{dt}=\frac{1}{2}vA_{cs}cp_{a,mt}\left(T_{a,in-}T_{a,ex}\right)+\dot{Q}_{evap}-\dot{Q}_{conv}, \tag{83}$$

Where $\dot{Q}_{evap}$ and $\dot{Q}_{conv}$ (kW) are the evaporation and convection heat transfer rates between the drying air and wood, which can be calculated as follows:

$$\dot{Q}_{evap}=r\dot{m}_{wv,s}A_{surf}, \tag{84}$$

$$\dot{Q}_{conv}=hA\left(T_a-T_{SO}\right), \tag{85}$$

The specific water vapor mass flow rate ($\dot{m}_{wv,surf}$)(kg/m^2 s) to the drying air can be calculated as follows:

$$\dot{m}_{wv,surf}=\frac{h_D}{R_{wv}T_{SO}}\left(P_{wv,surf}-P_{wv,a}\right), \tag{86}$$

The vapor pressure on the wood surface can be determined from the sorption isotherms of wood. The mass transfer coefficient (h_D)(m/s) can be calculated from the convection heat transfer coefficient (h)(kW/m^2 K) as follows:

$$h_D=h\frac{1}{\rho_{a,mt}cp_{a,mt}Le^{0.58}}/\left(1-\frac{\rho_{wv,m}}{P}\right), \tag{87}$$

Water transfer in wood involves liquid free water and water vapor flow while MC of lumber is above the F.S.P.

According to Darcy's law the liquid free water flux is in proportion to pressure gradient and permeability. So Darcy's law for liquid free water may be written as:

$$J_f=\frac{K_l\rho_l}{\mu_l}.\frac{\partial P_c}{\partial\chi} \tag{88}$$

where:

J_f = liquid free water flow flux, kg/m^2·s,

K_l = specific permeability of liquid water, $m^3(liquid)/m$,

ρ_l = density of liquid water, kg/m^3,

μ_l = viscosity of liquid water, $p_a.s$,

p_c = capillary pressure, p_a,

χ = water transfer distance, m,

$\partial p_c / \partial \chi$ = capillary pressure gradient, p_a / m.

The water vapor flow flux is also proportional to pressure gradient and permeability as follows:

$$J_{vf} = \frac{K_V \rho_v}{\mu_V} \cdot \frac{\partial P_V}{\partial \chi} \tag{89}$$

where:

J_{vf} = water vapor flow flux, $kg/m^2 \cdot s$,

K_V = specific permeability of water vapor, $m^3 (vapor)/m$,

ρ_v, μ_v = density and viscosity of water vapor respectively, kg/m^3 and $p_a \cdot s$,

$\partial p_V / \partial \chi$ =vapor partial pressure gradient, p_a / m.

Therefore, the water transfer equation above F.S.P. during high temperature drying can be written as:

$$\rho_s \frac{\partial (MC)}{\partial t} = \frac{\partial}{\partial x}\left(J_f + J_{vf}\right), \tag{90}$$

where:

ρ_S = basic density of wood, kg/m^3,

MC = moisture content of wood, %,

t = time, s,

$\partial (MC) / \partial t$ = the rate of moisture content change, %/s

x = water transfer distance, m.

Water transfer in wood below F.S.P. involves bound water diffusion and water vapor diffusion. The bound water diffusion in lumber usually is unsteady diffusion; the diffusion equation follows Fick's second law as follows:

$$\frac{\partial (MC)}{\partial t} = \frac{\partial}{\partial x}\left(D_b \frac{\partial (MC)}{\partial x}\right), \tag{91}$$

Where D_b is bound water diffusion coefficient, m^2/s, $\partial(MC)/\partial x$ is MC gradient of lumber, %/m.

The bound water diffusion flux J_b can be expressed as:

$$J_b = D_b \rho_s \frac{\partial(MC)}{\partial x}, \tag{92}$$

where: ρ_s is basic density of wood, kg/m^3.

The water vapor diffusion equation is similar to bound water diffusion equation as follows:

$$\frac{\partial(MC)}{\partial t} = \frac{\partial}{\partial(MC)}\left(D_V \frac{\partial(MC)}{\partial x}\right), \tag{93}$$

where D_V is water vapor diffusion coefficient, m^2/s.

The water vapor diffusion flux can be expressed as:

$$J_V = D_V \rho_s \frac{\partial(MC)}{\partial x}, \tag{94}$$

Therefore, the water transfer equation below F.S.P. during high temperature drying can be expressed as:

$$\rho_s \frac{\partial(MC)}{\partial t} = \frac{\partial}{\partial x}\left(J_b + J_V\right), \tag{95}$$

Two types of wood samples (namely; Guilan spruce and pine) were selected for drying investigation. Natural defects such as knots, checks, splits, etc., which would reduce strength of wood are avoided. All wood samples were dried to a moisture content of approximately 30%.The effect of drying temperature and drying modes on the surface roughness, hardness and color development of wood samples are evaluated.

The average roughness is the area between the roughness profile and its mean line, or the integral of the absolute value of the roughness profile height over the evaluation length:

$$R_a = \frac{1}{L}\int_0^L \left|r(x)dx\right| \tag{96}$$

When evaluated from digital data, the integral is normally approximated by a trapezoidal rule:

$$R_a = \frac{1}{N}\sum_{n=1}^{N}|r_n|, \tag{97}$$

The root-mean-square (rms) average roughness of a surface is calculated from another integral of the roughness profile:

$$R_q = \sqrt{\frac{1}{L}\int_0^L r^2(x)dx}, \tag{98}$$

The digital equivalent normally used is:

$$R_q = \sqrt{\frac{1}{N}\sum_{n=1}^{N} r_n^2}, \tag{99}$$

R_z (ISO) is a parameter that averages the height of the five highest peaks plus the depth of the five deepest valleys over the evaluation length. These parameters, which are characterized by ISO 4287 were employed to evaluate influence of drying methods on the surface roughness of the samples.

We investigated the influence of drying temperatures on the surface roughness characteristics of veneer samples as well. The results showed that the effect of drying temperatures used in practice is not remarkable on surface roughness of the sliced veneer and maximum drying temperature (130°C) applied to sliced veneers did not affect significantly surface roughness of the veneers. Veneer sheets were classified into four groups and dried at 20, 110, 150, and 180°C. According to the results, the smoothest surfaces were obtained for 20°C drying temperature while the highest values of surface roughness were obtained for 180°C. Because some surface checks may develop in the oven-drying process. It was also found in a study that the surface roughness values of beech veneers dried at 110°C was higher than that of dried at 20°C.

In another experimental study, veneer sheets were oven-dried in a veneer dryer at 110 °C(normal drying temperature) and 180°C (high drying temperature) after peeling process. The surfaces of some veneers were then exposed at indoor laboratory conditions to obtain inactive wood surfaces for glue bonds, and some veneers were treated with borax, boric acid and ammonium acetate solutions. After these treatments, surface roughness measurements were made on veneer surfaces. Alder veneers were found to be smoother than beech veneers. It was concluded that the values mean roughness profile (R_a) decreased slightly or no clear changes were obtained in R_a values after the natural inactivation process. However, little increases were obtained for surface roughness parameters, no clear changes were found especially for beech veneers.

The changes created by weathering on impregnated wood with several different wood preservatives were investigated. The study was performed on the accelerated weathering test cycle, using UV irradiation and water spray in order to simulate natural weathering. Wood samples were treated with ammonium copper quat (ACQ 1900 and ACQ 2200), chromated copper arsenate (CCA), Tanalith E 3491 and Wolmanit CX-8 in accelerated weathering experiment. The changes on the surface of the weathered samples were characterized by roughness measurements on the samples with 0, 200, 400 and 600 h of total weathering. Generally the surface values of alder wood treated with copper-containing preservatives decreased with over the irradiation time except for treated Wolmanit CX-82% when comparing values. Surface values of pine treated samples generally increased with increasing irradiation time except for ACQ-1900 groups.

Because the stylus of detector was so sensitive first each sample was smoothened with emery paper then measurement test was performed before and after drying. The Mitutoyo Surface roughness tester SJ-201P instrument was employed for surface roughness measurements. Cut-off length was 2.5 mm, sampling length was 12.5 mm and detector tip radius was 5 μm in the surface roughness measurements. Table 3 and Table 2 displays the changes in surface roughness parameters (R_a, R_z and R_q) of the Pine and Guilan spruce at varying drying methods. In both cases the surface roughness becomes higher during microwave and infrared heating while surface smoothness of both pine and Guilan spruce increased during convection and combined drying. However, the roughness of wood is a complex phenomenon because wood is an anisotropic and heterogeneous material. Several factors such as anatomical differences, growing characteristics, machining properties of wood, pretreatments (e.g., steaming, drying, etc.) applied to wood before machining should be considered for the evaluation of the surface roughness of wood.

TABLE 5 Surface roughness (μm) for pine.

Drying methods	Drying conditions	R_a	R_z	R_q
Microwave	Before drying	4.52	24.68	5.39
	After drying	5.46	30.21	6.62
Infrared	Before drying	4.42	25.52	5.43
	After drying	4.87	26.55	5.69
Convection	Before drying	4.66	26.87	5.86
	After drying	4.08	24.64	5.12
Combined	Before drying	5.23	32.59	6.42
	After drying	3.41	21.7	4.27

TABLE 6 Surface roughness (μm) for Guilan spruce.

Drying methods	Drying conditions	R_a	R_z	R_q
Microwave	Before drying	6.44	34.18	7.85
	After drying	7.77	44.3	9.82
Infrared	Before drying	4.92	30.61	6.30
	After drying	6.42	38.93	8.17
Convection	Before drying	4.97	32.41	6.5
	After drying	4.78	32.27	6.34
Combined	Before drying	10.41	59.5	13.37
	After drying	9.11	54.31	11.5

Hardness represents the resistance of wood to indentation and marring. In order to measure the hardness of wood samples, the Brinell hardness method was applied. In this method a steel hemisphere of diameter 10 mm was forced into the surface under test. The Brinell method measures the diameter of the mark caused by the steel ball in the specimens. The specimens were loaded parallel and perpendicular to the direction of wood grains After applying the force the steel ball was kept on the surface for about 30 seconds. In both type of samples the hardness measured in longitudinal direction is reported to be higher than tangential. The amount of fibers and its stiffness carrying the load are expected to be lower when the load direction is angled to the grain. Results showed that hardness of wood increased in combined drying. The hardness of wood is proportional to its density. The hardness of wood varies, depending on the position of the measurement. Latewood is harder than early wood and the lower part of a stem is harder than the upper part. Increase in moisture content decreases the hardness of wood. It was observed the effect of different drying temperatures during air circulation drying. The result indicates no significant influence of temperature on hardness; still the specimens dried at higher temperature gave a hard and brittle impression. It was also investigated whether wood hardness is affected by temperature level during microwave drying and whether the response is different from that of conventionally dried wood. It was concluded that there is a significant difference in wood hardness parallel to the grain between methods when drying progresses to relatively lower level of moisture content that is, wood hardness becomes higher during microwave drying. Variables such as density and moisture content have a greater influence on wood hardness than does the drying method or the drying temperature.

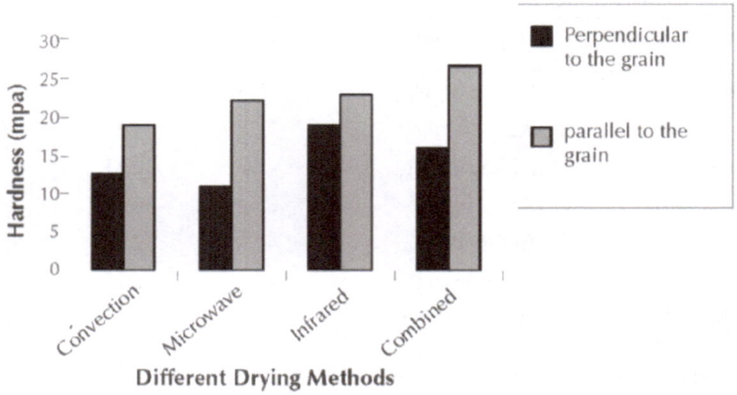

FIGURE 13 Brinell hardness for Guilan spruce.

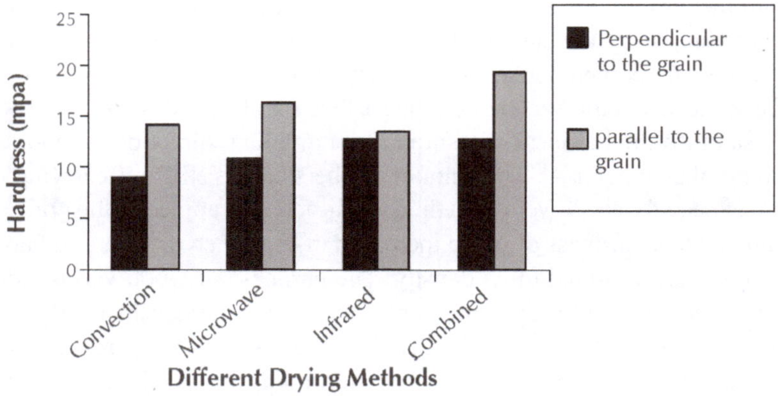

FIGURE 14 Brinell hardness for pine.

Color development of wood surfaces can be measured by using optical devices such as spectrophotometers. With optical measurement methods, the uniformity of color can be objectively evaluated and presented as L*, a* and b* coordinates named by CIEL*a*b* color space values. Measurements were made both on fresh and dried boards and always from the freshly planted surface. Three measurements in each sample board were made avoiding knots and other defects and averaged to one recording. The spectrum of reflected light in the visible region (400–750 nm) was measured and transformed to the CIEL*a*b* color scale using a 10° standard observer and D65 standard illuminant.

These color space values were used to calculate the total color change (ΔE^*) applied to samples according to the following equations:

$$\begin{aligned}\Delta L^* &= L_f^* - L_i^* \\ \Delta a^* &= a_f^* - a_i^* \\ \Delta b^* &= b_f^* - b_i^* \\ \Delta E^* &= \sqrt{(\Delta L^*)^2 + (\Delta a^*)^2 + (\Delta b^*)^2}\end{aligned} \quad , \qquad (100)$$

f and *i* are subscripts after and before drying, respectively.

In this three-dimensional coordinates, L* axis represents nonchromatic changes in lightness from an L* value of 0 (black) to an L* value of 100 (white), +a* represents red, –a* represents green, +b* represents yellow and –b* represents blue.

As can be seen from Figs. 4 and 5 color space values of both pine and Guilan spruce changed after drying.

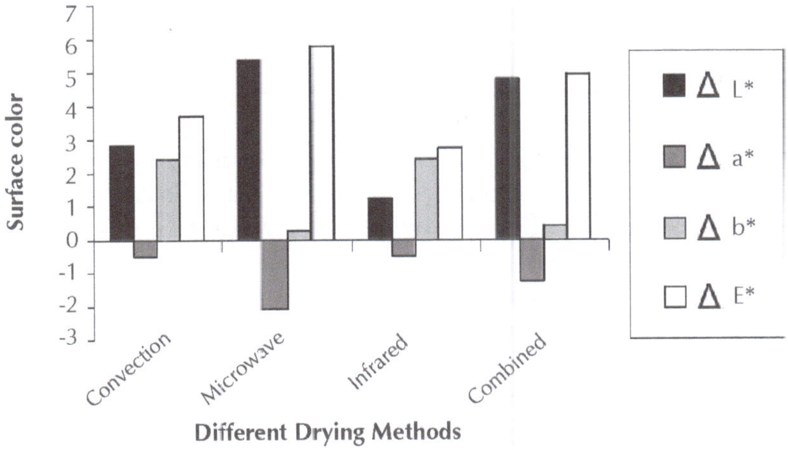

FIGURE 15 Surface color of Pine.

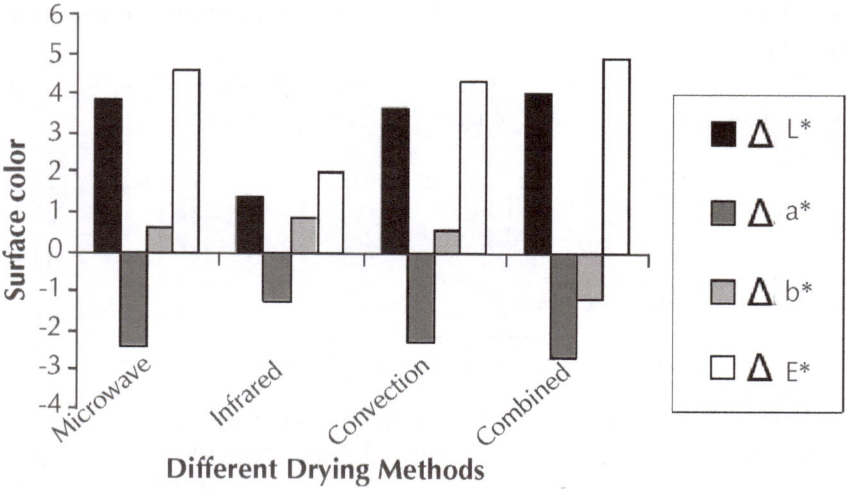

FIGURE 16 Surface color.

Results shows that Δa^*generally decreased but Δb^*increased for both pine and Guilan spruce wood samples except during combined heating. The lightness values ΔL^* increased during drying. The L^* of wood species such as tropical woods, which originally have dark color increases by exposure to light. This is due to the special species and climate condition of pines wood. Positive values of Δb^* indicate an increment of yellow color and negative values an increase of blue color. Negative values of Δa^* indicate a tendency of wood surface to greenish. A low ΔE^* corresponds to a low color change or a stable color. The biggest changes in color appeared in ΔE^* values of pine samples during infrared drying while for Guilan spruce it was reversed. Due to differences in composition of wood components, the color of fresh, untreated wood varies between different species, between different trees of the same species and even within a tree. Within a species wood color can vary due to the genetic factors and environmental conditions. In discoloration, chemical reactions take place in wood, changing the number and type of chromophores.

Discolorations caused by the drying process are those that actually occur during drying and are mainly caused by nonmicrobial factors. Many environmental factors such as solar radiation, moisture and temperature cause weathering or oxidative degradation of wooden products during their normal use; these ambient phenomena can eventually change the chemical, physical, optical and mechanical properties of wood surfaces.

A number of studies have been conducted that have attempted to find a solution to kiln brown stain, the majority of them being pretreatment processes. Biological treatment, compression rolling, sap displacement and chemical inhibitors have all been used as pretreatments.

In all cases these processes were successful in reducing or eliminating stain but were not considered economically viable. Vacuum drying and modified schedules have been tried as modified drying processes with only limited success. Within industry various schedules have been developed, though these are generally kept secret and it is difficult to gauge their success. Generally it seems that industry has adopted a postdrying process involving the mechanical removal of the kiln brown stain layer [555–108].

9.7 CONCLUSION

Microwave processing of materials is a relatively new technology that provides new approaches to improve the physical properties of materials. Microwave drying generate heat from within the grains by rapid movement of polar molecules causing molecular friction and help in faster and more uniform heating than does conventional heating. If wood is exposed to an electromagnetic field with such high frequency as is characteristic for microwaves, the water molecules, which are dipoles, begin to turn at the same frequency as the electromagnetic field. Wood is a complex composite material, which consists mainly of cellulose (40–45%), hemicelluloses (20–30%) and lignin (20–30%). These polymers are also polar molecules, and therefore even they are likely to be affected by the electromagnetic field. This could possibly cause degradation in terms wood hardness. For spruce the average of hardness is shown to be much higher than pine. From the experimental results it can be observed that in combined microwave dryer, the hardness was relatively improved in comparison to the other drying methods. Microwave and infrared drying can increase wood surface roughness while the smoothness of wood increases during convection and combined drying. The effect varies with the wood species. Thus this work suggests keeping the core temperature below the critical value until the wood has dried below fiber saturation as one way of ensuring that the dried wood is acceptably bright and light in color. In the model presented in this chapter, a simple method of predicting moisture distributions leads to prediction of drying times more rapid than those measured in experiments. From this point of view, the drying reveals many aspects which are not normally observed or measured and which may be of value in some application.

The derivation of the drying curves is an example. It is clear from the experiments over a range of air velocities that it is not possible to make accurate predictions and have the experimental curves coincide at all points with the predicted distributions simply by introducing a VCF into the calculations. This suggest that a close agreement between calculated and experimental curves over the entire drying period could be obtained by using a large value of VCF in the initial stages of drying and progressively decreasing it as drying proceeds.

KEYWORDS

- **conservation equations**
- **method of characteristics**
- **microwave**
- **moisture distributions**
- **spectrophotometers**

CHAPTER 10

DYNAMIC MODELING FOR MASS AND MOMENTUM TRANSPORT

CONTENTS

10.1 INTRODUCTION

In the two phases flow is extremely important to the concept of volume concentration. This is the relative volume fraction of one phase in the volume of the pipe. Such an environment typical fluid is a high density and little compressibility.

Two-phase flows exhibit various flow regimes, or flow patterns, depending on the relative concentration of the two phases and the flow rate. A simple but generally adequate set of descriptive phrases for most of the important liquid-vapor flow regimes consists of bubble flow, slug flow. Churn flow, annular flow, and droplet flow. Bubble flow describes the flow of distinct, roughly spherical vapor regions surrounded by continuous liquid. The bubble diameter is generally considerably smaller than that of the container through which they flow, Bubble flow usually occurs at low vapor concentrations.

If the vapor and liquid are flowing through a pipe, bubbles may coalesce into long vapor regions that have almost the same diameter as the pipe. This is called slug flow. At moderate to high flow velocities and roughly equal concentrations of vapor and liquid, the flow pattern is often very irregular and chaotic. If the flow contains no distinct entities with spherical or, in a pipe, cylindrical symmetry, it is said to be churn flow. At high vapor concentration, the liquid may exist as a thin film wetting the pipe wall (annular flow) or as small. Roughly spherical droplets in the vapor stream (droplet flow). If both a thin film and droplets exist, the flow is described as annular droplet flow.

Rapid transient pressure fluctuations can result in vibrations or resonance that can cause even flanged pipes and fittings (bend and elbows) to dislodge, resulting in a leak or rupture. In fact, the cavitations that commonly occur with water hammer can – as the phenomenon's name implies – release energy that sounds like someone pounding on the pipe with a hammer.

Thermal and species fluxes occur during the exchange of energy and partial mass through the interface between the vapor and liquid phases and must be suitably modeled. In certain systems, it is also important to describe also the dynamics of the surface phase that can be nontrivial, due to the flows in the surrounding bulk phases perturbing the interface mechanical equilibrium.

The real behavior of liquid mixture in heat pipe needs the integration of continuity, momentum, energy and species equations obeying the appropriate entropy production laws in a context of irreversible processes thermodynamics in the space-time. In this work the effect of braking of intensity of phase transitions was analyzed from experiments in binary bubbly mixtures.

At numerous experiments this effect is reached by selection and change of concentration of components of mixes. The offered parameter allows choosing a necessary binary mix theoretically.

Results of calculations for concrete binary mixes were resulted. The application of offered parameter in power engineering, chemical, cryogenic technology was discussed.

In various branches of manufacture there is a necessity of use of liquids for cooling of heating up surfaces. Thus the reasons of allocation of a considerable quantity of heat can be various. For example, in engines of cars in the course of a friction and fuel ignition, at machine tool work, influence of a rotating drill on metal details and etc.

In all resulted examples cooling surfaces is reached by influence of liquids. For more effective cooling binary mixtures (mix of two liquids) are used. In many cases mixtures should be also cold-resistant (i.e., not to freeze). In language of mechanics, it means that the liquid should have the minimum speed of phase transformations (for example, boiling up and freezing).

Structure and concentration of a component for binary solution are selected by practical consideration. Structure is varied when speed of phase transformations is become low. A lot of time and the expensive equipments are required. Thus the considerable amounts of reagents are spent.

Water is so much a part of our lives that we seldom question its role. Yet water possesses a unique combination of chemical, physical, and thermal properties that make it ideally suited for many purposes. In addition, although important regional shortages may exist, water is found in large quantities on the surface of the earth. For both these reasons, water plays a central role in both human activity and natural processes.

One surprising feature of the water molecule is its simplicity, formed as it is from two diatomic gases, hydrogen (H_2) and oxygen (O_2). Yet the range and variety of water's properties are remarkable. Some property values in the design tables-especially density and viscosity values-are used regularly by pipeline engineers. Other properties, such as compressibility and thermal values, are used indirectly, primarily to justify modeling assumptions, such as the flow being isothermal and incompressible. Many properties of water depend on intermolecular forces that create powerful attractions (cohesion) between water molecules. That is, although a water molecule is electrically neutral, the two hydrogen atoms are positioned to create a tetrahedral charge distribution on the water molecule, allowing water molecules to be held strongly together with the aid of electrostatic attractions. These strong internal forces-technically called "hydrogen bonds"-arise directly from the nonsymmetrical distribution of charge.

The chemical behavior of water also is unusual. Water molecules are slightly ionized, making water an excellent solvent for both electrolytes and no electrolytes. In fact, water is nearly a universal solvent, able to wear away mountains, transport solutes, and support the biochemistry of life. But the same properties that create so many benefits also create problems, many of which must be faced by the pipeline engineer.

Toxic chemicals, disinfection by-products, aggressive and corrosive compounds, and many other substances can be carried by water in a pipeline, possibly causing damage to the pipe and placing consumers at risk. Other challenges also arise. Water's almost unique property of expanding on freezing can easily burst pipes. As a result, the pipeline engineer either may have to bury a line or may need to supply expensive heat-tracing systems on lines exposed to freezing weather, particularly if there is a risk that standing water may sometimes occur. Furthermore, the selected properties of:

10.1.1 WATER PHYSICAL PROPERTIES

1. High density – 1000 (kg/m^3)
2. Density maximum at 4 (°C) – i.e., above freezing.
3. High surface tension – 0.073 (N/m)
4. High bulk modulus (usually assumed incompressible) – 2.07 (GPa)

10.1.2 THERMAL PROPERTIES

1. Specific heat-highest except for NH_3 – 4.187 (kJ/kg°C)
2. High heat of vaporization – 2.45 (MJ/kg)
3. High heat of fusion – 0.36 (MJ/kg)
4. High boiling point – c.f., H_2 20 (K), O_2 90 (K) and H_2O, 373(K)
5. Good conductor of heat relative to other liquids and nonmetal solids.

10.1.3 CHEMICAL AND OTHER PROPERTIES

1. Slightly ionized-water is a good solvent for electrolytes and no electrolytes.
2. Transparent to visible light; opaque to near infrared.
3. High dielectric constant-responds to microwaves and electromagnetic fields.

The values are approximate. All the properties listed are functions of temperature, pressure, water purity, and other factors that should be known if more exact values are to be assigned. For example, surface tension is greatly influenced by the presence of soap films, and the boiling point depends on water purity and confining pressure. The values are generally indicative of conditions near 100(°C) and one atmosphere of pressure. Combination of its high density and small compressibility creates potentially dramatic transient conditions. In two-phase flow, the concept of hold-up is important. It is the relative fraction of one phase in the pipe. This is not necessarily equal to the relative fraction of that phase in the entering fluid mixture. The usual question for the engineer is that of calculating the pressure drop required to achieve specified flow rates of the gas and the liquid through a pipe of a given diameter. To make design calculations involving two-phase flow, Perry's Handbook is a useful resource. Also, an informative chapter in Holland and Bragg is devoted to gas-liquid two-phase flow.

10.2 MATERIALS AND METHODS

When one uses solutions, one must takes into account a more complex modeling of the dynamics and thermodynamics of the fluid, vapor and surface phases than in the case of a simple fluid. In fact, the bulk phases have more specific degrees of freedom and there are also more equilibrium equations to satisfy one for each component of the solution/mixture governing the equilibrium of the bulk phases. Other aspects that one needs to take into account are the possibility that both the vapor and the liquid phases may have a non ideal behavior and finally that the latent heat of phase change depends not only on the properties of the pure components of the solution but also on composition and on the exchange of partial masses.

In general also the thermodynamics of the surface phase, when relevant, gets more complicated due to the presence of a larger number of components. It is clear that dealing with a transfer process one is certainly out of equilibrium conditions. Thermal and species fluxes occur during the exchange of energy and partial mass through the interface between the vapor and liquid phases and must be suitably modeled. In certain systems, it is also important to describe also the dynamics of the surface phase that can be nontrivial, due to the flows in the surrounding bulk phases perturbing the interface mechanical equilibrium.

Real effects in solution also affect the characteristics of the equilibrium between the phases strongly modifying the phase diagrams with respect to ideal systems. Also the flow establishing in the bulk phases must be suitably characterized taking into account the presence of more species. The real behavior of liquid mixture in heat pipe needs the integration of continuity, momentum, energy and species equations obeying the appropriate entropy production laws in a context of irreversible processes thermodynamics in the space-time. Also the heat pipe geometry is extremely important in order to determine its actual functioning performances and requires an appropriate description involving also the initial and boundary conditions for the field equations.

The thermodynamics of the vapor-liquid equilibrium "VLE" in presence of multicomponent systems requires the introduction of molar partial properties, fugacity, fugacity coefficients, activities and activity coefficients, involves Raoult's and/or Henry's laws, the determinations of the properties change of mixing for both the bulk phases, the expression of partial molar properties in terms of molar properties and the proper formulation of the local field equations in jump form at the interface between the bulk phases for the fields variables playing the (key) role of boundary conditions for the bulk balance equations in which the classical "VLE" relations occur.

During a hydraulic transient event, the hydraulic-grade line (HGL), or head, at some locations may drop low enough to reach the pipe's elevation, resulting in subatmospheric pressures or even full-vacuum pressures. Some of the water may flash from liquid to vapor while vacuum pressures persist, resulting in a temporary water-column separation. When system pressures increase again, the vapor condenses to liquid as the water columns accelerate toward each other (with nothing to slow them down unless air entered the system at a vacuum breaker valve) until they collapse the vapor pocket; this is the most violent and damaging water hammer phenomenon possible.

Two-phase flow is a difficult subject principally because of the complexity of the form in which the two fluids exist inside the pipe, known as the flow regime.

It is difficult to construct a model from first principles in all but the most elementary situations. The usual question for the engineer is that of calculating the pressure drop required to achieve specified flow rates of the gas and the liquid through a pipe of a given diameter.

The experimental results show that the bubble flow usually occurs at low concentrations of vapor. It includes three main types of flow regimes in microgravity bubble, slug and an annular.

The Two-Fluid Model Analysis of two-phase flow begins with the most general principles governing the behavior of all matter, namely, conservation of mass, momentum, and energy. These principles can be expressed mathematically at every point in space and time by local, instantaneous field equations.

10.2.1 LAWS OF CONSERVATION

Although the implications of the characteristics of water are enormous, no mere list of its properties will describe a physical problem completely. Whether we are concerned with water quality in a reservoir or with transient conditions in a pipe, natural phenomena also obey a set of physical laws that contributes to the character and nature of a system's response.

If engineers are to make quantitative predictions, they must first understand the physical problem and the mathematical laws that model its behavior. Basic physical laws must be understood and be applied to a wide variety of applications and in a great many different environments: from flow through a pump to transient conditions in a channel or pipeline. The derivations of these equations are not provided, however, because they are widely available and take considerable time and effort to do properly. Instead, the laws are presented, summarized, and discussed in the pipeline context.

More precisely, a quantitative description of fluid behavior requires the application of three essential relations:

(1) A kinematics relation obtained from the law of mass conservation in a control volume,
(2) Equations of motion provided by both Newton's second law and the energy equation, and
(3) An equation of state adapted from compressibility considerations, leading to a wave speed relation in transient flow and justifying the assumption of an incompressible fluid in most steady flow applications.

10.2.2 CONTINUITY EQUATION FOR UNSTEADY FLOW

The continuity equation for a fluid is based on the principle of conservation of mass. The general form of the continuity equation for unsteady fluid flow is as follows:

$$(\partial H / \partial t) + (V dH / \partial x) + \left(a^2 \partial V / g \partial x\right) = 0,$$

The second term on the left-hand side of the preceding equation is small relative to other terms and is typically neglected, yielding the following simplified continuity equation, as used in the majority of unsteady models:

$$(\partial H / \partial t) + \left(a^2 \partial V / g \partial x\right) = 0,$$

10.2.3 CONSTITUTIVE RELATIONS

Any change in flow or pressure, at any point in the system, can trigger hydraulic transients. If the change is gradual, the resulting transient pressures may not be severe. However, if the change of flow is rapid or sudden, the resulting transient pressure can cause surges or water hammer. Since each system has a different characteristic time, the qualitative adjectives gradual and rapid correspond to different quantitative time intervals for each system.

There are many possible causes for rapid or sudden changes in a pipe system, including power failures, or a rapid valve opening or closure. These can result from natural causes, equipment malfunction, or even operator error. It is therefore important to consider the several ways in which hydraulic transients can occur in a system and to model them. If identifying, modeling, and protecting against several possible hydraulic transient events seems to take a lot of time and resources, remember that it is far safer and less expensive to learn about system's vulnerabilities in a computer model and far easier to clean up – than from expensive service interruptions and field repairs.

The purpose of this type of transient analysis is to ensure the system and its components can withstand the resulting transient pressures and determine how long to wait for the transient energy to dissipate. For many systems, starting backup pumps before the transient energy has decayed sufficiently can cause worse surge pressures than those caused by the initial power failure.

Conversely, relying on rapid backup systems to prevent transient pressures may not be realistic given that most transient events occur within seconds of the power failure while isolating the electrical load, bringing the generator on-line, and restarting pumps (if they have not timed out) can take several minutes.

10.2.4 INTERFACIAL MASS AND ENERGY EXCHANGE

10.2.4.1 CONSERVATION OF MASS

The central concept is that of conservation of mass, and its key expression is the continuity or mass conservation equation. One remarkable fact about changes in a physical system is that not everything changes. In fact, most physical laws are conservation laws: They are generalized statements about regularities that occur in the midst of change.

10.2.4.2 CONSERVATION OF MASS AT STEADY STATE

At any node in a system containing incompressible fluid, the total volumetric or mass flows in must equal the flows out, less the change in storage. Separating these into flows from connecting pipes, demands, and storage, gives the continuity equation:

$$\sum Q_{IN}\Delta t = \sum Q_{OUT}\Delta t + \Delta Vs,$$

where, Q_{IN} – total flow into the node (m^3/s), Q_{OUT} – total demand at the node (m^3/s), ΔV_S – change in storage volume (m^3), Δt – change in time (s).

10.2.4.3 CONSERVATION OF ENERGY

The first law of thermodynamics states that for any given system and time interval, the change in total energy is equal to the difference between the heat transferred to the system and the work done by the system on its surroundings. In hydraulic terms, changes in the total energy of a fluid do not consider changes in its internal (molecular) forms of energy, such as electrical and chemical energy, because these are usually relatively small. In hydraulic terms, energy is often represented as energy per unit weight, resulting in units of length. At any point in a hydraulic system, the total energy of a fluid consists of three components that can be expressed as an equivalent elevation, or head:

Pressure Head: p/γ Elevation Head: z Velocity Head: $V^2/2g$

where: p-pressure (N/m^2), γ-specific weight (N/m^3), z-elevation (m), V-velocity (m/s), g- gravitational acceleration constant (m/s^2, ft/s^2).

Converting the total energy to an equivalent head allows it to be plotted on the same scale as elevation for any point in the system, either on pipeline profiles or maps, allowing engineers to visualize changes as slopes or contour lines, respectively. This gives a better feel for the resulting behavior of the system, especially when reviewing the results of an EPS or transient analysis. Further, the difference between this energy level and the pipeline elevation is equal to the total gauge pressure.

10.2.4.4 CONSERVATION OF ENERGY AT STEADY STATE

The conservation of energy principle states that the head losses through the system must balance at each point. For pressure networks, this means that the total head loss between any two nodes in the system must be the same regardless of what path is taken between the two points. The sign of the head loss must be consistent with the assumed flow direction (i.e., gain head when proceeding opposite the flow direction and lose head when proceeding in the flow direction). The same basic principle can be applied to any path between two points. The combined head loss around a loop must be zero to achieve the same hydraulic grade as at the beginning. Governing Equations for Unsteady (or Transient) Flow Hydraulic transient flow is also known as unsteady fluid flow. During a transient analysis, the fluid and system boundaries can be either elastic or inelastic:

- Elastic theory describes unsteady flow of a compressible liquid in an elastic system (e.g., where pipes can expand and contract).
- Rigid-column theory describes unsteady flow of an incompressible liquid in a rigid system. It is only applicable to slower transient phenomena. Both branches of transient theory stem from the same governing equations.

The continuity equation and the momentum equation are needed to determine V and p in a one-dimensional flow system. Solving these two equations produces a theoretical result that usually corresponds quite closely to actual system measurements if the data and assumptions used to build the numerical model are valid.

Transient analysis results that are not comparable with actual system measurements are generally caused by inappropriate system data (especially boundary conditions) and inappropriate assumptions.

10.2.5 INTERFACIAL MOMENTUM EXCHANGE

10.2.5.1 MOMENTUM EQUATION FOR UNSTEADY FLOW

The equations of motion for a fluid can be derived from the consideration of the forces acting on a small element, or control volume, including the shear stresses generated by the fluid motion and viscosity. The three-dimensional momentum equations of a real fluid system are known as the Navier-Stokes equations. Since flow perpendicular to pipe walls is approximately zero, flow in a pipe can be considered one-dimensional, for which the continuity equation reduces to:

$$(\partial V / \partial t)+(V\partial V / \partial x)+(gdH / \partial x)+(fV \ |V| / 2D)=0,$$

The last term on the left-hand side represents friction losses in the direction of flow:

$$fV|V|/2D$$

where: f – Darcy-Weisbach friction coefficient, D – inside diameter of the pipe (or equivalent Dimension),V – velocity of fluid, γ – specific weight of the fluid.

The first term on the left-hand side is the local acceleration term, while the second term represents the convective acceleration, proportional to the spatial change of velocity at a point in the fluid, which is often neglected to yield the following simplified equation:

$$(\partial V/\partial t)+(gdH/\partial x)+(fV|V\ |/2D)=0,$$

Equations though rigorous and explicit, incorporate the following assumptions, which are often not strictly valid in real water systems:

- Fluid is homogeneous – water typically incorporates a small amount of dissolved and/or entrained air whose exact percentage changes along the system.
- Fluid and pipe wall are linearly elastic – in aging water pipes whose shape has become noncircular and whose integrity may be compromised by cracks virtually every water system leaks), fluid may escape the system rather than being compressed and deformations imposed on piping may not be entirely recovered.
- Flow is one-dimensional – this assumption has been shown to be inaccurate at tees in suction lines. Minor losses result from three-dimensional vortices.
- Pipe flows full – even in pressurized systems, air or vapor can accumulate at local high points, forcing the water to accelerate and pass underneath it. In extreme cases, this phenomenon can significantly diminish pumping efficiency (e.g., vapor lock).
- Average velocity is used – experiments show that the velocity distribution changes across a cross section during transient events, even for flow in straight pipes.
- Viscous losses similar to steady state – emerging research in transient or unsteady friction is challenging this assumption. Nevertheless, these assumptions are essentially valid for the majority of the time in the majority of water systems. Solving these equations yields accurate numerical simulation results in most cases.

The representative system length, L, can be approximated for a network by taking the longest path connecting a pump to a storage element, such as a tank or reservoir. Flow-control operations that occur over an interval longer than the characteristic time are designated "gradual" or "slow". The characteristic time is significant in transient flow analysis because it dictates which method is applicable for evaluating a particular flow-control operation in a given system.

The rigid model provides accurate results only for surge transients generated by slow flow-control operations that do not cause significant liquid compression or pipe deformation. Instantaneous, rapid, and gradual changes must be analyzed with the elastic model.

In addition to the equations describing transient flow, it is important to know about the effect of boundaries – such as tanks, dead ends, and pipe branches – that modify the effects of hydraulic transient phenomena.

Hydraulic systems commonly have interconnected pipelines with differing characteristics, such as material and diameter. These pipeline segments and connection points (nodes) define a system's topology. When a wave traveling in a pipe and defined by a head pulse H_o comes to a node, it is transmitted with a head value H_s to all other connected pipes and reflects back to the initial pipe with a head value H_r. The wave reflection occurring at a node changes the head and flow conditions in each of the pipes connected to the node. If the distances between the pipe connections are small, the head at all connections can be assumed to be the same (that is, the head loss through the node is negligible), and the transmission factor (s) can be defined as:

$$S = \left(\Delta Hs / \Delta_O\right) = \left(2A_0 / a_0\right) / \sum_{i=0}^{n} A_i / a_i \,,$$

Where :s – transmission factor (dimensionless),H_o – incident head pulse (m), H_s – head of transmitted wave (m), A_o – incoming pipe area (m^2), a_o – incoming wave speed (m/s), A_i – area of i-th pipe (m^2), a_i – wave speed of i-th pipe (m/s), n – number of outgoing pipes, i – pipe number index.

In a closed system without friction to dampen transients, transients would persist indefinitely. However, viscous and friction effects typically cause transients to attenuate within seconds to minutes.

As an essential tool to keep track of the transient pressure wave reflections and the friction and elastic effects during the simulation, as follows:

- Because friction does exist in an actual system, the potential head change calculated using the Joukowski equation underestimates the actual head rise. This underestimation is due to packing – an additional increase in head occurring at the valve as the pressure wave travels upstream.

- The small velocity behind the wave front means that the velocity difference across the wave front is less than V_o, so the pressure change is progressively less than the potential surge as the wave travels upstream. This effect, which is concurrent with line packing, is called attenuation or reduction.
- Transient pressure waves are partially transmitted and simultaneously reflected back at every junction with other pipes, depending on their wave speed and diameter.

10.2.5.2 MODELING CAPABILITIES

The differences between computer model results and actual system measurements were caused by several factors, including the following difficulties:

1. Precise determination of the pressure wave speed for the piping system was difficult, if not impossible. This was especially true for buried pipelines, whose wave speeds were influenced by bedding conditions and the compaction of the surrounding soil.
2. Precise modeling of dynamic system elements (such as valves, pumps, and protection devices) was difficult because they were subject to deterioration with age and adjustments made during maintenance activities. Measurement equipment may also be inaccurate.
3. Unsteady or transient friction coefficients and losses depend on fluid velocities and accelerations. These were difficult to predict and calibrate even in laboratory conditions.
4. Prediction of the presence of free gases in the system liquid was sometimes impossible.

These gases can significantly affect the pressure wave speed. In addition, the exact timing of vapor-pocket formation and column separation was difficult to simulate. Calibrating model parameters based on field data minimized the first source of error.

Unsteady or transient friction coefficients and the effects of free gases are more challenging to account for. Fortunately, friction effects are usually minor in most water systems and vaporization can be avoided by specifying protection devices and/or stronger pipes and fittings able to withstand sub atmospheric or vacuum conditions, which are usually short-lived.

For present research with free gas systems and the potential for water-column separation, the numerical simulation of hydraulic transients was more complex. Small pressure spikes caused by the type of tiny vapor pockets that was difficult to simulate accurately seldom result in a significant change to the transient envelops. Larger vapor-pocket collapse events resulting in significant upsurge pressures were simulated with enough accuracy to support definitive conclusions.

TABLE 1 Simulators, models and problems.

Cases	Range of problems
Steady or gradually varying turbulent flow	Rapidly varying or transient flow
Incompressible, Newtonian, single-phase fluids	Slightly compressible, two-phase fluids (vapor and liquid) and two-fluid systems (air and liquid)
Full pipes	Closed-conduit pressurized systems with air intake and release at discrete points

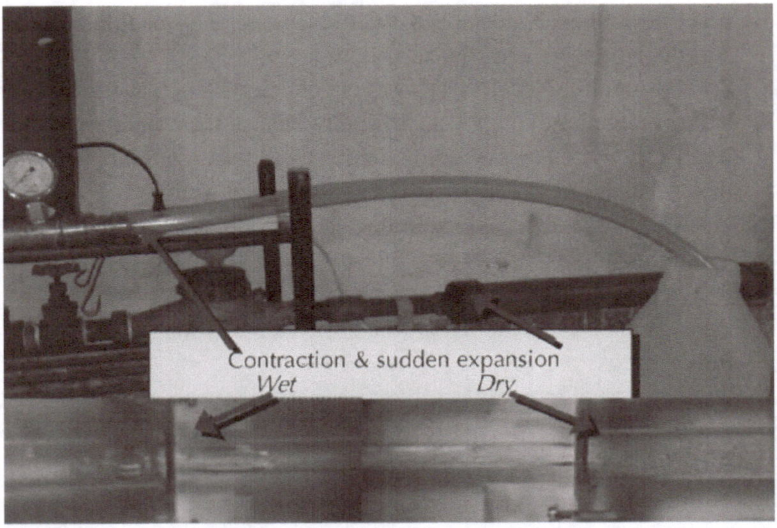

FIGURE 1 Snapshot laboratory setup for studying the structure of flow.

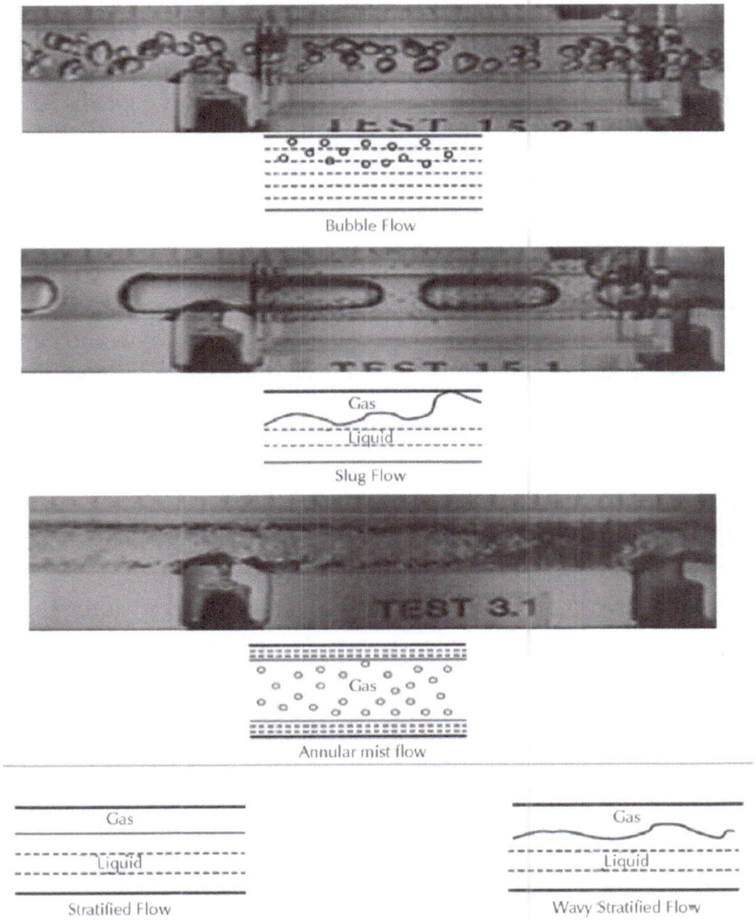

FIGURE 2 Types of flows.

Figure 1 shows an experimental setup, which investigated the formation of different modes of fluid flow with gas bubbles and steam. Fig. 2 shows snapshots of the flow pattern for various configurations of the tube [4–5]. The flow enters the tee at the bottom of the picture, and then is divided at the entrance to the tube. The inner diameter of the tee is 1.27 cm (Fig.1). The narrowing of the flow is achieved by reducing the diameter of the hose.

Within the reduction is a liquid recirculation zone, called the "vena contraction" Wet picture that, when the liquid is recirculated to the "vena contraction." However, there are conditions, whereby the gas phase of the contract is caught vena. Fig. 3 shows the different flow regimes observed in the experimental setup [6].

Increased flow is achieved by increasing the size of the pipe. Again, there is an area of the liquid recirculation near the "corner" a sudden expansion. Depending on the level of consumption of bubbly liquid or gas, it falls into the trap in this area.

In the inlet fluid moves out of the pipe diameter of 12.7 (*mm*) in the pipe diameter of 25 (*mm*).

Normal extension occurs at the beginning of the flow. Soon comes the expansion section, and the flow rate continues to increase. Two-phase jet stream created, ultimately, with areas of air flowing above and below the bubble region [6]. The behavior of gas-liquid mixture in the expansion is proportional to increasing the diameter of the pipe [7–8].

It is shown in Figs.1–3 that in place of the sudden expansion of a transition flow regime of the turbulent flow. Depending on the flow or gas bubble mixture it falls into the trap in this area.

Experiments were conducted on the pipe, whose diameter is suddenly doubled [9]. In this case the region of turbulence of fluid are observed around the "corner" a sudden expansion. The expansion is observed at the beginning of the flow. As a result of turbulence flow gap expansion increases and the flow rate continues to increase. In the end, creates a stream of two-phase flow, with airfields, the current above and below the bubble area [10–11].

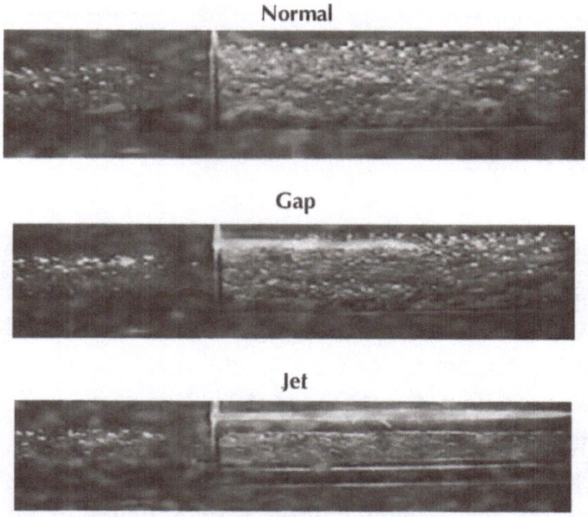

FIGURE 3 Narrowing and sudden expansion flow level.

With this experimental setup is shown that the formation of different modes of two-phase flow depends on the relative concentration of these phases and the flow rate. Fig. 1 and Fig.2 shows a diagram of core flow of vapor-liquid flow regimes, in particular, the bubble, stratified, slug, stratified, and the wave dispersion circular flow.

In these experiments the mode of vapor-liquid flow in a pipe, when the bubbles are connected in long steam field, whose dimensions are commensurate with the diameter of the pipe [12–13].This flow is called the flow of air from the tube. In the transition from moderate to high-speed flow, when the concentration of vapor and liquid are approximately equal, the flow regime is often irregular and even chaotic. By the simulated conditions, It is assumed that the electricity suddenly power off without warning (i.e., no time to turn the diesel generators or pumps) [14–16].

In this chapter, mathematical model described existence processes. Theoretical method for make decision suggested suitable solution for problem. By means of the received results of the offered model and without resorting to experiments, necessary binary mixtures were restricted.

10.3 RESULTS

Such situations are the strong reason of the installation of pressure sensors, equipped with high-speed data loggers. Therefore, the following items are consequences, which may result in these situations.

Hydraulic transients can lead to the following physical phenomena. High or low transient pressures arise in the piping and connections. They often alternate from highest to lowest levels and vice versa. High pressures are a consequence of the collapse of steam bubbles or cavities is similar to steam pump cavitations. It can yield the tensile strength of the pipes. It can also penetrate the groundwater into the pipeline [17–18].

High-speed flows are also very fast pulse pressure. It leads to temporary but very significant transient forces in the bends and other devices that can make a connection to deform. Even strain buried pipes under the influence of cyclical pressures may lead to deterioration of joints. In the low-pressure pumping stations at downstream a very rapid closing of the valve, known as shut off valve, may lead to high pressure transient flows.

Water column, usually are separated with sharp changes in the profile or the local high points. It is because of the excess of atmospheric pressure. The spaces between the columns are filled with water or the formation of steam (e.g., steam at ambient temperature) or air, if allowed admission into the pipe through the valve. Collapse of cavitation bubbles or steam can cause the dramatic impact of rising pressure on the transition process.

If the water column is divided very quickly, it could in turn lead to rupture of the pipeline. Vapors cavitation may also lead to curvature of the pipe. High-pressure wave can also be caused by the rapid removal of air from the pipeline.

Steam bubbles or cavities are generated during the hydraulic transition. The level of hydraulic pressure (EGD) or pressure in some areas could fall low enough to reach the top of the pipe. It leads to sub-atmospheric pressure or even full-vacuum pressures. Part of the water may undergo a phase transition, changing from liquid to steam, while maintaining the vacuum pressure.

This leads to a temporary separation of the water column. When the system pressure increases, the columns of water rapidly approach to each other. The pair reverts to the liquid until vapor cavity completely dissolved. This is the most powerful and destructive power of water hammer phenomenon.

If system pressure drops to vapor pressure of the liquid, the fluid passes into the vapor, leading to the separation of liquid columns. Consequently, the vapor pressure is a fundamental parameter for hydraulic transient modeling. The vapor pressure varies considerably at high temperature or altitude.

When the source of the pressure fluctuations is the so-called, "Acceleration factor", one can say that in order to accelerate the mass of fluid in the system until the new rate of additional efforts. The source of the pressure fluctuations can be related to dynamics of heat and mass transfer.

10.3.1 DYNAMICS OF HEAT AND MASS TRANSFER

The dynamics of heat and mass transfer of vapor bubble in a binary solution of liquids, in Ref. [8], was studied for significant thermal, diffusion and inertial effect.

It was assumed that binary mixture with a density ρ_l, consisting of components 1 and 2, respectively, the density ρ_1 and ρ_2.

Moreover:

$$\rho_1 + \rho_2 = \rho_l ,$$

where, the mass concentration of component 1 of the mixture.

Also Ref. [8] considers a two-temperature model of interphase heat exchange for the bubble liquid. This model assumes homogeneity of the temperature in phases.

The intensity of heat transfer for one of the dispersed particles with an endless stream of carrier phase will be set by the dimensionless parameter of Nusselt Nu_l.

Bubble dynamics described by the Rayleigh equation:

$$R\dot{w}_l + \frac{3}{2}w_l^2 = \frac{p_1 + p_2 - p_\infty - 2\sigma / R}{\rho_l} - 4\nu_1 \frac{w_l}{R} \tag{1}$$

where p_1 and p_2 – the pressure component of vapor in the bubble, p_∞ – the pressure of the liquid away from the bubble, σ and ν_1 – surface tension coefficient of kinematic viscosity for the liquid.

Consider the condition of mass conservation at the interface.

Mass flow j_i^{TH} component $(i = 1,2)$ of the interface $r = R(t)$ in j_i^{TH} phase per unit area and per unit of time and characterizes the intensity of the phase transition is given by:

$$j_i = \rho_i\left(\dot{R} - w_l - w_i\right), \quad (i = 1,2), \tag{2}$$

where: w_i – the diffusion velocity component on the surface of the bubble. The relative motion of the components of the solution near the interface is determined by Fick's law:

$$\rho_1 w_1 = -\rho_2 w_2 = -\rho_l D \left.\frac{\partial k}{\partial r}\right|_R \tag{3}$$

If we add Eq. (2), while considering that:

$\rho_1 + \rho_2 = \rho_l$ and draw the Eq. (3), we obtain,

$$\dot{R} = w_l + \frac{j_1 + j_2}{\rho_l}, \tag{4}$$

Multiplying the first Eq.(2) on ρ_2, the second in ρ_1 and subtract the second equation from the first. In view of Eq. (3) we obtain

$$k_R j_2 - (1 - k_R) j_1 = -\rho_l D \left.\frac{\partial k}{\partial r}\right|_R$$

where k_R – the concentration of the first component at the interface. With the assumption of homogeneity of parameters inside the bubble changes in the mass of each component due to phase transformations can be written as $\frac{d}{dt}\left(\frac{4}{3}\pi R^3 \rho_i^{/}\right) = 4\pi R^2 j_i$ or

$$\frac{R}{3}\dot{\rho}_i^{/} + \dot{R}\rho_i^{/} = j_i, \ (i = 1,2), \tag{5}$$

Express the composition of a binary mixture in mole fractions of the component relative to the total amount of substance in liquid phase

$$N = \frac{n_1}{n_1 + n_2}, \tag{6}$$

The number of moles i^{TH} component n_i, which occupies the volume V, expressed in terms of its density:

$$n_i = \frac{\rho_i V}{\mu_i}, \tag{7}$$

Substituting Eq. (7) in Eq. (6), we obtain

$$N_1(k) = \frac{\mu_2 k}{\mu_2 k + \mu_1 (1-k)}, \tag{8}$$

By law, Raul partial pressure of the component above the solution is proportional to its molar fraction in the liquid phase, that is,

$$p_1 = p_{S1}(T_v) N_1(k_R), \ \ p_2 = p_{S2}(T_v)\left[1 - N_1(k_R)\right], \tag{9}$$

Equations of state phases have the form:

$$p_i = BT_v \rho_i^{/} / \mu_i, (i = 1,2), \tag{10}$$

where: B – Gas constant, T_v – the temperature of steam, $\rho_i^{/}$ – the density of the mixture components in the vapor bubble, μ_i – molecular weight, p_{si} – saturation pressure.

The boundary conditions $r = \infty$ and on a moving boundary can be written as

$$k\big|_{r=\infty} = k_0, k\big|_{r=R} = k_R, T_l\big|_{r=\infty} = T_0, T_l\big|_{r=R} = T_v, \tag{11}$$

$$j_1 l_1 + j_2 l_2 = \lambda_l D \frac{\partial T_l}{\partial r}\bigg|_{r=R}, \tag{12}$$

where: l_i – specific heat of vaporization.

By the definition of Nusselt parameter – the dimensionless parameter characterizing the ratio of particle size and the thickness of thermal boundary layer in the phase around the phase boundary and determined from additional considerations or from experience.

The heat of the bubble's intensity with the flow of the carrier phase will be further specified as:

$$\left(\lambda_l \frac{\partial T_l}{\partial r}\right)_{r=R} = Nu_l \cdot \frac{\lambda_l (T_0 - T_v)}{2R}, \tag{13}$$

In Ref. [16] obtained an analytical expression for the Nusselt parameter

$$Nu_l = 2\sqrt{\frac{\omega R_0^2}{a_l}} = 2\sqrt{\frac{R_0}{a_l}\sqrt{\frac{3\gamma p_0}{\rho_l}}} = 2\sqrt{\sqrt{3\gamma} \cdot Pe_l}, \tag{14}$$

where: $a_l = \frac{\lambda_l}{\rho_l c_l}$ – thermal diffusivity of fluid, $Pe_l = \frac{R_0}{a_l}\sqrt{\frac{p_0}{\rho_l}}$ – Peclet number.

The intensity of mass transfer of the bubble with the flow of the carrier phase will continue to ask by using the dimensionless parameter Sherwood *Sh*:

$$\left(D \frac{\partial k}{\partial r}\right)_{r=R} = Sh \cdot \frac{D(k_0 - k_R)}{2R}$$

where: D – diffusion coefficient, k – the concentration of dissolved gas in liquid,

The subscripts 0 and R refer to the parameters in an undisturbed state and at the interface.

We define a parameter in the form of Sherwood [16]

$$Sh = 2\sqrt{\frac{\omega R_0^2}{D}} = 2\sqrt{\frac{R_0}{D}\sqrt{\frac{3\gamma p_0}{\rho_l}}} = 2\sqrt{\sqrt{3\gamma} \cdot Pe_D}, \tag{15}$$

where $Pe_D = \frac{R_0}{D}\sqrt{\frac{p_0}{\rho_l}}$ – diffusion Peclet number.

The system of Eqs. (1)–(15) is a closed system of equations describing the dynamics and heat transfer of insoluble gas bubbles with liquid [19–58].

10.3.2 DYNAMIC MODELING FOR MASS AND MOMENTUM TRANSPORT

If we use Eqs. (7)–(9), we obtain relations for the initial concentration of component 1:

$$k_0 = \frac{1-\chi_2^0}{1-\chi_2^0+\mu\left(\chi_1^0-1\right)},\ \mu = \mu_2/\mu_1,\ \chi_i^0 = p_{si0}/p_0,\ i=1,2, \tag{16}$$

where: μ_2, μ_1 – Molecular weight of the liquid components of the mixture, p_{si0} – saturated vapor pressure of the components of the mixture at an initial temperature of the mixture T_0, which were determined by integrating the Clausius-Clapeyron relation. The parameter χ_i^0 is equal to

$$\chi_i^0 = \exp\left[\frac{l_i\mu_i}{B}\left(\frac{1}{T_{ki}}-\frac{1}{T_0}\right)\right], \tag{17}$$

Gas-phase liquid components in the derivation of Eq. (2) seemed perfect gas equations of state:

$$p_i = \rho_i B T_i / \mu_i.$$

where B – universal gas constant, p_i – The vapor pressure inside the bubble T_i to the temperature in the ratio of Eq. (2), T_{ki} – Temperature evaporating the liquid components of binary solution at an initial pressure p_0, l_i – specific heat of vaporization.

The initial concentration of the vapor pressure of component p_0 is determined from the relation:

$$c_0 = \frac{k_0\chi_1^0}{k_0\chi_1^0+\left(1-k_0\right)\chi_2^0}, \tag{18}$$

In this paper the problem of radial motions of a vapor bubble in binary solution was solved. It was investigated at various pressure drops in the liquid for different initial radii R_0 for a bubble. It is of great practical interest of aqueous solutions of ethanol and ethylene glycol.

It was revealed an interesting effect. The parameters characterized the dynamics of bubbles in aqueous ethyl alcohol. It was studied in the field of variable pressure lie between the limiting values of the parameter p_0 for pure components [59–68].

The pressure drops and consequently the role of diffusion are assumed unimportant. The pressure drop along with the heat dissipation is included diffusion dissipation. The rate of growth and collapse of the bubble is much higher than in the corresponding pure components of the solution under the same conditions. A completely different situation existed during the growth and collapse of vapor bubble in aqueous solutions of ethylene glycol.

In this case, the effect of diffusion resistance, leaded to inhibition of the rate of phase transformations. The growth rate and the collapse of the bubble is much smaller than the corresponding values, but for the pure components of the solution. Further research and calculations have to give a physical explanation for the observed effect. The influence of heat transfer and diffusion on damping of free oscillations of a vapor bubble binary solution.

It was found that the dependence of the damping rate of oscillations of a bubble of water solutions of ethanol, methanol, and toluene monotonic on k_0.

It was mentioned for the aqueous solution of ethylene glycol similar dependence with a characteristic minimum at: $k_0 \approx 0.02$.

Moreover, for: $0.01 \leq k_0 \leq 0.3$
decrement, binary solution has less damping rates for pulsations of a bubble in pure (one-component) water and ethylene glycol.

This means that in the range of concentrations of water: $0.01 \leq k_0 \leq 0.3$

Pulsations of the bubble (for water solution of ethylene glycol) decay much more slowly and there are inhibition of the process of phase transformations. A similar process was revealed and forced oscillations of bubbles in an acoustic field.

In this book the influence of nonstationary heat and mass transfer processes was investigated in the propagation of waves in a binary solution of liquids with bubbles. The influence of component composition and concentration of binary solution was investigated on the dispersion, dissipation and attenuation of monochromatic waves in two-phase, two-component media.

The aqueous solution of ethyl alcohol in aqueous ethylene glycol decrements showed perturbations less relevant characteristics of pure components of the solution.

Unsteady interphase heat transfer revealed in calculation, the structure of stationary shock waves in bubbly binary solutions. The problem signifies on effect a violation of monotonicity behavior of the calculated curves for concentration, indicating the presence of diffusion resistance.

In some of binary mixtures, it is seen the effect of diffusion resistance. It is led to inhibition of the intensity of phase transformations.

The physical explanation revealed the reason for an aqueous solution of ethylene glycol. Pronounced effect of diffusion resistance is related to the solution with limited ability. It diffuses through the components of $D = 10^{-9}$ (m^2/sec),

D– Diffusion coefficient volatility of the components is very different, and thus greatly different concentrations of the components in the solution and vapor phase.

In the case of aqueous solution of ethanol volatility component are roughly the same $\chi_1^0 \approx \chi_2^0$.

In accordance with Eq. (3) $c_0 \approx k_0$, so the finiteness of the diffusion coefficient does not lead to significant effects in violation of the thermal and mechanical equilibrium phases.

Figures 4 and 5 show the dependence $k_0(c_0)$ of ethyl alcohol and ethylene glycol's aqueous solutions. From Fig. 24 it is clear that almost the entire range of k_0, $k_0 \approx c_0$.

At the same time for an aqueous solution of ethylene glycol, by the calculations and Fig. 2, $0.01 \leq k_0 \leq 0.3$, $k_0 \leq c_0$, and when $k_0 > 0.3$, $k_0 \approx c_0$.

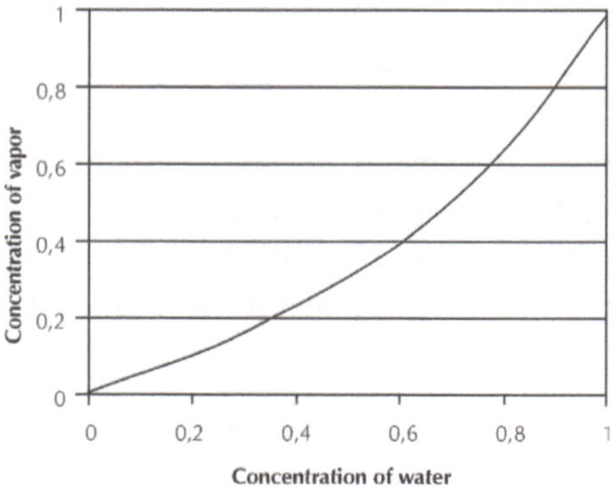

FIGURE 4 The dependence ($k_0(c_0)$) for an aqueous solution of ethanol.

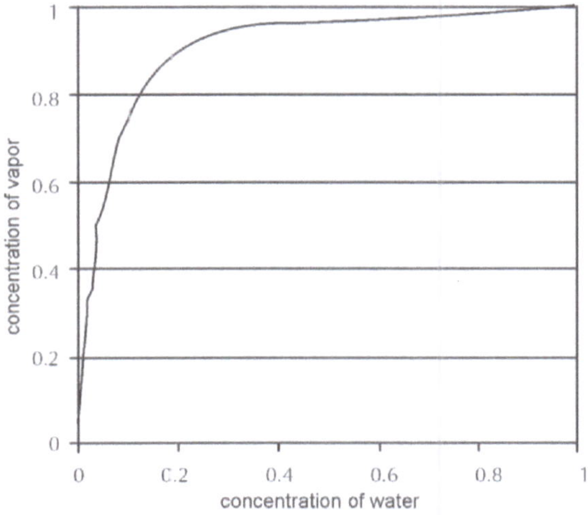

FIGURE 5 Dependence of $(k_0(c_0))$ for an aqueous solution of ethylene glycol.

In Figs. 6 and 7 show the boiling point of the concentration for the solution of two systems.

When $k_0 = 1$, $c_0 = 1$ and get clean water to steam bubbles. It is for boiling of a liquid at $T_0 = 373°K$.

If $k_0 = 0$, $c_0 = 0$ and have correspondingly pure bubble ethanol $(T_0 = 350^{\circ} K)$ and ethylene glycol $(T_0 = 470^{\circ} K)$.

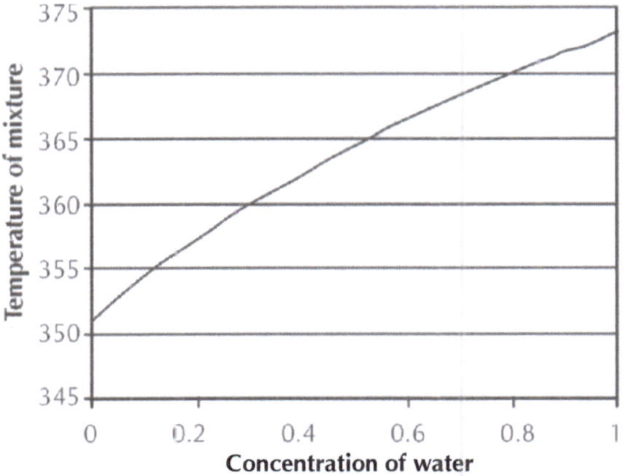

FIGURE 6 The dependence of the boiling temperature of the concentration of the solution to an aqueous solution of ethanol.

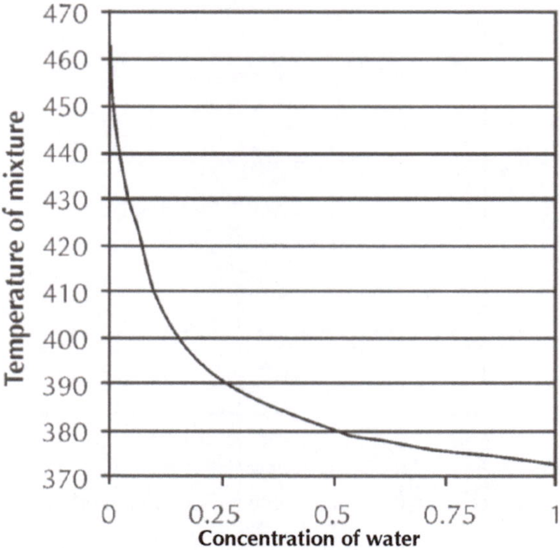

FIGURE 7 The dependence of the boiling point of the concentration of the solution to an aqueous solution of ethylene glycol.

It should be noted that in all the above works regardless of the above problems in the mathematical description of the cardinal effects of component composition of the solution shows the value of the parameter β equal

$$\beta=\left(1-\frac{1}{\gamma}\right)\frac{\left(c_0-k_0\right)\left(N_{c_0}-N_{k_0}\right)}{k_0\left(1-k_0\right)}\frac{c_l}{c_{pv}}\left(\frac{c_{pv}T_0}{L}\right)^2\sqrt{\frac{a_l}{D}}, \tag{19}$$

where: N_{k_0}, N_{c_0} – molar concentration of 1-th component in the liquid and steam.

$$N_{k_0}=\frac{\mu k_0}{\mu k_0+1-k_0},$$

$$N_{c_0}=\frac{\mu c_0}{\mu c_0+1-c_0}$$

where: γ – Adiabatic index, c_l and c_{pv}, respectively, the specific heats of liquid and vapor at constant pressure, a_l – Thermal diffusivity.

$L = l_1c_0 + l_2(1 - c_0)$

We also note that option Eq. (4) is a self-similar solution describing the growth of a bubble in a superheated solution. This solution has the form:

$$R = 2\sqrt{\frac{3}{\pi}\frac{\lambda_l \Delta T \sqrt{t}}{L\rho_v\sqrt{a_l}(1+\beta)}}, \tag{20}$$

Here ρ_v – vapor density, t – time, R – radius of the bubble, λ_l – the coefficient of thermal conductivity, ΔT – overheating of the liquid.

Figures 8 and 9 show the dependence $\beta(k_0)$ for the above binary solutions. For aqueous ethanol β is negative for any value of concentration and dependence on k_0 is monotonic.

For an aqueous solution of ethylene glycol β – is positive and has a pronounced maximum at $k_0 = 0.02$.

As a result of this chapter, at low-pressure drops (superheating and super cooling of the liquid, respectively), diffusion does not occur in aqueous solutions of ethyl alcohol. By approximate equality of $k_0$and c_0 all calculated dependence lie between the limiting curves for the case of one-component constituents of the solution.

They are included dependence of pressure, temperature, vapor bubble radius, the intensity of phase transformations, and so from time to time).

The pressure difference becomes important diffusion processes. Mass transfer between bubble and liquid is in a more intensive mode than in single-component constituents of the solution.

In particular, the growth rate of the bubble in a superheated solution is higher than in pure water and ethyl alcohol. It is because of the negative β according to Eq. (5).

In an aqueous solution of ethylene glycol, there are the same perturbations due to significant differences between $k_0$andc_0.It is especially when$0.01 \leq k_0 \leq 0.3$, the effect of diffusion inhibition contributes to a significant intensity of mass transfer. In particular, during the growth of the bubble, the rate of growth in solution is much lower than in pure water and ethylene glycol. It is because of the positive β by Eq. (5).

Moreover, the maximum braking effect is achieved at the maximum value of β, when$k_0 = 0.02$.

A similar pattern is observed at the pulsations and the collapse of the bubble. Dependence of the damping rate of fluctuations in an aqueous solution of ethyl alcohol from the water concentration is monotonic .Aqueous solution of ethylene glycol dependence of the damping rate has a minimum at $k_0 = 0.02$, $0.01 \leq k_0 \leq 0.3$.

The function decrement is small respectively large difference between k_0 and c_0 and β takes a large value. These ranges of concentrations in the solution have significant effect of diffusion inhibition.

For aqueous solutions of glycerin, methanol, toluene, etc., calculations are performed. Comparison with experimental data confirms the possibility of theoretical prediction of the braking of heat and mass transfer.

It was analyzed the dependence of the parameter β, decrement of oscillations of a bubble from the equilibrium concentration of the mixture components. Therefore, in every solutions, it was determined the concentration of the components of a binary mixture.

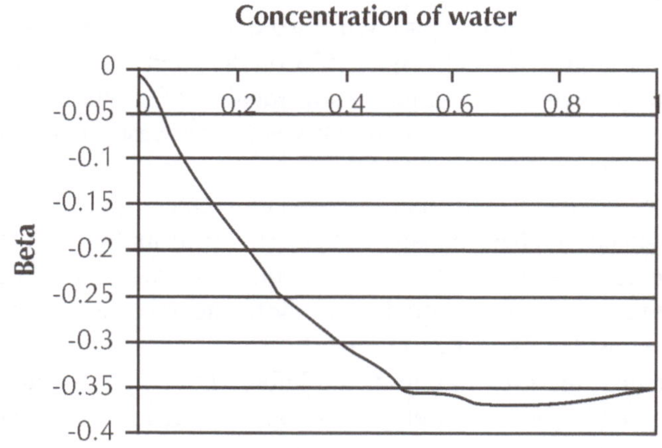

FIGURE 8 The dependence $\beta(k_0)$ for an aqueous solution of ethanol.

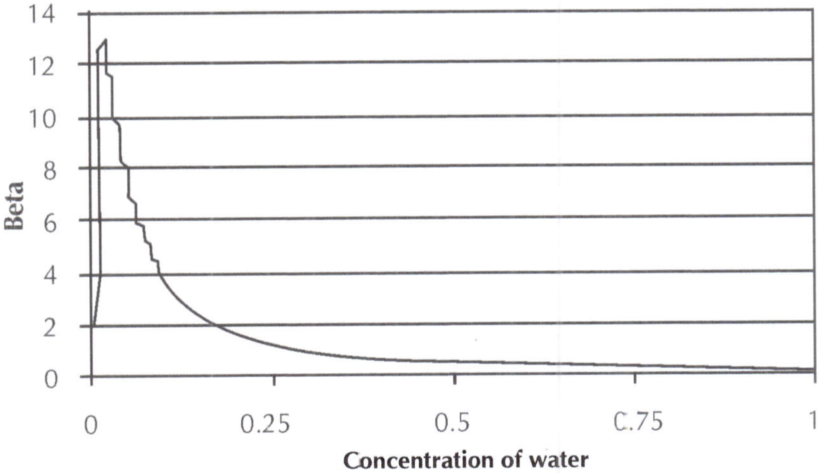

FIGURE 9 Dependence of $\beta(k_0)$ for an aqueous solution of ethylene glycol.

Figures 10 and 11 are illustrated by theoretical calculations. These figures defined on the example of aqueous solutions of ethyl alcohol and ethylene glycol (antifreeze used in car radiators). It is evident that the first solution is not suitable to the task.

The aqueous solution of ethylene glycol with a certain concentration is theoretically much more slowly boils over with clean water and ethylene gly-col. This confirms the reliability of the method.

Calculations show that such a solution is almost never freezes. The same method can offer concrete solutions for cooling of hot parts and components of various machines and mechanisms.

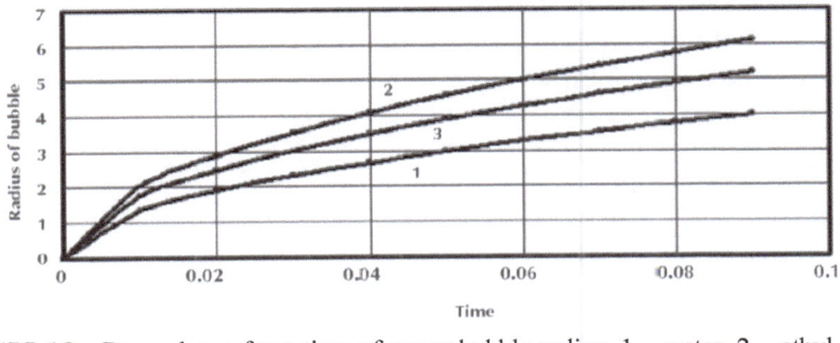

FIGURE 10 Dependence from time of vapor bubble radius. 1 – water, 2 – ethyl spirit, 3 – water mixtures of ethyl spirit.

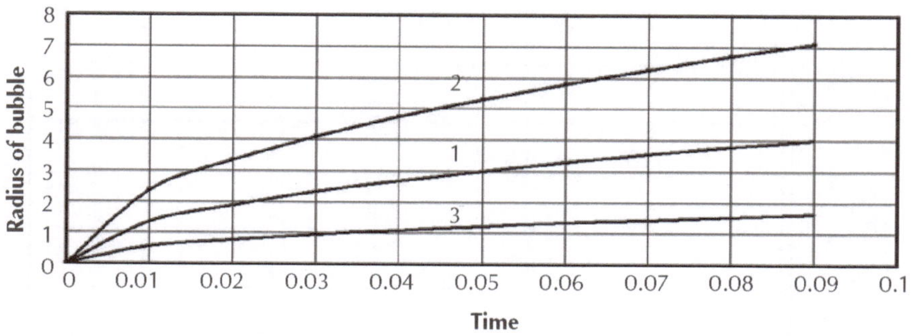

FIGURE 11 Dependence from time of vapor bubble radius. 1 – water, 2 – ethylene glycol, 3 – water mixtures of ethylene glycol.

10.4 CONCLUSION

The solution of the reduced system of equations revealed an interesting effect. The parameters were characterized the dynamics of bubbles in aqueous ethyl alcohol in the field of variable pressure.

They lied between the limiting values of relevant parameters for the pure components. It was for the case, which pressure drops and consequently the role of diffusion was unimportant.

A completely different situation is observed during the growth and collapse of vapor bubble in aqueous solutions of ethylene glycol. The effect of diffusion resistance, leads to inhibition of the rate of phase transformations. For pure components of the solution, the growth and the collapse rate of the bubble is much smaller than the corresponding values.

KEYWORDS

- **bubble binary solution**
- **hydraulic transition**
- **Nusselt parameter**
- **vapor bubble**
- **vena contraction**
- **water hammer**

REFERENCES AND FURTHER READING

REFERENCES FOR CHAPTER 1

1. Joukowski, N., Paper to Polytechnic Soc. Moscow, spring of 1898, English translation by Miss, O. Simin. Proc. AWWA, 1904, 57–58.
2. Allievi, L., "General Theory of Pressure Variation in Pipes", Ann. D. Ing., 1982, 166–171.
3. Wood, F. M., "History of Water Hammer", Civil Engineering Research Report, #65, Queens University, Canada, 1970, 66–70.
4. Parmakian, J., "Water Hammer Design Criteria", J. of Power Div., ASCE, Sept., 1957, 456–460.
5. Parmakian, J., "Water Hammer Analysis", Dover Publications, Inc., New York, New York, 1963, 51–58.
6. Streeter, V. L., Lai, C., "Water Hammer Analysis Including Fluid Friction." Journal of Hydraulics Division, ASCE, 88, 1962, 79 p.
7. Streeter, V. L., Wylie, E. B., "Fluid Mechanics", McGraw-Hill Ltd., USA, 1979, 492–505.
8. Streeter, V. L., Wylie, E. B., "Fluid Mechanics", McGraw-Hill Ltd., USA, 1981, 398–420.
9. Wylie, E. B., Streeter, V. L., Talpin, L. B., Matched impedance to control fluid transients. Trans. ASME 105, 2, 1983, 219–224.
10. Wylie, E. B., Streeter, V. L., Fluid Transients in Systems, Prentice-Hall, Englewood Cliffs, New Jersey, 1993, 4
11. Wylie, E. B., Streeter, V. L., Fluid Transients, Feb Press, Ann Arbor, MI, 1982, corrected copy, 1983, 158
12. Brunone, B., Karney, B. W., Mecarelli, M., Ferrante, M. "Velocity Profiles and Unsteady Pipe Friction in Transient Flow" Journal of Water Resources Planning and Management, ASCE, 126(4), Jul., 2000, 236–244.
13. Koelle, E., Luvizotto Jr. E., Andrade, J.P. G., "Personality Investigation of Hydraulic Networks using MOC – Method of Characteristics" Proceedings of the 7th International Conference on Pressure Surges and Fluid Transients, Harrogate Durham, United Kingdom, 1996, 1–8.
14. Filion, Y., Karney, B. W., "A Numerical Exploration of Transient Decay Mechanisms in Water Distribution Systems", Proceedings of the ASCE Environmental Water Resources Institute Conference, American Society of Civil Engineers, Roanoke, Virginia, 2002, 30.
15. Hamam, M. A., Mc Corquodale, J. A., "Transient Conditions in the Transition from Gravity to Surcharged Sewer Flow", Canadian, J. of Civil Eng., Canada, Sep., 1982, p.65–98.
16. Savic, D. A., Walters, G. A., "Genetic Algorithms Techniques for Calibrating Network Models", Report No. 95/12, Centre for Systems and Control Engineering, 1995, 137–146.
17. Savic, D. A., Walters, G. A., Genetic Algorithms Techniques for Calibrating Network Models, University of Exeter, Exeter, United Kingdom, 1995, 41–77.
18. Walski, T. M., Lutes, T. L., "Hydraulic Transients Cause Low-Pressure Problems", Journal of the American Water Works Association, 75(2), 1994, 58.
19. Wu, Z. Y., Simpson, A. R., Competent genetic-evolutionary optimization of water distribution systems. J. Comput. Civ. Eng. 15(2), 2001, 89–101.
20. Gerasimov Yu.I., The course of physical chemistry. V.1. Goskhimizdat, M., 1963, 736.
21. Dikarevsky, M., Impactprotection opositelnyh closed systems. Moscow: Kolos, 1981, 80.
22. Nigmatulin, R. I., Nagiyev, F. B., Khabeev, N. S., Destruction and collapse of vapor

bubbles and strengthening shock waves in a liquid with vapor bubbles. Assembly. "Gas and wave dynamics, " No.3, "MSU", 1979, 124–129.
23. Allievi, L., "General Theory of Pressure Variation in Pipes", Ann. D. Ing., 1982, 166–171.
24. Joukowski, N., Paper to Polytechnic Soc. Moscow, spring of 1898, English translation by Miss, O. Simin. Proc. AWWA, 1904, 57–58.
25. Wisniewski, K. P., Design of pumping stations closed irrigation systems: Right. / K.P. Vishnevsky, Podlasov A.V. – Moscow: Agropromizdat, 1990, 93.
26. Nigmatulin, R. I., Khabeev, N. S., Nagiyev, F. B., Dynamics, heat and mass transfer of vapor-gas bubbles in a liquid. Int. J. Heat Mass Transfer,vol.24, N6, Printed in Great Britain, 1981, 1033–1044.
27. Vargaftik, N. B., Handbook of thermo-physical properties of gases and liquids. Oxford: Pergamon Press, 1972, 98.
28. Laman, B. F., Hydro pumps and installation, 1988, 278.
29. Nagiyev, F. B., Kadyrov, B. A., Heat transfer and the dynamics of vapor bubbles in a liquid binary solution. DAN Azerbaijani SSR, 1986, № 4, 10–13.
30. Alyshev, V. M., Hydraulic calculations of open channels on your PC. – Part 1 Tutorial. – Moscow: MSUE, 2003, 185.
31. Streeter, V. L., Wylie, E. B., "Fluid Mechanics", McGraw-Hill Ltd., USA, 1979, 492–505.
32. Sharp, B., "Water hammer Problems and Solutions", Edward Arnold Ltd., London, 1981, 43–55.
33. Skousen, P., "Valve Handbook", McGraw Hill, New York, HAMMER Theory and Practice, 1998, 687–721.
34. Shaking, N. I., Water hammer to break the continuity of the flow in pressure conduits pumping stations: Dis. on Kharkov, 1988, 225.
35. Tijsseling, Alan E Vardy "Time scales and FSI in unsteady liquid-filled pipe flow", 1993, 5–12.
36. Wu, P. Y., Little, W. A., measurement of friction factor for flow of gases in very fine channels used for micro miniature, Joule Thompson refrigerators, Cryogenics 24 (8), 1983, 273–277.
37. Song, C. C. et al.,"Transient Mixed-Flow Models for Storm Sewers", J. of Hyd. Div., Nov., 1983, Vol. 109, 458–530.
38. Stephenson, D., "Pipe Flow Analysis", Elsevier, Vol. 19, S. A., 1984, 670–788.
39. Chaudhry, M. H., "Applied Hydraulic Transients", Van Nostrand Reinhold Co., N. Y., 1979, 1322–1324.
40. Chaudhry, M. H., Yevjevich, V. "Closed Conduit Flow", Water Resources Publication, USA, 1981, 255–278.
41. Chaudhry, M. H., Applied Hydraulic Transients, Van Nostrand Reinhold, New York, USA, 1987, p.165–167.
42. Kerr, S. L., "Minimizing service interruptions due to transmission line failures: Discussion," Journal of the American Water Works Association, 41, 634, July 1949, 266–268.
43. Kerr, S. L., "Water hammer control," Journal of the American Water Works Association, 43, December 1951, 985–999.
44. Apoloniusz Kodura, Katarzyna Weinerowska," Some Aspects of Physical and Numerical Modeling of Water Hammer in Pipelines", 2005, 126–132.
45. Anuchina, N. N., Volkov, V. I., Gordeychuk, V. A., Es'kov, N. S., Ilyutina, O. S., Kozyrev, O.M. "Numerical simulations of Rayleigh-Taylor and Richtmyer-Meshkov instability using mah-3 code," J. Comput. Appl. Math. 168, 2004, 11.
46. Fox, J. A., "Hydraulic Analysis of Unsteady Flow in Pipe Network", Wiley, NY, 1977, p.78–89.
47. Karassik, I. J., "Pump Handbook – Third Edition", McGraw-Hill, 2001, 19–22.
48. Fok, A., "Design Charts for Air Chamber on Pump Pipelines", J. of Hyd. Div., ASCE, Sept., 1978, p.15–74.
49. Fok, A., Ashamalla, A., Aldworth, G., "Considerations in Optimizing Air Chamber for Pumping Plants", Symposium on Fluid Transients and Acoustics in the Power Industry, San Francisco, U.S.A. Dec., 1978, 112–114.
50. Fok, A., "Design Charts for Surge Tanks on Pump Discharge Lines", BHRA 3rd Int. Conference on Pressure Surges, Bedford, England, Mar., 1980, 23–34.
51. Fok, A., "Water hammer and Its Protection in Pumping Systems", Hydro technical

Conference, CSCE, Edmonton, May, 1982, 45–55.
52. Fok, A., "A Contribution to the Analysis of Energy Losses in Transient Pipe Flow", PhD. Thesis, University of Ottawa, 1987, 176–182.
53. Hariri Asli, K., Nagiyev, F. B., Water Hammer and Fluid Condition, Ministry of Energy, Gilan Water and Wastewater Co., Research Week Exhibition, Tehran, Iran, December, 2007, p. 132–148, http://isrc.nww.co.ir.
54. Hariri Asli, K., Nagiyev, F. B., Water Hammer Analysis and Formulation, Ministry of Energy, Gilan Water and Wastewater Co., Research Week Exhibition, Tehran, Iran, December, 2007, 111–131, http://isrc.nww.co.ir.
55. Hariri Asli, K., Nagiyev, F. B., Water Hammer and hydrodynamics instabilities, Interpenetration of two fluids at parallel between plates and turbulent moving in pipe, Ministry of Energy, Guilan Water and Wastewater Co., Research Week Exhibition, Tehran, Iran, December, 2007, 90–110, http://isrc.nww.co.ir.
56. Hariri Asli, K., Nagiyev, F. B., Water Hammer and pump pulsation, Ministry of Energy, Guilan Water and Wastewater Co., Research Week Exhibition, Tehran, Iran, December, 2007, 51–72, http://isrc.nww.co.ir.
57. Hariri Asli, K., Nagiyev, F. B., Reynolds number and hydrodynamics' instability", Ministry of Energy, Guilan Water and Wastewater Co., Research Week Exhibition, Tehran, Iran, December, 2007, 31–50, http://isrc.nww.co.ir.
58. Hariri Asli, K., Nagiyev, F. B., Water Hammer and valves, Ministry of Energy, Guilan Water and Wastewater Co., Research Week Exhibition, Tehran, Iran, December, 2007, 20–30, http://isrc.nww.co.ir.
59. Hariri Asli, K., Nagiyev, F. B., "Interpenetration of two fluids at parallel between plates and turbulent moving in pipe", Ministry of Energy, Guilan Water and Wastewater Co., Research Week Exhibition, Tehran, Iran, December, 2007, 73–89, http://isrc.nww.co.ir.
60. Hariri Asli, K., Nagiyev, F. B., Decreasing of Unaccounted For Water "UFW" by Geographic Information System "GIS" in Rasht urban water system, civil engineering organization of Guilan, Technical and Art Journal, 2007, 3–7,http://www.art-of-music.net/.
61. Hariri Asli, K., Portable Flow meter Tester Machine Apparatus, Certificate on registration of invention, Tehran, Iran, #010757, Series a/82, 24/11/2007, 1–3.
62. Hariri Asli, K., Nagiyev, F. B., Haghi, A. K., "Interpenetration of two fluids at parallel between plates and turbulent moving in pipe", 9th Conference on Ministry of Energetic works at research week, Tehran, Iran, 2008, 73–89, http://isrc.nww.co.ir.
63. Hariri Asli, K., Nagiyev, F. B., Haghi, A. K., "Water hammer and valves", 9th Conference on Ministry of Energetic works at research week, Tehran, Iran, 2008, 20–30, http://isrc.nww.co.ir.
64. Hariri Asli, K., Nagiyev, F. B., Haghi, A. K., "Water hammer and hydrodynamics instability", 9th Conference on Ministry of Energetic works at research week, Tehran, Iran, 2008, 90–110, http://isrc.nww.co.ir.
65. Hariri Asli, K., Nagiyev, F. B. Haghi, A. K., "Water hammer analysis and formulation", 9th Conference on Ministry of Energetic works at research week, Tehran, Iran, 2008, 27–42, http://isrc.nww.co.ir.
66. Hariri Asli, K., Nagiyev, F. B., Haghi, A. K., "Water hammer &fluid condition", 9th Conference on Ministry of Energetic works at research week, Tehran, Iran, 2008, 27–43, http://isrc.nww.co.ir.
67. Hariri Asli, K., Nagiyev, F. B., Haghi, A. K., "Water hammer and pump pulsation", 9th Conference on Ministry of Energetic works at research week, Tehran, Iran, 2008, 27–44, http://isrc.nww.co.ir.
68. Hariri Asli, K., Nagiyev, F. B., Haghi, A. K.,"Reynolds number and hydrodynamics instability", 9th Conference on Ministry of Energetic works at research week, Tehran, Iran, 2008, 27–45, http://isrc.nww.co.ir.
69. Hariri Asli, K., Nagiyev, F. B., Haghi, A. K., "Water hammer and fluid Interpenetration", 9th Conference on Ministry of Energetic works at research week, Tehran, Iran, 2008, 27–47, http://isrc.nww.co.ir.
70. Hariri Asli, K., GIS and water hammer disaster at earthquake in Rasht water pipeline, civil engineering organization of Guilan,

Technical and Art Journal, 2008, 14–17, http://www.art-of-music.net/.
71. Hariri Asli, K., GIS and water hammer disaster at earthquake in Rasht water pipeline, 3rd International Conference on Integrated Natural Disaster Management, Tehran university, ISSN: 1735–5540, 18–19 Feb., INDM, Tehran, Iran, 2008, №13,53/1–12, http://www.civilica.com/Paper-INDM03-INDM03_001.html
72. Hariri Asli, K., Nagiyev, F. B., Bubbles characteristics and convective effects in the binary mixtures. Transactions issue mathematics and mechanics series of physical-technical and mathematics science, ISSN: 0002–3108, Azerbaijan, Baku, 2009, 68–74, http://www.imm.science.az/journals.html.
73. Hariri Asli, K., Nagiyev, F. B., Haghi, A. K., Aliyev, S. A., Three-Dimensional conjugate heat transfer in porous media, 1st Festival on Water and Wastewater Research and Technology, Tehran, Iran, 12–17 Dec.2009, 26–28, http://isrc.nww.co.ir.
74. Hariri Asli, K., Nagiyev, F. B., Haghi, A. K., Aliyev, S. A., Some Aspects of Physical and Numerical Modeling of water hammer in pipelines,1st Festival on Water and Wastewater Research and Technology, Tehran, Iran, 12–17 Dec.2009, 26–29, http://isrc.nww.co.ir
75. Hariri Asli, K., Nagiyev, F. B., Haghi, A. K., Aliyev, S. A., Modeling for Water Hammer due to valves: From theory to practice, 1st Festival on Water and Wastewater Research and Technology, Tehran, Iran, 12–17 Dec.2009, 26,30, http://isrc.nww.co.ir.
76. Hariri Asli, K., Nagiyev, F. B., Haghi, A. K., Aliyev, S. A., Water hammer and hydrodynamics instabilities modeling: From Theory to Practice, 1st Festival on Water and Wastewater Research and Technology, Tehran, Iran, 12–17 Dec.2009, 26–31, http://isrc.nww.co.ir
77. Hariri Asli, K., Nagiyev, F. B., Haghi, A. K., Aliyev, S. A., A computational approach to study fluid movement, 1st Festival on Water and Wastewater Research and Technology, Tehran, Iran, 12–17 Dec.2009, 27–32, http://isrc.nww.co.ir.
78. Hariri Asli, K., Nagiyev, F. B., Haghi, A. K., Aliyev, S. A., Water Hammer Analysis: Some Computational Aspects and practical hints, 1st Festival on Water and Wastewater Research and Technology, Tehran, Iran, 12–17 Dec.2009, 27–33, http://isrc.nww.co.ir
79. Hariri Asli, K., Nagiyev, F. B., Haghi, A. K., Aliyev, S. A., Water Hammer and Fluid condition: A computational approach, 1st Festival on Water and Wastewater Research and Technology, Tehran, Iran, 12–17 Dec.,2009, 27–34, http://isrc.nww.co.ir.
80. Hariri Asli, K., Nagiyev, F. B., Haghi, A. K., Aliyev, S. A., A computational Method to Study Transient Flow in Binary Mixtures, 1st Festival on Water and Wastewater Research and Technology, Tehran, Iran, 12–17 Dec.2009, 27–35, http://isrc.nww.co.ir.
81. Hariri Asli, K., Nagiyev, F. B., Haghi, A. K., Physical modeling of fluid movement in pipelines, 1st Festival on Water and Wastewater Research and Technology, Tehran, Iran, 12–17 Dec.2009, 27–36, http://isrc.nww.co.ir.
82. Hariri Asli, K., Nagiyev, F. B., Haghi, A. K., Aliyev, S. A., Interpenetration of two fluids at parallel between plates and turbulent moving, 1st Festival on Water and Wastewater Research and Technology, Tehran, Iran, 12–17 Dec. 2009, 27–37, http://isrc.nww.co.ir.
83. Hariri Asli, K., Nagiyev, F. B., Haghi, A. K., Aliyev, S. A., Modeling of fluid interaction produced by water hammer, 1st Festival on Water and Wastewater Research and Technology, Tehran, Iran, 12–17 Dec.2009, 27–38, http://isrc.nww.co.ir.
84. Hariri Asli, K., Nagiyev, F. B., Haghi, A. K., Aliyev, S. A., GIS and water hammer disaster at earthquake in Rasht pipeline, 1st Festival on Water and Wastewater Research and Technology, Tehran, Iran, 12–17 Dec.2009, 27–39, http://isrc.nww.co.ir.
85. Hariri Asli, K., Nagiyev, F. B., Haghi, A. K., Aliyev, S. A., Interpenetration of two fluids at parallel between plates and turbulent moving,1st Festival on Water and Wastewater Research and Technology, Tehran, Iran, 12–17 Dec.2009, 27–40, http://isrc.nww.co.ir.
86. Hariri Asli, K., Nagiyev, F. B., Haghi, A. K., Aliyev, S. A., Water hammer and hydrodynamics' instability, 1st Festival on Water and Wastewater Research and Technol-

ogy, Tehran, Iran, 12–17 Dec.2009, 27–41, http://isrc.nww.co.ir.

87. Hariri Asli, K., Nagiyev, F. B., Haghi, A. K., Aliyev, S. A., Water hammer analysis and formulation, 1st Festival on Water and Wastewater Research and Technology, Tehran, Iran, 12–17 Dec.2009, 27–42, http://isrc.nww.co.ir.
88. Hariri Asli, K., Nagiyev, F. B., Haghi, A. K., Aliyev, S. A., Water hammer and fluid condition, 1st Festival on Water and Wastewater Research and Technology, Tehran, Iran, 12–17 Dec.2009, 27–43, http://isrc.nww.co.ir.
89. Hariri Asli, K., Nagiyev, F. B., Haghi, A. K., Aliyev, S. A., Water hammer and pump pulsation,1st Festival on Water and Wastewater Research and Technology, Tehran, Iran, 12–17 Dec.2009, 27–44, http://isrc.nww.co.ir.
90. Hariri Asli, K., Nagiyev, F. B., Haghi, A. K., Aliyev, S. A., Reynolds number and hydrodynamics instabilities, 1st Festival on Water and Wastewater Research and Technology, Tehran, Iran, 12–17 Dec.2009, 27–45, http://isrc.nww.co.ir.
91. Hariri Asli, K., Nagiyev, F. B., Haghi, A. K., Aliyev, S. A., water hammer and valves, 1st Festival on Water and Wastewater Research and Technology, Tehran, Iran, 12–17 Dec.2009, 27–46, http://isrc.nww.co.ir.
92. Hariri Asli, K., Nagiyev, F. B., Haghi, A. K., Aliyev, S. A., "Water hammer and fluid Interpenetration", 1st Festival on Water and Wastewater Research and Technology, Tehran, Iran, 12–17 Dec.2009, 27–47, http://isrc.nww.co.ir.
93. Hariri Asli, K., Nagiyev, F. B., Modeling of fluid interaction produced by water hammer, International Journal of Chemoinformatics and Chemical Engineering, IGI, ISSN: 2155–4110, EISSN: 2155–4129, USA, 2010, 29–41, http://www.igi-global.com/journals/details.asp?ID=34654.
94. Hariri Asli, K., Nagiyev, F. B., Haghi, A. K., Water hammer and fluid condition; a computational approach, Computational Methods in Applied Science and Engineering, USA, Chapter 5, Nova Science Publications, ISBN:978-1-60876-052-7, USA, 2010, 73–94, https://www.novapublishers.com/catalog/.
95. Hariri Asli, K., Nagiyev, F. B., Haghi, A. K., Some aspects of physical and numerical modeling of water hammer in pipelines. Computational Methods in Applied Science and Engineering, USA, Chapter 23, Nova Science Publications, ISBN:978-1-60876-052-7, USA, 2010, 365–387, https://www.novapublishers.com/catalog/
96. Hariri Asli, K., Nagiyev, F. B., Haghi, A. K., Modeling for water hammer due to valves; from theory to practice, Computational Methods in Applied Science and Engineering, USA, Chapter 11, Nova Science Publications ISBN:978-1-60876-052-7, USA, 2010, 229–236, https://www.novapublishers.com/catalog/
97. Hariri Asli, K., Nagiyev, F. B., Haghi, A. K., A computational method to Study transient flow in binary mixtures, Computational Methods in Applied Science and Engineering, USA, Chapter 13, Nova Science Publications ISBN:978-1-60876-052-7, USA, 2010, 229–236, https://www.novapublishers.com/catalog/
98. Hariri Asli, K., Nagiyev, F. B., Haghi, A. K., Water hammer analysis; some computational aspects and practical hints, Computational Methods in Applied Science and Engineering, USA, Chapter 16, Nova Science Publications ISBN:978-1-60876-052-7, USA, 2010, 263–281, https://www.novapublishers.com/catalog/
99. Hariri Asli, K., Nagiyev, F. B., Haghi, A. K., Water hammer and hydrodynamics instabilities modeling, Computational Methods in Applied Science and Engineering, USA, Chapter 17, From Theory to Practice, Nova Science Publications ISBN:978—1-60876-052-7, USA, 2010, 283–301, https://www.novapublishers.com/catalog/
100. Hariri Asli, K., Nagiyev, F. B., Haghi, A. K., A computational approach to study water hammer and pump pulsation phenomena, Computational Methods in Applied Science and Engineering, USA, Chapter 22, Nova Science Publications, ISBN:978-1-60876-052-7, USA, 2010, 349–363, https://www.novapublishers.com/catalog/
101. Hariri Asli, K., Nagiyev, F. B., Haghi, A. K., A computational approach to study fluid movement, Nanomaterials Yearbook – 2009, From Nanostructures,

Nanomaterials and Nanotechnologies to Nanoindustry, Chapter 16, Nova Science Publications, USA, ISBN: 978–1-60876–451–8, USA, 2010, 181–196, https://www.novapublishers.com/catalog/product_info.php?products_id=11587
102. Hariri Asli, K., Nagiyev, F. B., Haghi, A. K., Physical modeling of fluid movement in pipelines, Nanomaterials Yearbook – 2009, From Nanostructures, Nanomaterials and Nanotechnologies to Nanoindustry, Chapter 17, Nova Science Publications, USA, ISBN: 978–1-60876–451–8, USA, 2010, 197–214, https://www.novapublishers.com/catalog/product_info.php?products_id=11587
103. Hariri Asli, K., Nagiyev, F. B., Haghi, A. K., "Some Aspects of Physical and Numerical Modeling of water hammer in pipelines", Nonlinear Dynamics An International Journal of Nonlinear Dynamics and Chaos in Engineering Systems, ISSN: 1573–269X (electronic version) Journal no. 11071 Springer, Netherlands, 2009, ISSN: 0924–090X (print version), Springer, Heidelberg, Germany, Number 4 / June, 2010, Volume 60, 677–701, http://www.springerlink.com/openurl.aspgenre=article&id=doi:10.1007/s11071-009-9624-7.
104. Hariri Asli, K., Nagiyev, F. B., Haghi, A. K., Interpenetration of two fluids at parallel between plates and turbulent moving in pipe; a case study, Computational Methods in Applied Science and Engineering, USA, Chapter 7, Nova Science Publications, ISBN:978-1-60876-052-7, USA, 2010, 107–133, https://www.novapublishers.com/catalog/
105. Hariri Asli, K., Nagiyev, F. B., Beglou, M. J., Haghi, A. K., Kinetic analysis of convective drying, International Journal of the Balkan Tribological Association, ISSN: 1310-4772, Sofia, Bulgaria, 2009, Vol. 15, No 4, 546–556, jbalkta@gmail.com
106. Hariri Asli, K., Nagiyev, F. B., Haghi, A. K., Three-dimensional Conjugate Heat Transfer in Porous Media, International Journal of the Balkan Tribological Association, ISSN: 1310-4772, Sofia, Bulgaria, 2009, Vol. 15, No 3, 336–346, jbalkta@gmail.com
107. Hariri Asli, K., Nagiyev, F. B., Haghi, A. K., Aliyev, S. A., Pure Oxygen penetration in wastewater flow, Recent Progress in Research in Chemistry and Chemical Engineering, Nova Science Publications, ISBN: 978-1-61668-501-0, Nova Science Publications, USA, 2010, 17–27, https://www.novapublishers.com/catalog/product_info.php?products_id=13174110100.
108. Hariri Asli, K., Nagiyev, F. B., Haghi, A. K., Aliyev, S. A., Improved modeling for prediction of water transmission failure, Recent Progress in Research in Chemistry and Chemical Engineering, Nova Science Publications, ISBN: 978-1-61668-501-0, Nova Science Publications, USA, 2010, 28–36, https://www.novapublishers.com/catalog/product_info.php?products_id=13174

REFERENCES FOR CHAPTER 2

1. Hariri Asli, K., GIS and Water hammer disaster at earthquake in Rasht water pipeline, 3rd International Conference on Integrated Natural Disaster Management, INDM2008, http://www.civilica.com/Paper-INDM03-INDM03_001.html.
2. Hariri Asli, K., Nagiyev, F.B., Beglou, M. J., Haghi, A.K., Kinetic analysis of convective drying, International Journal of the Balkan Tribological Association, ISSN: 1310–4772, Sofia, Bulgaria, Vol. 15, No 4, 546–556, 2009, jbalkta@gmail.com
3. Hariri Asli K, Nagiyev, F.B., Bubbles characteristics and convective effects in the binary mixtures. Transactions issue mathematics and mechanics series of physical-technical and mathematics science, ISSN 0002–3108, Azerbaijan, Baku, 215–220, 2008, www.imm.science.az/journals/AMEA_xeberleri/…/215–220.pdf
4. Hariri Asli, K., Nagiyev, F.B., Haghi, A.K.,Three-dimensional Conjugate Heat Transfer in Porous Media, International Journal of the Balkan Tribological Association, ISSN: 1310–4772, Sofia, Bulgaria, Vol. 15, No 3, 336–346, 2009, jbalkta@gmail.com
5. Hariri Asli, K., Nagiyev, F.B., Haghi, A.K., Water hammer and fluid condition; a computational approach, Computational Meth-

ods in Applied Science and Engineering, USA, Chapter 5, Nova Science Publications, ISBN: 978–1-60876–052–7, USA, 73–94, 2010, https://www.novapublishers.com/catalog/

6. Hariri Asli, K., Nagiyev, F.B., Haghi, A.K.,Interpenetration of two fluids at parallel between plates and turbulent moving in pipe; a case study, Computational Methods in Applied Science and Engineering, USA, Chapter 7, Nova Science Publications, ISBN: 978–1-60876–052–7, USA, 107–133, 2010, https://www.novapublishers.com/catalog/
7. Hariri Asli, K., Nagiyev, F.B., Haghi, A.K., Modeling for water hammer due to valves; from theory to practice, Computational Methods in Applied Science and Engineering, USA, Chapter11, Nova Science Publications ISBN: 978–1-60876–052–7, USA, 229–236, 2010, https://www.novapublishers.com/catalog/
8. Hariri Asli, K., Nagiyev, F.B., Haghi, A.K., A computational method to Study transient flow in binary mixtures, Computational Methods in Applied Science and Engineering, USA, Chapter13, Nova Science Publications ISBN: 978–1-60876–052–7, USA, 229–236, 2010, https://www.novapublishers.com/catalog/
9. Hariri Asli, K., Nagiyev, F.B., Haghi, A.K., Water hammer analysis; some computational aspects and practical hints, Computational Methods in Applied Science and Engineering, USA, Chapter16, Nova Science Publications ISBN: 978–1-60876–052–7, USA, 263–281, 2010, https://www.novapublishers.com/catalog/
10. Hariri Asli, K., Nagiyev, F.B., Haghi, A.K., Water hammer and hydrodynamics instabilities modeling, Computational Methods in Applied Science and Engineering, USA, Chapter17, From Theory to Practice, Nova Science Publications ISBN: 978–1-60876–052–7, USA, 283–301, 2010, https://www.novapublishers.com/catalog/
11. Hariri Asli, K., Nagiyev, F.B., Haghi, A.K., A computational approach to study water hammer and pump pulsation phenomena, Computational Methods in Applied Science and Engineering, USA, Chapter22, Nova Science Publications, ISBN: 978–1-60876–052–7, USA, 349–363, 2010, https://www.novapublishers.com/catalog/
12. Hariri Asli, K., Nagiyev, F.B., Haghi, A.K., Some aspects of physical and numerical modeling of water hammer in pipelines. Computational Methods in Applied Science and Engineering, USA, Chapter23, Nova Science Publications, ISBN 978–1-60876–052–7, USA, 365–387, 2010, https://www.novapublishers.com/catalog/
13. Hariri Asli, K., Nagiyev F.B., Haghi, A.K., A computational approach to study fluid movement,Nanomaterials Yearbook – 2009, From Nanostructures, Nanomaterials and Nanotechnologies to Nanoindustry, Chapter16, Nova Science Publications, USA, ISBN:978–1-60876–451–8, USA, 181–196,2010. https://www.novapublishers.com/catalog/product_info.php?products_id=11587
14. Hariri Asli, K., Nagiyev, F.B., Haghi, A.K., Physical modeling of fluid movement in pipelines,Nanomaterials Yearbook – 2009, From Nanostructures, Nanomaterials and Nanotechnologies to Nanoindustry, Chapter17,Nova Science Publications, USA, ISBN:978–1-60876–451–8, USA, 197–214,2010. https://www.novapublishers.com/catalog/product_info.php?products_id=11587
15. Hariri Asli, K., Nagiyev, F.B.,Haghi, A.K., Aliyev, S.A., Improved modeling for prediction of water transmission failure, Recent Progress in Research in Chemistry and Chemical Engineering, Chapter2, Nova Science Publications, ISBN: 978–1-61668–501–0, Nova Science Publications, USA, 28–36, 2010. https://www.novapublishers.com/catalog/product_info.php?products_id=13174
16. Hariri Asli, K., Nagiyev, F.B.,Haghi, A.K., Aliyev, S.A. Pure Oxygen penetration in wastewater flow, Recent Progress in Research in Chemistry and Chemical Engineering, Chapter3, Nova Science Publications, ISBN: 978–1-61668–501–0, Nova Science Publications, USA, 17–27, 2010, https://www.novapublishers.com/catalog/product_info.php?products_id=13174
17. Hariri Asli, K., Mathematics and numerical modeling Technology, Journal of Mathematics and Technology, ISSN: 2078–0257,

No.3, August, Baku, Azerbaijan, 68–74, 2010, https://www.International%20Journal%20of%20Academic%20Research-IJAR.htm
18. Hariri Asli, K., Nagiyev, F.B.,Haghi, A.K., Aliyev, S.A., Physical and Numerical Modeling of Fluid Flow in Pipelines: A computational approach, International Journal of the Balkan Tribological Association, ISSN: 1310–4772, Vol.16, No 1, Sofia, Bulgaria, 20–34, 2010, jbalkta@gmail.com
19. Hariri Asli, K., Nagiyev, F.B.,Haghi, A.K., Aliyev, S.A., A Numerical Study on heat transfer in Microtubes, International Journal of the Balkan Tribological Association, ISSN: 1310–4772, Vol.16, No 1, Sofia, Bulgaria, 9–19, 2010, jbalkta@gmail.com
20. Hariri Asli, K., Nagiyev, F.B.,Haghi, A.K.,A numerical study on fluid dynamics, Material Science Synthesis, Properties, Applicators,ISBN: 978–1-60876–872–1, Chapter15, Nova Science Publications, USA, 101–110, 2010. https://www.novapublishers.com/catalog/product_info.php?products_id=12129
21. Hariri Asli, K., Nagiyev, F.B.,Haghi, A.K.,Some interpenetration for turbulent moving of fluid in pipe, Material Science Synthesis, Properties, Applicators,ISBN: 978–1-60876–872–1, Chapter16, Nova Science Publications, USA,111–117, 2010. https://www.novapublishers.com/catalog/product_info.php?products_id=12129
22. Hariri Asli, K., Nagiyev, F.B.,Haghi, A.K.,Fluid flow analysis due to water hammer, Material Science Synthesis, Properties, Applicators,ISBN: 978–1-60876–872–1, Chapter17, Nova Science Publications, USA, 120–128, 2010. https://www.novapublishers.com/catalog/product_info.php?products_id=12129
23. Hariri Asli, K., Nagiyev, F.B.,Haghi, A.K.,Transient flow in binary mixtures, Material Science Synthesis, Properties, Applicators,ISBN: 978–1-60876–872–1, Chapter19, Nova Science Publications, USA, 164–176, 2010. https://www.novapublishers.com/catalog/product_info.php?products_id=12129
24. Hariri Asli, K., Nagiyev, F.B.,Haghi, A.K.,Hydrodynamics instabilities modeling, Material Science Synthesis, Properties, Applicators,ISBN: 978–1-60876–872–1, Chapter20 Nova Science Publications, USA, 140–146, 2010. https://www.novapublishers.com/catalog/product_info.php?products_id=12129
25. Hariri Asli, K., Nagiyev, F.B.,Haghi, A.K.,Fluid dynamics and pump pulsation, Material Science Synthesis, Properties, Applicators, ISBN: 978–1-60876–872–1, Chapter21, Nova Science Publications, USA, 147–155, 2010. https://www.novapublishers.com/catalog/product_info.php?products_id=12129
26. Hariri Asli, K., Nagiyev, F.B.,Haghi, A.K., Aliyev, S.A., Hariri Asli, H., Flow in water pipeline: A computational approach, International Journal of Academic Research, ISSN: 1310–4772, ISSN: 2075–4124, Vol. 2, Issue 5, September 30, Baku, Azerbaijan, 164–176, 2010, https://www.International%20Journal%20of%20Academic%20Research-IJAR.htm
27. Hariri Asli, K., Nagiyev, F.B.,Haghi, A.K., Aliyev, S.A., Nonlinear Heterogeneous Model for Water Hammer Disaster, International Journal of the Balkan Tribological Association, ISSN: 1310–4772, Vol. 16, No 2, Sofia, Bulgaria, 209–222, 2010, jbalkta@gmail.com
28. Hariri Asli, K., Nagiyev, F.B.,Haghi, A.K.,Heat flow and mass transfer in capillary Porous body, Journal of the Balkan Tribological Association, Vol. 16, No 3, Tribotechnics and tribomechanics, Sofia, Bulgaria, 353–361, 2010, jbalkta@gmail.com
29. Hariri Asli, K., Nagiyev, F.B.,Haghi, A.K.,A Numerical Study on thermal drying of Porous solid, Journal of the Balkan Tribological Association, Vol. 16, No 3, Tribotechnics – thermal drying, Sofia, Bulgaria, 373–381,2010, jbalkta@gmail.com
30. Hariri Asli, K., Haghi, A.K.,A Numerical Study on Fluid Flow and Pressure drop in Microtubes, Journal of the Balkan Tribological Association, Vol. 16, No 3, Tribotechnics and tribomechanics, Sofia, Bulgaria, 382–392, 2010, jbalkta@gmail.com
31. Hariri Asli, K., Nagiyev, F.B.,Haghi, A.K., Aliyev, S.A., Hariri Asli, H., Improved Nonlinear Heterogeneous Model for Wastewater Treatment, International Journal on "Technical and Physical Problems of Engi-

neering", (IJTPE), Published by the International Organization on TPE (IOTPE), ISSN: 2077–3528, Baku, Azerbaijan, 30–36, 2010, http://www.iotpe.com/TPE-Journal/PublicationPolicy.html

32. Hariri Asli, K., GIS and Nonlinear Dynamics Model: Some Computational Aspects and Practical Hints, International Journal on "Technical and Physical Problems of Engineering", (IJTPE), Published by the International Organization on TPE (IOTPE), ISSN: 2077–3528, 1–5, Baku, Azerbaijan, 2010, http://www.iotpe.com/TPE-Journal/PublicationPolicy.html
33. Skousen, P., "Valve Handbook", McGraw Hill, New York, HAMMER Theory and Practice, 1998, 687–721.
34. Shaking, N.I., Water hammer to break the continuity of the flow in pressure conduits pumping stations: Dis. on Kharkov, 1988, 225.
35. Tijsseling, "Alan E Vardy Time scales and FSI in unsteady liquid-filled pipe flow", 1993, 5–12.
36. Wu, P.Y., Little, W.A., Measurement of friction factor for flow of gases in very fine channels used for micro miniature, Joule Thompson refrigerators, Cryogenics 24 (8), 1983, 273–277.
37. Song, C.C. et al., "Transient Mixed-Flow Models for Storm Sewers", J. of Hyd. Div., Nov., 1983, Vol. 109, 458–530.
38. Stephenson, D., "Pipe Flow Analysis", Elsevier, Vol. 19, S.A., 1984, 670–788.
39. Chaudhry, M.H., "Applied Hydraulic Transients", Van Nostrand Reinhold Co., N.Y., 1979, 1322–1324.
40. Chaudhry, M.H., Yevjevich, V. "Closed Conduit Flow", Water Resources Publication, USA, 1981, 255–278.
41. Chaudhry, M. H., Applied Hydraulic Transients, Van Nostrand Reinhold, New York, USA, 1987, p.165–167.
42. Kerr, S.L., "Minimizing service interruptions due to transmission line failures: Discussion, " Journal of the American Water Works Association, 41, 634, July 1949, 266–268.
43. Kerr, S.L., "Water hammer control, " Journal of the American Water Works Association, 43, December 1951, 985–999.
44. Apoloniusz Kodura, Katarzyna Weinerowska," Some Aspects of Physical and Numerical Modeling of Water Hammer in Pipelines", 2005, 126–132.
45. Anuchina, N.N., Volkov, V.I., Gordeychuk, V.A., Es'kov, N.S. Ilyutina, O.S., Kozyrev, O.M. "Numerical simulations of Rayleigh-Taylor and Richtmyer-Meshkov instability using mah-3 code, ' J. Comput. Appl. Math. 168, 2004, 11.
46. Fox, J.A., "Hydraulic Analysis of Unsteady Flow in Pipe Network", Wiley, N.Y., 1977, p.78–89.
47. Karassik, I.J., "Pump Handbook – Third Edition", McGraw-Hill, 2001 19–22.
48. Fok, A., "Design Charts for Air Chamber on Pump Pipelines", J. of Hyd. Div., ASCE, Sept., 1978, p.15–74.
49. Fok, A., Ashamalla, A., Aldworth, G., "Considerations in Optimizing Air Chamber for Pumping Plants", Symposium on Fluid Transients and Acoustics in the Power Industry, San Francisco, U.S.A. Dec., 1978, p.112–114.
50. Fok, A., "Design Charts for Surge Tanks on Pump Discharge Lines", BHRA 3rd Int. Conference on Pressure Surges, Bedford, England, Mar., 1980, p.23–34.
51. Fok, A., "Water hammer and Its Protection in Pumping Systems", Hydro technical Conference, CSCE, Edmonton, May, 1982, p.45–55.
52. Fok, A., "A contribution to the Analysis of Energy Losses in Transient Pipe Flow", PhD. Thesis, University of Ottawa, 1987, p.176–182.
53. Hariri Asli, K., Nagiyev, F.B , Water Hammer and fluid condition, Ministry of Energy, Gilan Water and Wastewater Co., Research Week Exhibition, Tehran, Iran, December, 2007, p. 132–148, http://isrc.nww.co.ir.
54. Hariri Asli, K., Nagiyev, F.B., Water Hammer analysis and formulation, Ministry of Energy, Gilan Water and Wastewater Co., Research Week Exhibition, Tehran, Iran, December, 2007, p. 111–131, http://isrc.nww.co.ir.
55. Hariri Asli, K., Nagiyev, F.B., Water Hammer and hydrodynamics instabilities, Interpenetration of two fluids at parallel between plates and turbulent moving in pipe, Ministry of Energy, Guilan Water and Wastewater

Co., Research Week Exhibition, Tehran, Iran, December, 2007, p.90–110, http://isrc.nww.co.ir.
56. Hariri Asli, K., Nagiyev, F.B., Water Hammer and pump pulsation, Ministry of Energy, Guilan Water and Wastewater Co., Research Week Exhibition, Tehran, Iran, December, 2007, p. 51–72, http://isrc.nww.co.ir.
57. Hariri Asli, K., Nagiyev, F.B., Reynolds number and hydrodynamics' instability", Ministry of Energy, Guilan Water and Wastewater Co., Research Week Exhibition, Tehran, Iran, December, 2007, p.31–50, http://isrc.nww.co.ir.
58. Hariri Asli, K., Nagiyev, F.B., Water Hammer and valves, Ministry of Energy, Guilan Water and Wastewater Co., Research Week Exhibition, Tehran, Iran, December, 2007, p.20–30, http://isrc.nww.co.ir.
59. Hariri Asli, K., Nagiyev, F.B., "Interpenetration of two fluids at parallel between plates and turbulent moving in pipe", Ministry of Energy, Guilan Water and Wastewater Co., Research Week Exhibition, Tehran, Iran, December, 2007, p.73–89, http://isrc.nww.co.ir.
60. Hariri Asli, K., Nagiyev, F.B., Decreasing of Unaccounted For Water "UFW" by Geographic Information System"GIS" in Rasht urban water system, civil engineering organization of Guilan, Technical and Art Journal, 2007, p.3–7, http://www.art-of-music.net/.
61. Hariri Asli, K., Portable Flow meter Tester Machine Apparatus, Certificate on registration of invention, Tehran, Iran, #010757, Series a/82, 24/11/2007, 1–3
62. Hariri Asli, K., Nagiyev, F.B., Haghi, A.K., "Interpenetration of two fluids at parallel between plates and turbulent moving in pipe", 9th Conference on Ministry of Energetic works at research week, Tehran, Iran, 2008, p.73–89, http://isrc.nww.co.ir.
63. Hariri Asli, K., Nagiyev, F.B., Haghi, A.K., "Water hammer and valves", 9th Conference on Ministry of Energetic works at research week, Tehran, Iran, 2008, p.20–30, http://isrc.nww.co.ir.
64. Hariri Asli, K., Nagiyev, F.B., Haghi, A.K., "Water hammer and hydrodynamics instability", 9th Conference on Ministry of Energetic works at research week, Tehran, Iran, 2008, p.90–110, http://isrc.nww.co.ir.
65. Hariri Asli, K., Nagiyev, F.B., Haghi, A.K., "Water hammer analysis and formulation", 9th Conference on Ministry of Energetic works at research week, Tehran, Iran, 2008, p. 27–42, http://isrc.nww.co.ir.
66. Hariri Asli, K., Nagiyev, F.B., Haghi, A.K., "Water hammer &fluid condition", 9th Conference on Ministry of Energetic works at research week, Tehran, Iran, 2008, p.27–43, http://isrc.nww.co.ir.
67. Hariri Asli, K., Nagiyev, F.B., Haghi, A.K., "Water hammer and pump pulsation", 9th Conference on Ministry of Energetic works at research week, Tehran, Iran, 2008, p.27–44, http://isrc.nww.co.ir.
68. Hariri Asli, K., Nagiyev, F.B., Haghi, A.K., "Reynolds number and hydrodynamics instability", 9th Conference on Ministry of Energetic works at research week, Tehran, Iran, 2008, p. 27–45, http://isrc.nww.co.ir.
69. Hariri Asli, K., Nagiyev, F.B., Haghi, A.K., "Water hammer and fluid Interpenetration", 9th Conference on Ministry of Energetic works at research week, Tehran, Iran, 2008, p.27–47, http://isrc.nww.co.ir.
70. Hariri Asli, K., GIS and water hammer disaster at earthquake in Rasht water pipeline, civil engineering organization of Guilan, Technical and Art Journal, 2008, 14–17, http://www.art-of-music.net/.
71. Hariri Asli, K., GIS and water hammer disaster at earthquake in Rasht water pipeline, 3rd International Conference on Integrated Natural Disaster Management, Tehran university, ISSN: 1735–5540, 18–19 Feb., INDM, Tehran, Iran, 2008, №13, 53/1–12, http://www.civilica.com/Paper-INDM03-INDM03_001.html
72. Hariri Asli, K., Nagiyev, F.B., Bubbles characteristics and convective effects in the binary mixtures. Transactions issue mathematics and mechanics series of physical-technical and mathematics science, ISSN: 0002–3108, Azerbaijan, Baku, 2009, 68–74, http://www.imm.science.az/journals.html.
73. Hariri Asli, K., Nagiyev, F.B., Haghi, A.K., Aliyev, S.A., Three-Dimensional conjugate heat transfer in porous media, 1st Festival on Water and Wastewater Research and

Technology, Tehran, Iran, 12–17 Dec.2009, 26–28, http://isrc.nww.co.ir.

74. Hariri Asli, K., Nagiyev, F.B., Haghi, A.K., Aliyev, S.A., Some Aspects of Physical and Numerical Modeling of water hammer in pipelines, 1st Festival on Water and Wastewater Research and Technology, Tehran, Iran, 12–17 Dec.2009, 26–29, http://isrc.nww.co.ir
75. Hariri Asli, K., Nagiyev, F.B., Haghi, A.K., Aliyev, S.A., Modeling for Water Hammer due to valves: From theory to practice, 1st Festival on Water and Wastewater Research and Technology, Tehran, Iran, 12–17 Dec.2009, 26,30, http://isrc.nww.co.ir.
76. Hariri Asli, K., Nagiyev, F.B., Haghi, A.K., Aliyev, S.A., Water hammer and hydrodynamics instabilities modeling: From Theory to Practice, 1st Festival on Water and Wastewater Research and Technology, Tehran, Iran, 12–17 Dec. 2009, 26–31, http://isrc.nww.co.ir
77. Hariri Asli, K., Nagiyev, F.B., Haghi, A.K., Aliyev, S.A., A computational approach to study fluid movement, 1st Festival on Water and Wastewater Research and Technology, Tehran, Iran, 12–17 Dec.2009, 27–32, http://isrc.nww.co.ir.
78. Hariri Asli, K., Nagiyev, F.B., Haghi, A.K., Aliyev, S.A., Water Hammer Analysis: Some Computational Aspects and practical hints, 1st Festival on Water and Wastewater Research and Technology, Tehran, Iran, 12–17 Dec.2009, 27–33, http://isrc.nww.co.ir
79. Hariri Asli, K., Nagiyev, F.B., Haghi, A.K., Aliyev, S.A., Water Hammer and Fluid condition: A computational approach, 1st Festival on Water and Wastewater Research and Technology, Tehran, Iran, 12–17 Dec.,2009, 27–34, http://isrc.nww.co.ir.
80. Hariri Asli, K., Nagiyev, F.B., Haghi, A.K., Aliyev, S.A., A computational Method to Study Transient Flow in Binary Mixtures, 1st Festival on Water and Wastewater Research and Technology, Tehran, Iran, 12–17 Dec.2009, 27–35, http://isrc.nww.co.ir.
81. Hariri Asli, K., Nagiyev, F.B., Haghi, A.K., Aliyev, S.A., Physical modeling of fluid movement in pipelines, 1st Festival on Water and Wastewater Research and Technology, Tehran, Iran, 12–17 Dec.2009, 27–36, http://isrc.nww.co.ir.
82. Hariri Asli, K., Nagiyev, F.B., Haghi, A.K., Aliyev, S.A., Interpenetration of two fluids at parallel between plates and turbulent moving, 1st Festival on Water and Wastewater Research and Technology, Tehran, Iran, 12–17 Dec. 2009, 27–37, http://isrc.nww.co.ir.
83. Hariri Asli, K., Nagiyev, F.B., Haghi, A.K., Aliyev, S.A., Modeling of fluid interaction produced by water hammer, 1st Festival on Water and Wastewater Research and Technology, Tehran, Iran, 12–17 Dec.2009, 27–38, http://isrc.nww.co.ir.
84. Hariri Asli, K., Nagiyev, F.B., Haghi, A.K., Aliyev, S.A., GIS and water hammer disaster at earthquake in Rasht pipeline, 1st Festival on Water and Wastewater Research and Technology, Tehran, Iran, 12–17 Dec.2009, 27–39, http://isrc.nww.co.ir.
85. Hariri Asli, K., Nagiyev, F.B., Haghi, A.K., Aliyev, S.A., Interpenetration of two fluids at parallel between plates and turbulent moving, 1st Festival on Water and Wastewater Research and Technology, Tehran, Iran, 12–17 Dec 2009, 27–40, http://isrc.nww.co.ir.
86. Hariri Asli, K., Nagiyev, F.B, Haghi, A.K., Aliyev, S.A., Water hammer and hydrodynamics' instability, 1st Festival on Water and Wastewater Research and Technology, Tehran, Iran, 12–17 Dec.2009, 27–41, http://isrc.nww.co.ir.
87. Hariri Asli, K., Nagiyev, F.B., Haghi, A.K., Aliyev, S.A., Water hammer analysis and formulation, 1st Festival on Water and Wastewater Research and Technology, Tehran, Iran, 12–17 Dec.2009, 27–42, http://isrc.nww.co.ir.
88. Hariri Asli, K., Nagiyev, F.B., Haghi, A.K., Aliyev, S.A., Water hammer &fluid condition, 1st Festival on Water and Wastewater Research and Technology, Tehran, Iran, 12–17 Dec.2009, 27–43, http://isrc.nww.co.ir.
89. Hariri Asli, K., Nagiyev, F.B., Haghi, A.K., Aliyev, S.A., Water hammer and pump pulsation, 1st Festival on Water and Wastewater Research and Technology, Tehran, Iran, 12–17 Dec.2009, 27–44, http://isrc.nww.co.ir.
90. Hariri Asli, K., Nagiyev, F.B., Haghi, A.K., Aliyev, S.A., Reynolds number and hydrodynamics instabilities, 1st Festival on Wa-

ter and Wastewater Research and Technology, Tehran, Iran, 12–17 Dec.2009, 27–45, http://isrc.nww.co.ir.

91. Hariri Asli, K., Nagiyev, F.B., Haghi, A.K., Aliyev, S.A., water hammer and valves, 1st Festival on Water and Wastewater Research and Technology, Tehran, Iran, 12–17 Dec.2009, 27–46, http://isrc.nww.co.ir.
92. Hariri Asli, K., Nagiyev, F.B., Haghi, A.K., Aliyev, S.A., "Water hammer and fluid Interpenetration", 1st Festival on Water and Wastewater Research and Technology, Tehran, Iran, 12–17 Dec.2009, 27–47, http://isrc.nww.co.ir.
93. Chaudhry, M.H., "Applied Hydraulic Transients", Van Nostrand Reinhold Co., N.Y., 1979, 1322–1324.
94. Parmakian, J., "Water hammer Analysis", Dover Publications, Inc., New York, New York, 1963, 51–58.
95. Streeter, V.L., Wylie, E.B., "Fluid Mechanics", McGraw-Hill Ltd., USA, 1979, 492–505.
96. Leon Arturo, S., "An efficient second-order accurate shock-capturing scheme for modeling one and two-phase water hammer flows" PhD Thesis, March 29, 2007, 4–44.
97. Wu, P.Y., Little, W.A. Measurement of friction factor for flow of gases in very fine channels used for micro miniature, Joule Thompson refrigerators, Cryogenics 24 (8), 1983, 273–277.
98. Harms, T.M., Kazmierczak, M.J., Cerner, F.M., Holke, A., Henderson, H.T., Pilchowski, H.T., Baker, K., Experimental Investigation of Heat Transfer and Pressure Drop through Deep Micro channels in a (100) Silicon Substrate, in: Proceedings of the ASME. Heat Transfer Division, HTD 351, 1997, 347–357.
99. Fedorov, A.G., Viskanta, R., Three-dimensional Conjugate Heat Transfer into Microchannel Heat Sink for Electronic Packaging, Int. J. Heat Mass Transfer 43, 2000, 399–415.
100. Qu, W., Mala, G.M., Li, D., Heat Transfer for Water Flow in Trapezoidal Silicon Microchannels, 1993, 399–404.
101. Choi, S.B., Barren, R.R., Warrington, R.O., Fluid Flow and Heat Transfer in Microtubes, ASME DSC 40, 1991, 89–93.
102. Adams, T.M., Abdel-Khalik, S.I., Jeter, S.M., Qureshi, Z. H. AN Experimental investigation of single-phase Forced Convection in Microchannels, Int. J. Heat Mass Transfer, 41, 1998, 851–857.
103. Peng, X.F., Peterson, G.P., Convective Heat Transfer and Flow Friction for Water Flow in Microchannel Structure, Int. J. Heat Mass Transfer 36, 1996, 2599–2608.
104. Mala, G., Li, D., Dale, J.D., Heat Transfer and Fluid Flow in Microchannels, J. Heat Transfer, 40, 1997, 3079–3088.
105. Xu, B., Ooi, K.T., Mavriplis, C., Zaghloul, M. E., Viscous dissipation effects for liquid flow in microchannels, Micorsystems, 2002, 53–57.
106. Tuckerman, D.B., R.F.W Pease, high performance heat sinking for VLSI, IEEE Electron device letter, DEL-2, 1981, 126–129.

REFERENCES FOR CHAPTER 3

1. Qu, W.; Mala, G.M.; Li, D.; Heat Transfer for Water Flow in Trapezoidal Silicon Microchannels, 1993, 399–404.
2. Hariri Asli, K.; Nagiyev, F.B.; Haghi, A.K.; Aliyev, S.A.; A computational approach to study fluid movement, 1st Festival on Water and Wastewater Research and Technology, Tehran, Iran, 12–17 Dec.2009, 27–32, http://isrc.nww.co.ir.
3. Peng, X.F.; Peterson, G.P.; Convective Heat Transfer and Flow Friction for Water Flow in Microchannel Structure, Int. J. Heat Mass Transfer 36, 1996, 2599–2608.
4. Bergant Anton, Discrete Vapour Cavity Model with Improved Timing of Opening and Collapse of Cavities, 1980, 1–11.
5. Ishii, M.; Thermo-Fluid Dynamic Theory of Two-Phase Flow, Collection de, D.R.; Liles and, W.H.; Reed, "A Sern-Implict Method for Two-Phase Fluid la Direction des Etudes et. Recherché d'Electricite de France, Vol. 22 Dynamics," Journal of Computational Physics 26, Paris, 1975, 390–407.
6. Hariri Asli, K.; Nagiyev, F.B.; Haghi, A.K.; Aliyev, S.A.; A computational approach to study fluid movement, 1st Festival on Water and Wastewater Research and Technology, Tehran, Iran, 12–17 Dec.2009, 27–32, http://isrc.nww.co.ir.

7. Pickford, J.; "Analysis of Surge", Macmillian, London, 1969, 153–156.
8. Pipeline Design for Water and Wastewater, American Society of Civil Engineers, New York, 1975, 54 p.
9. Xu, B.; Ooi, K.T.; Mavriplis, C.; Zaghloul, M.E.; Viscous dissipation effects for liquid flow in microchannels, Micorsystems, 2002, 53–57.
10. Fedorov, A.G.; Viskanta, R.; Three-dimensional Conjugate Heat Transfer into Microchannel Heat Sink for Electronic Packaging, Int. J. Heat Mass Transfer 43, 2000, 399–415.
11. Tuckerman, D.B.; Heat transfer microstructures for integrated circuits, Ph.D. thesis, Stanford University, 1984, 10–120.
12. Harms, T.M.; Kazmierczak, M.J.; Cerner, F.M.; Holke, A.; Henderson, H.T.; Pilchowski, H.T.; Baker, K.; Experimental Investigation of Heat Transfer and Pressure Drop through Deep Micro channels in a (100) Silicon Substrate, in: Proceedings of the ASME. Heat Transfer Division, HTD 351, 1997, 347–357.
13. Holland, F.A.; Bragg, R.; Fluid Flow for Chemical Engineers, Edward Arnold Publishers, London, 1995, 1–3.
14. Lee TS, Pejovic S (1996) Air influence on similarity of hydraulic transients and vibrations. ASME J. Fluid Eng. 118(4), 706–709.
15. Li, J.; McCorquodale, A.; "Modeling Mixed Flow in Storm Sewers", Journal of Hydraulic Engineering, ASCE, Vol. 125, No. 11, 1999, 1170–1180.
16. Minnaert, M.; on musical air bubbles and the sounds of running water. Phil. Mag., 1933, v. 16, №7, 235–248.
17. Moeng, C.H.; McWilliams, J.C.; Rotunno, R.; Sullivan, P.P.; Weil, J.; "Investigating 2D modeling of atmospheric convection in the PBL," J. Atm. Sci. 61, 2004, 889–903.
18. Tuckerman, D.B.; R.F.W Pease, high performance heat sinking for VLSI, IEEE Electron device letter, DEL-2, 1981, 126–129.
19. Nagiyev, F.B.; Khabeev, N.S, Bubble dynamics of binary solutions. High Temperature, 1988, v. 27, № 3, 528–533.
20. Shvarts, D.; Oron, D.; Kartoon, D.; Rikanati, A.; Sadot, O.; "Scaling laws of nonlinear Rayleigh-Taylor and Richtmyer-Meshkov instabilities in two and three dimensions," C. R. Acad. Sci. Paris IV, 719, 2000, 312 p.
21. Cabot, W.H.; Cook, A.W.; Miller, P.L.; Laney, D.E.; Miller, M.C.; Childs, H.R.; "Large eddy simulation of Rayleigh-Taylor instability," Phys. Fluids, September, 2005, vol. 17, p. 91–106
22. Cabot, W.; University of California, Lawrence Livermore National laboratory, Livermore, CA, Physics of Fluids, 2006, p. 94–550.
23. Goncharov, V.N.; "Analytical model of nonlinear, single-mode, classical Rayleigh-Taylor instability at arbitrary Atwood numbers", Phys. Rev. Letters 88, 134502, 2002, 10–15.
24. Ramaprabhu, P.; Andrews, M.J.; "Experimental investigation of Rayleigh-Taylor mixing at small Atwood numbers," J. Fluid Mech. 502, 2004, 233 p.
25. Clark, T.T.; "A numerical study of the statistics of a two-dimensional Rayleigh-Taylor mixing layer", Phys. Fluids 15, 2003, 2413.
26. Cook, A.W.; Cabot, W.; Miller, P.L.; "The mixing transition in Rayleigh-Taylor instability", J. Fluid Mech. 511,2004, 333.
27. Waddell, J.T.; Niederhaus, C.E.; Jacobs, J.W.; "Experimental study of Rayleigh-Taylor instability: Low Atwood number liquid systems with single-mode initial perturbations," Phys. Fluids 13, 2001, 1263–1273.
28. Weber, S.V.; Dimonte, G.; Marinak, M.M.; "Arbitrary Lagrange-Eulerian code simulations of turbulent Rayleigh-Taylor instability in two and three dimensions," Laser and Particle Beams 21, 2003, 455 p.
29. Dimonte, G.; Youngs, D.; Dimits, A.; Weber, S.; Marinak, M. "A comparative study of the Rayleigh-Taylor instability using high-resolution three-dimensional numerical simulations: the Alpha group collaboration," Phys. Fluids 16,2004, 1668.
30. Young, Y.N.; Tufo, H.; Dubey, A.; Rosner, R.; "On the miscible Rayleigh-Taylor instability: two and three dimensions," J. Fluid Mech. 447, 377, 2001, 2003–2500.
31. George, E.; Glimm, J.; "Self-similarity of Rayleigh-Taylor mixing rates," Phys. Fluids 17, 054101, 2005, 1–3.
32. Oron. D., Arazi, L.; Kartoon, D.; Rikanati, A.; Alon, U.; Shwarts, D.; "Dimensionality dependence of the Rayleigh-Taylor and

Richtmyer-Meshkov instability late-time scaling laws", Phys. Plasmas 8, 2001, 2883 p.
33. Nigmatulin, R.I.; Nagiyev, F.B.; Khabeev, N.S.; Effective heat transfer coefficients of the bubbles in the liquid radial pulse. Mater. Second-Union. Conf. Heat Mass Transfer, "Heat massoob-men in the biphasic. with. ". Minsk, 1980, v.5, 111–115.
34. Nagiyev, F.B.; Khabeev, N.S, Bubble dynamics of binary solutions. High Temperature, 1988, v. 27, № 3, 528–533.
35. Nagiyev, F.B.; Damping of the oscillations of bubbles boiling binary solutions. Mater. VIII Resp. Conf. mathematics and mechanics. Baku, October 26–29, 1988, 177–178.
36. Nagyiev, F.B.; Kadyrov, B.A.; Small oscillations of the bubbles in a binary mixture in the acoustic field. Math. AN Az.SSR Ser. Physicotech. and mate. Science, 1986, № 1, 23–26.
37. Nagiyev, F.B.; Dynamics, heat and mass transfer of vapor-gas bubbles in a two-component liquid. Turkey-Azerbaijan petrol semin., Ankara, Turkey, 1993, 32–40.
38. Nagiyev, F.B.; The method of creation effective coolness liquids, Third Baku international Congress. Baku, Azerbaijan Republic, 1995, 19–22.
39. Nagiyev, F.B.; The linear theory of disturbances in binary liquids bubble solution. Dep. In VINITI, 1986, № 405, in 86, 76–79.
40. Nagiyev, F.B.; Structure of stationary shock waves in boiling binary solutions. Math. USSR, Fluid Dynamics, 1989, № 1, 81–87.
41. Rayleigh, On the pressure developed in a liquid during the collapse of a spherical cavity. Philos. Mag. Ser. 6, v. 34, N 200, 1917, 94–98.
42. Perry. R.H., Green, D.W.; Maloney, J.O.; Perry's Chemical Engineers Handbook, 7th Edition, McGraw-Hill, New York, 1997, 1–61.
43. Nigmatulin, R.I.; Dynamics of multiphase media., Moscow, "Nauka", 1987, v. 1, 2, 12–14.
44. Kodura, A.; Weinerowska, K.; the influence of the local pipeline leak on water hammer properties, Materials of the II Polish Congress of Environmental Engineering, Lublin, 2005, 125–133.
45. Kane, J.; Arnett, D.; Remington, B.A.; Glendinning, S.G.; Baz'an, G.; "Two-dimensional versus three-dimensional supernova hydrodynamic instability growth", Astrophys. J., 2000, 528–989.
46. Quick, R.S.; "Comparison and Limitations of Various Water hammer Theories", J. Hyd. Div., ASME, May, 1933, 43–45.
47. Jaeger, C.; "Fluid Transients in Hydro-Electric Engineering Practice", Blackie and Son Ltd., 1977, 87–88.
48. Jaime Suárez, A.; "Generalized water hammer algorithm for piping systems with unsteady friction" 2005, 72–77.
49. Fok, A.; Ashamalla, A.; Aldworth, G.; "Considerations in Optimizing Air Chamber for Pumping Plants", Symposium on Fluid Transients and Acoustics in the Power Industry, San Francisco, U.S.A. Dec., 1978, 112–114.
50. Fok, A.; "Design Charts for Surge Tanks on Pump Discharge Lines", BHRA 3rd Int. Conference on Pressure Surges, Bedford, England, Mar., 1980, 23–34.
51. Fok, A.; "Water hammer and Its Protection in Pumping Systems", Hydro technical Conference, CSCE, Edmonton, May, 1982, 45–55.
52. Fok, A.; "A contribution to the Analysis of Energy Losses in Transient Pipe Flow", PhD. Thesis, University of Ottawa, 1987, 176–182.
53. Hariri Asli, K.; Nagiyev, F.B.; Water Hammer and fluid condition, Ministry of Energy, Gilan Water and Wastewater Co., Research Week Exhibition, Tehran, Iran, December, 2007, 132–148, http://isrc.nww.co.ir.
54. Hariri Asli, K.; Nagiyev, F.B.; Water Hammer analysis and formulation, Ministry of Energy, Gilan Water and Wastewater Co., Research Week Exhibition, Tehran, Iran, December, 2007, 111–131, http://isrc.nww.co.ir.
55. Hariri Asli, K.; Nagiyev, F.B.; Water Hammer and hydrodynamics instabilities, Interpenetration of two fluids at parallel between plates and turbulent moving in pipe, Ministry of Energy, Guilan Water and Wastewater Co., Research Week Exhibition, Tehran, Iran, December, 2007, 90–110, http://isrc.nww.co.ir.

56. Hariri Asli, K.; Nagiyev, F.B.; Water Hammer and pump pulsation, Ministry of Energy, Guilan Water and Wastewater Co., Research Week Exhibition, Tehran, Iran, December, 2007, 51–72, http://isrc.nww.co.ir.
57. Hariri Asli, K.; Nagiyev, F.B.; Reynolds number and hydrodynamics' instability", Ministry of Energy, Guilan Water and Wastewater Co., Research Week Exhibition, Tehran, Iran, December, 2007, 31–50, http://isrc.nww.co.ir.
58. Hariri Asli, K.; Nagiyev, F.B.; Water Hammer and valves, Ministry of Energy, Guilan Water and Wastewater Co., Research Week Exhibition, Tehran, Iran, December, 2007, 20–30, http://isrc.nww.co.ir.
59. Hariri Asli, K.; Nagiyev, F.B.; "Interpenetration of two fluids at parallel between plates and turbulent moving in pipe", Ministry of Energy, Guilan Water and Wastewater Co., Research Week Exhibition, Tehran, Iran, December, 2007, 73–89, http://isrc.nww.co.ir.
60. Hariri Asli, K.; Nagiyev, F.B.; Decreasing of Unaccounted For Water "UFW" by Geographic Information System"GIS" in Rasht urban water system, civil engineering organization of Guilan, Technical and Art Journal, 2007, 3–7, http://www.art-of-music.net/.
61. Hariri Asli, K.; Portable Flow meter Tester Machine Apparatus, Certificate on registration of invention, Tehran, Iran, #010757, Series a/82, 24/11/2007, 1–3
62. Hariri Asli, K.; Nagiyev, F.B.; Haghi, A.K.; "Interpenetration of two fluids at parallel between plates and turbulent moving in pipe", 9th Conference on Ministry of Energetic works at research week, Tehran, Iran, 2008, 73–89, http://isrc.nww.co.ir.
63. Hariri Asli, K.; Nagiyev, F.B.; Haghi, A.K.; "Water hammer and valves", 9th Conference on Ministry of Energetic works at research week, Tehran, Iran, 2008, 20–30, http://isrc.nww.co.ir.
64. Hariri Asli, K.; Nagiyev, F.B.; Haghi, A.K.; "Water hammer and hydrodynamics instability", 9th Conference on Ministry of Energetic works at research week, Tehran, Iran, 2008, 90–110, http://isrc.nww.co.ir.
65. Hariri Asli, K.; Nagiyev, F.B.; Haghi, A.K.; "Water hammer analysis and formulation", 9th Conference on Ministry of Energetic works at research week, Tehran, Iran, 2008, 27–42, http://isrc.nww.co.ir.
66. Hariri Asli, K.; Nagiyev, F.B.; Haghi, A.K.; "Water hammer & fluid condition", 9th Conference on Ministry of Energetic works at research week, Tehran, Iran, 2008, 27–43, http://isrc.nww.co.ir.
67. Hariri Asli, K.; Nagiyev, F.B.; Haghi, A.K.; "Water hammer and pump pulsation", 9th Conference on Ministry of Energetic works at research week, Tehran, Iran, 2008, 27–44, http://isrc.nww.co.ir.
68. Hariri Asli, K.; Nagiyev, F.B.; Haghi, A.K.; "Reynolds number and hydrodynamics instability", 9th Conference on Ministry of Energetic works at research week, Tehran, Iran, 2008, 27–45, http://isrc.nww.co.ir.
69. Hariri Asli, K.; Nagiyev, F.B. Haghi, A.K.; "Water hammer and fluid Interpenetration", 9th Conference on Ministry of Energetic works at research week, Tehran, Iran, 2008, 27–47, http://isrc.nww.co.ir.
70. Hariri Asli, K.; GIS and water hammer disaster at earthquake in Rasht water pipeline, civil engineering organization of Guilan, Technical and Art Journal, 2008, 14–17, http://www.art-of-music.net/.
71. Hariri Asli, K.; GIS and water hammer disaster at earthquake in Rasht water pipeline, 3rd International Conference on Integrated Natural Disaster Management, Tehran university, ISSN: 1735–5540, 18–19 Feb., INDM, Tehran, Iran, 2008, №13, 53/1–12, http://www.civilica.com/Paper-INDM03-INDM03_001.html
72. Hariri Asli, K.; Nagiyev, F.B.; Bubbles characteristics and convective effects in the binary mixtures. Transactions issue mathematics and mechanics series of physical-technical and mathematics science, ISSN: 0002–3108, Azerbaijan, Baku, 2009, 68–74, http://www.imm.science.az/journals.html
73. Hariri Asli, K.; Nagiyev, F.B.; Haghi, A.K.; Aliyev, S.A.; Three-Dimensional conjugate heat transfer in porous media, 1st Festival on Water and Wastewater Research and Technology, Tehran, Iran, 12–17 Dec.2009, 26–28, http://isrc.nww.co.ir.

74. Hariri Asli, K.; Nagiyev, F.B.; Haghi, A.K.; Aliyev, S.A.; Some Aspects of Physical and Numerical Modeling of water hammer in pipelines, 1st Festival on Water and Wastewater Research and Technology, Tehran, Iran, 12–17 Dec.2009, 26–29, http://isrc.nww.co.ir
75. Hariri Asli, K.; Nagiyev, F.B.; Haghi, A.K.; Aliyev, S.A.; Modeling for Water Hammer due to valves: From theory to practice, 1st Festival on Water and Wastewater Research and Technology, Tehran, Iran, 12–17 Dec.2009, 26,30, http://isrc.nww.co.ir.
76. Hariri Asli, K.; Nagiyev, F.B.; Haghi, A.K.; Aliyev, S.A.; Water hammer and hydrodynamics instabilities modeling: From Theory to Practice, 1st Festival on Water and Wastewater Research and Technology, Tehran, Iran, 12–17 Dec.2009, 26–31, http://isrc.nww.co.ir
77. Hariri Asli, K.; Nagiyev, F.B.; Haghi, A.K.; Aliyev, S.A.; A computational approach to study fluid movement, 1st Festival on Water and Wastewater Research and Technology, Tehran, Iran, 12–17 Dec.2009, 27–32, http://isrc.nww.co.ir.
78. Hariri Asli, K.; Nagiyev, F.B.; Haghi, A.K.; Aliyev, S.A.; Water Hammer Analysis: Some Computational Aspects and practical hints, 1st Festival on Water and Wastewater Research and Technology, Tehran, Iran, 12–17 Dec.2009, 27–33, http://isrc.nww.co.ir
79. Hariri Asli, K.; Nagiyev, F.B.; Haghi, A.K.; Aliyev, S.A.; Water Hammer and Fluid condition: A computational approach, 1st Festival on Water and Wastewater Research and Technology, Tehran, Iran, 12–17 Dec.,2009, 27–34, http://isrc.nww.co.ir.
80. Hariri Asli, K.; Nagiyev, F.B.; Haghi, A.K.; Aliyev, S.A.; A computational Method to Study Transient Flow in Binary Mixtures, 1st Festival on Water and Wastewater Research and Technology, Tehran, Iran, 12–17 Dec.2009, 27–35, http://isrc.nww.co.ir.
81. Hariri Asli, K.; Nagiyev, F.B.; Haghi, A.K.; Physical modeling of fluid movement in pipelines, 1st Festival on Water and Wastewater Research and Technology, Tehran, Iran, 12–17 Dec.2009, 27–36, http://isrc.nww.co.ir.
82. Hariri Asli, K.; Nagiyev, F.B.; Haghi, A.K.; Aliyev, S.A.; Interpenetration of two fluids at parallel between plates and turbulent moving, 1st Festival on Water and Wastewater Research and Technology, Tehran, Iran, 12–17 Dec. 2009, 27–37, http://isrc.nww.co.ir.
83. Hariri Asli, K.; Nagiyev, F.B.; Haghi, A.K.; Aliyev, S.A.; Modeling of fluid interaction produced by water hammer, 1st Festival on Water and Wastewater Research and Technology, Tehran, Iran, 12–17 Dec.2009, 27–38, http://isrc.nww.co.ir.
84. Hariri Asli, K.; Nagiyev, F.B.; Haghi, A.K.; Aliyev, S.A.; GIS and water hammer disaster at earthquake in Rasht pipeline, 1st Festival on Water and Wastewater Research and Technology, Tehran, Iran, 12–17 Dec.2009, 27–39, http://isrc.nww.co.ir.
85. Hariri Asli, K.; Nagiyev, F.B.; Haghi, A.K.; Aliyev, S.A.; Interpenetration of two fluids at parallel between plates and turbulent moving, 1st Festival on Water and Wastewater Research and Technology, Tehran, Iran, 12–17 Dec.2009, 27–40, http://isrc.nww.co.ir.
86. Hariri Asli, K.; Nagiyev, F.B.; Haghi, A.K.; Aliyev, S.A.; Water hammer and hydrodynamics' instability, 1st Festival on Water and Wastewater Research and Technology, Tehran, Iran, 12–17 Dec.2009, 27–41, http://isrc.nww.co.ir.
87. Hariri Asli, K.; Nagiyev, F.B.; Haghi, A.K.; Aliyev, S.A.; Water hammer analysis and formulation, 1st Festival on Water and Wastewater Research and Technology, Tehran, Iran, 12–17 Dec.2009, 27–42, http://isrc.nww.co.ir.
88. Hariri Asli, K.; Nagiyev, F.B.; Haghi, A.K.; Aliyev, S.A.; Water hammer &fluid condition, 1st Festival on Water and Wastewater Research and Technology, Tehran, Iran, 12–17 Dec.2009, 27–43, http://isrc.nww.co.ir.
89. Hariri Asli, K.; Nagiyev, F.B.; Haghi, A.K.; Aliyev, S.A.; Water hammer and pump pulsation, 1st Festival on Water and Wastewater Research and Technology, Tehran, Iran, 12–17 Dec.2009, 27–44, http://isrc.nww.co.ir.
90. Hariri Asli, K.; Nagiyev, F.B.; Haghi, A.K.; Aliyev, S.A.; Reynolds number and hydrodynamics instabilities, 1st Festival on Water and Wastewater Research and Technol-

ogy, Tehran, Iran, 12–17 Dec.2009, 27–45, http://isrc.nww.co.ir.
91. Brunone, B.; Karney, B.W.; Mecarelli, M.; Ferrante, M. "Velocity Profiles and Unsteady Pipe Friction in Transient Flow" Journal of Water Resources Planning and Management, ASCE, 126(4), Jul., 2000, 236–244.
92. Koelle, E.; Luvizotto Jr. E., Andrade, J.P.G., "Personality Investigation of Hydraulic Networks using MOC – Method of Characteristics" Proceedings of the 7th International Conference on Pressure Surges and Fluid Transients, Harrogate Durham, United Kingdom, 1996, 1–8.
93. Filion, Y.; Karney, B.W.; "A Numerical Exploration of Transient Decay Mechanisms in Water Distribution Systems", Proceedings of the ASCE Environmental Water Resources Institute Conference, American Society of Civil Engineers, Roanoke, Virginia, 2002, 30.
94. Hamam, M.A.; Mc Corquodale, J.A.; "Transient Conditions in the Transition from Gravity to Surcharged Sewer Flow", Canadian, J. of Civil Eng., Canada, Sep., 1982, 65–98.
95. Savic, D.A.; Walters, G.A.; "Genetic Algorithms Techniques for Calibrating Network Models", Report No. 95/12, Centre for Systems and Control Engineering, 1995, 137–146.
96. Walski, T.M.; Lutes, T.L.; "Hydraulic Transients Cause Low-Pressure Problems", Journal of the American Water Works Association, 75(2), 1994, 58.
97. Lee TS, Pejovic S (1996) Air influence on similarity of hydraulic transients and vibrations. ASME J. Fluid Eng. 118(4), 706–709.
98. Chaudhry, M.H.; "Applied Hydraulic Transients", Van Nostrand Reinhold Co., N.Y., 1979, 1322–1324.
99. Parmakian, J.; "Water hammer Analysis", Dover Publications, Inc., New York, New York, 1963, 51–58.
100. Streeter, V.L.; Wylie, E.B.; 'Fluid Mechanics", McGraw-Hill Ltd., USA, 1979, 492–505.
101. Leon Arturo, S.; "An efficient second-order accurate shock-capturing scheme for modeling one and two-phase water hammer flows" PhD Thesis, March 29, 2007, 4–44.
102. Adams, T.M.; Abdel-Khalik, S.I.; Jeter, S.M.; Qureshi, Z.H.; AN Experimental investigation of single-phase Forced Convection in Microchannels, Int. J. Heat Mass Transfer, 41, 1998, 851–857.
103. Peng, X.F.; Peterson, G.P.; Convective Heat Transfer and Flow Friction for Water Flow in Microchannel Structure, Int. J. Heat Mass Transfer 36, 1996, 2599–2608.
104. Mala, G.; Li, D.; Dale, J.D ; Heat Transfer and Fluid Flow in Microchannels, J. Heat Transfer, 40, 1997, 3079–3088.
105. Xu, B.; Ooi, K.T.; Mavriplis, C.; Zaghloul, M.E.; Viscous dissipation effects for liquid flow in microchannels, Micorsystems, 2002, 53–57.
106. Tuckerman, D.B.; R.F.W Pease, high performance heat sinking for VLSI, IEEE Electron device letter, DEL-2, 1981, 126–129.

REFERENCES FOR CHAPTER 4

1. Qu, W.; Mala, G.M.; Li, D.; Heat Transfer for Water Flow in Trapezoidal Silicon Microchannels, 1993, 399–404.
2. Hariri Asli, K.; Nagiyev, F.B.; Haghi, A.K.; Aliyev, S.A.; A computational approach to study fluid movement, 1st Festival on Water and Wastewater Research and Technology, Tehran, Iran, 12–17 Dec.2009, 27–32, http://isrc.nww.co.ir.
3. Peng, X.F.; Peterson, G.P.; Convective Heat Transfer and Flow Friction for Water Flow in Microchannel Structure, Int. J. Heat Mass Transfer 36, 1995, 2599–2608.
4. Bergant Anton, Discrete Vapour Cavity Model with Improved Timing of Opening and Collapse of Cavities, 1980, 1–11.
5. Ishii, M.; Thermo-Fluid Dynamic Theory of Two-Phase Flow, Collection de, D. R. Liles and, W. H. Reed, "A Sem-Implict Method for Two-Phase Fluid la Direction des Etudes et. Recherché d'Electricite de France, Vol. 22 Dynamics," Journal of Computational Physics 26, Paris, 1975, 390–407.
6. Hariri Asli, K.; Nagiyev, F.B.; Haghi, A.K.; Aliyev, S.A.; A computational approach to study fluid movement, 1st Festival on Water and Wastewater Research and Technology, Tehran, Iran, 12–17 Dec.2009, 27–32, http://isrc.nww.co.ir.

7. Pickford, J.; "Analysis of Surge", Macmillian, London, 1969, 153–156.
8. Pipeline Design for Water and Wastewater, American Society of Civil Engineers, New York, 1975, 54
9. Xu, B.; Ooi, K.T.; Mavriplis, C.; Zaghloul, M. E.; Viscous dissipation effects for liquid flow in microchannels, Micorsystems, 2002, 53–57.
10. Fedorov, A.G.; Viskanta, R.; Three-dimensional Conjugate Heat Transfer into Microchannel Heat Sink for Electronic Packaging, Int. J. Heat Mass Transfer 43, 2000, 399–415.
11. Tuckerman, D.B.; Heat transfer microstructures for integrated circuits, Ph.D. thesis, Stanford University, 1984, 10–120.
12. Harms, T.M.; Kazmierczak, M.J.; Cerner, F.M.; Holke, A.; Henderson, H.T.; Pilchowski, H.T.; Baker, K.; Experimental Investigation of Heat Transfer and Pressure Drop through Deep Micro channels in a (100) Silicon Substrate, in: Proceedings of the ASME. Heat Transfer Division, HTD 351, 1997, 347–357.
13. Holland, F.A.; Bragg, R.; Fluid Flow for Chemical Engineers, Edward Arnold Publishers, London, 1995, 1–3.
14. Lee TS, Pejovic S (1996) Air influence on similarity of hydraulic transients and vibrations. ASME J. Fluid Eng. 118(4), 706–709.
15. Li, J.; McCorquodale, A.; "Modeling Mixed Flow in Storm Sewers", Journal of Hydraulic Engineering, ASCE, Vol. 125, No. 11, 1999, 1170–1180.
16. Minnaert, M.; on musical air bubbles and the sounds of running water. Phil. Mag., 1933, v. 16, №7, 235–248.
17. Moeng, C.H.; McWilliams, J.C.; Rotunno, R.; Sullivan, P.; Weil, J.; "Investigating 2D modeling of atmospheric convection in the PBL," J. Atm. Sci. 61, 2004, 889–903.
18. Tuckerman, D.B.; R.F.W Pease, high performance heat sinking for VLSI, IEEE Electron device letter, DEL-2, 1981, 126–129.
19. Nagiyev, F.B.; Khabeev, N.S, Bubble dynamics of binary solutions. High Temperature, 1988, v. 27, № 3, 528–533.
20. Shvarts, D.; Oron, D.; Kartoon, D.; Rikanati, A.; Sadot, O.; "Scaling laws of nonlinear Rayleigh-Taylor and Richtmyer-Meshkov instabilities in two and three dimensions," C. R. Acad. Sci. Paris IV, 719, 2000, 312
21. Cabot, W.H.; Cook, A.W.; Miller, L.; Laney, D.E.; Miller, M.C.; Childs, H.R.; "Large eddy simulation of Rayleigh-Taylor instability," Phys. Fluids, September, 2005, vol. 17, 91–106.
22. Cabot, W.; University of California, Lawrence Livermore National laboratory, Livermore, CA, Physics of Fluids, 2006, 94–550.
23. Goncharov, V.N.; "Analytical model of nonlinear, single-mode, classical Rayleigh-Taylor instability at arbitrary Atwood numbers", Phys. Rev. Letters 88, 134502, 2002, 10–15.
24. Ramaprabhu , Andrews, M.J.; "Experimental investigation of Rayleigh-Taylor mixing at small Atwood numbers," J. Fluid Mech. 502, 2004, 233
25. Clark, T.T.; "A numerical study of the statistics of a two-dimensional Rayleigh-Taylor mixing layer", Phys. Fluids 15, 2003, 2413.
26. Cook, A.W.; Cabot, W.; Miller, L.; "The mixing transition in Rayleigh-Taylor instability", J. Fluid Mech. 511, 2004, 333.
27. Waddell, J.T.; Niederhaus, C.E.; Jacobs, J.W.; "Experimental study of Rayleigh-Taylor instability: Low Atwood number liquid systems with single-mode initial perturbations," Phys. Fluids 13, 2001, 1263–1273.
28. Weber, S.V.; Dimonte, G.; Marinak, M.M.; "Arbitrary Lagrange-Eulerian code simulations of turbulent Rayleigh-Taylor instability in two and three dimensions," Laser and Particle Beams 21, 2003, 455
29. Dimonte, G.; Youngs, D.; Dimits, A.; Weber, S.; Marinak, M. "A comparative study of the Rayleigh-Taylor instability using high-resolution three-dimensional numerical simulations: the Alpha group collaboration," Phys. Fluids 16, 2004, 1668.
30. Young, Y.N.; Tufo, H.; Dubey, A.; Rosner, R.; "On the miscible Rayleigh-Taylor instability: two and three dimensions," J. Fluid Mech. 447, 377, 2001, 2003–2500.
31. George, E.; Glimm, J.; "Self-similarity of Rayleigh-Taylor mixing rates," Phys. Fluids 17, 054101, 2005, 1–3.
32. Oron. D.; Arazi, L.; Kartoon, D.; Rikanati, A.; Alon, U.; Shvarts, D.; "Dimensionality dependence of the Rayleigh-Taylor and

Richtmyer-Meshkov instability late-time scaling laws", Phys. Plasmas 8, 2001, 2883
33. Nigmatulin, R.I.; Nagiyev, F.B.; Khabeev, N.S.; Effective heat transfer coefficients of the bubbles in the liquid radial pulse. Mater. Second-Union. Conf. Heat Mass Transfer, "Heat massoob-men in the biphasic. with. ". Minsk, 1980, v.5, 111–115.
34. Nagiyev, F.B.; Khabeev, N.S, Bubble dynamics of binary solutions. High Temperature, 1988, v. 27, № 3, 528–533
35. Nagiyev, F.B.; Damping of the oscillations of bubbles boiling binary solutions. Mater. VIII Resp. Conf. mathematics and mechanics. Baku, October 26–29, 1988. 177–178.
36. Nagyiev, F.B.; Kadyrov, B.A.; Small oscillations of the bubbles in a binary mixture in the acoustic field. Math. AN Az.SSR Ser. Physicotech. and Mate. Science, 1986, № 1, 23–26.
37. Nagiyev, F.B.; Dynamics, heat and mass transfer of vapor-gas bubbles in a two-component liquid. Turkey-Azerbaijan petrol semin., Ankara, Turkey, 1993, 32–40.
38. Nagiyev, F.B.; The method of creation effective coolness liquids, Third Baku international Congress. Baku, Azerbaijan Republic, 1995, 19–22.
39. Nagiyev, F.B.; The linear theory of disturbances in binary liquids bubble solution. Dep. In VINITI, 1986, № 405, in 86, 76–79.
40. Nagiyev, F.B.; Structure of stationary shock waves in boiling binary solutions. Math. USSR, Fluid Dynamics, 1989, № 1, 81–87.
41. Rayleigh, On the pressure developed in a liquid during the collapse of a spherical cavity. Philos. Mag. Ser. 6, v. 34, N 200, 1917, 94–98.
42. Perry. R.H.; Green, D.W.; Maloney, J.O.; Perry's Chemical Engineers Handbook, 7th Edition, McGraw-Hill, New York, 1997, 1–61.
43. Nigmatulin, R.I.; Dynamics of multiphase media., Moscow, "Nauka", 1987, v. 1, 2, 12–14.
44. Kodura, A.; Weinerowska, K.; the influence of the local pipeline leak on water hammer properties, Materials of the II Polish Congress of Environmental Engineering, Lublin, 2005, 125–133.
45. Kane, J.; Arnett, D.; Remington, B.A.; Glendinning, S.G.; Bazan, G.; "Two-dimensional versus three-dimensional supernova hydrodynamic instability growth", Astrophys. J.; 2000, 528–989.
46. Quick, R.S.; "Comparison and Limitations of Various Water hammer Theories", J. of Hyd. Div., ASME, May, 1933, 43–45.
47. Jaeger, C.; "Fluid Transients in Hydro-Electric Engineering Practice", Blackie and Son Ltd., 1977, 87–88.
48. Jaime Suarez, A.; "Generalized water hammer algorithm for piping systems with unsteady friction" 2005, 72–77.

REFERENCES FOR CHAPTER 5

1. Incropera, F.P.; Dewitt, D.P.; Fundamentals of Heat and Mass Transfer, second ed.; Wiley, New York, 1985.
2. Ghali, K.; Jones, B.; Tracy, E.; Modeling Heat and Mass Transfer in Fabrics, Int. J. Heat Mass Transfer 38(1), 13–21 (1995).
3. Flory, P.J.; Statistical Mechanics of Chain Molecules, Interscience Pub. NY, 1969.
4. Hadley, G. R.; Numerical Modeling of the Drying of Porous Materials, in: Proceedings of The Fourth International Drying Symposium, vol. 1, 151–158 (1984).
5. Hong, K.; Hollies, N.R.S.; Spivak, S.M.; Dynamic Moisture Vapour Transfer Through Textiles, Part I: Clothing Hygrometry and the Influence of Fiber Type, Textile Res. J. 58(12), 697–706 (1988).
6. Chen, C.S.; Johnson, W. H.; Kinetics of Moisture Movement in Hygroscopic Materials, In: Theoretical Considerations of Drying Phenomenon, Trans. ASAE.; 12, 109–113 (1969).
7. Barnes, J.; Holcombe, B.; Moisture Sorption and Transport in Clothing During Wear, Textile Res. J. 66 (12), 777–786 (1996).
8. Chen, P.; Pei, D.; A Mathematical Model of Drying Process, Int. J. Heat Mass Transfer 31 (12), 2517–2562 (1988).
9. Davis, A.; James, D.; Slow Flow Through a Model Fibrous Porous Medium, Int. J. Multiphase Flow 22, 969–989 (1996).
10. Jirsak, O.; Gok T.; Ozipek, B.; Pau, N.; Comparing Dynamic and Static Methods for Measuring Thermal Conductive Properties of Textiles, Textile Res. J. 68(1), 47–56 (1998).

11. Kaviany, M.; "Principle of Heat Transfer in Porous Media", Springer, New York, 1991.
12. Jackson, J.; James, D.; The Permeability of Fibrous Porous Media, Can. J. Chem. Eng. 64, 364–374 (1986).
13. Dietl, C.; George, O.P.; Bansal, N.K.; Modeling of Diffusion in Capillary Porous Materials During the Drying Process, Drying Technol. 13 (1&2), 267–293 (1995).
14. Ea, J.Y.; Water Vapour Transfer in Breathable Fabrics for Clothing, PhD thesis, University of Leeds, 1988.
15. Haghi, A.K.; Moisture permeation of clothing, JTAC 76, 1035–1055 (2004).
16. Haghi, A.K.; Thermal analysis of drying process, JTAC 74, 827–842 (2003).
17. Haghi, A.K.; Some Aspects of Microwave Drying, The Annals of Stefan cel Mare University, Year VII, No. 14, 22–25 (2000).
18. Haghi, A.K.; A Thermal Imaging Technique for Measuring Transient Temperature Field—An Experimental Approach, The Annals of Stefan cel Mare University, Year VI, No. 12, 73–76 (2000).
19. Haghi, A.K.; Experimental Investigations on Drying of Porous Media using Infrared Radiation, Acta Polytechnica,41(1), 55–57 (2001).
20. Haghi, A.K.;A Mathematical Model of the Drying Process, Acta Polytechnica 41(3) 20–23 (2001).
21. Haghi, A.K.;Simultaneous Moisture and Heat Transfer in Porous System, Journal of Computational and Applied Mechanics 2(2),195–204 (2001).
22. Haghi, A.K.; A Detailed Study on Moisture Sorption of Hygroscopic Fiber, Journal of Theoretical and Applied Mechanics 32(2) 47–62 (2002).
23. Dietl, C.; George, O.P.; Bansal, N.K.; Modeling of Diffusion in Capillary Porous Materials During the Drying Process, Drying Technol. 13 (1&2), 267–293 (1995).
24. Ea, J.Y.; Water Vapour Transfer in Breathable Fabrics for Clothing, PhD thesis, University of Leeds, 1988.
25. Flory, P.J.; Statistical Mechanics of Chain Molecules, Interscience Pub. NY, 1969.

REFERENCES FOR CHAPTER 6

1. Hariri Asli, K.; GIS and Water hammer disaster at earthquake in Rasht water pipeline, 3rd International Conference on Integrated Natural Disaster Management, INDM2008, http://www.civilica.com/Paper-INDM03-INDM03_001.html.
2. Hariri Asli, K.; Nagiyev, F.B.; Beglou, M.J.; Haghi, A.K.; Kinetic analysis of convective drying, International Journal of the Balkan Tribological Association, ISSN: 1310–4772, Sofia, Bulgaria, Vol. 15, No 4, 546–556, 2009, jbalkta@gmail.com
3. Hariri Asli K, Nagiyev, F.B.; Bubbles characteristics and convective effects in the binary mixtures. Transactions issue mathematics and mechanics series of physical-technical and mathematics science, ISSN 0002–3108, Azerbaijan, Baku, 215–220, 2008, www.imm.science.az/journals/AMEA_xeberleri/.../215–220.pdf
4. Hariri Asli, K.; Nagiyev, F.B.; Haghi, A.K.;Three-dimensional Conjugate Heat Transfer in Porous Media, International Journal of the Balkan Tribological Association, ISSN: 1310–4772, Sofia, Bulgaria, Vol. 15, No 3, 336–346, 2009, jbalkta@gmail.com
5. Hariri Asli, K.; Nagiyev, F.B.; Haghi, A.K.;Water hammer and fluid condition; a computational approach, Computational Methods in Applied Science and Engineering, USA, Chapter 5, Nova Science Publ ications,ISBN:978-1-60876-052-7, USA, 73–94, 2010, https://www.novapublishers.com/catalog/
6. Hariri Asli, K.; Nagiyev, F.B.; Haghi, A.K.;Interpenetration of two fluids at parallel between plates and turbulent moving in pipe; a case study, Computational Methods in Applied Science and Engineering, USA, Chapter 7, Nova Science Publications, ISBN:978-1-60876-052-7, USA, 107–133,2010, https://www.novapublishers.com/catalog/
7. Hariri Asli, K.; Nagiyev, F.B.; Haghi, A.K.; Modeling for water hammer due to valves; from theory to practice, Computational Methods in Applied Science and Engineering, USA, Chapter 11, Nova Science Publications ISBN:978-1-60876-052-7, USA,

229–236, 2010, https://www.novapublishers.com/catalog/

8. Hariri Asli, K.; Nagiyev, F.B.; Haghi, A.K.; A computational method to Study transient flow in binary mixtures, Computational Methods in Applied Science and Engineering, USA, Chapter 13, Nova Science Publications ISBN:978-1-60876-052-7, USA, 229–236, 2010, https://www.novapublishers.com/catalog/
9. Hariri Asli, K.; Nagiyev, F.B.; Haghi, A.K.;Water hammer analysis; some computational aspects and practical hints, Computational Methods in Applied Science and Engineering, USA, Chapter 16, Nova Science Publications ISBN:978-1-60876-052-7, USA, 263–281, 2010, https://www.novapublishers.com/catalog/
10. Hariri Asli, K.; Nagiyev, F.B.; Haghi, A.K.;Water hammer and hydrodynamics instabilities modeling, Computational Methods in Applied Science and Engineering, USA, Chapter 17, From Theory to Practice, Nova Science Publications ISBN:978-1-60876-052-7, USA, 283–301, 2010, https://www.novapublishers.com/catalog/
11. Hariri Asli, K.; Nagiyev, F.B.; Haghi, A.K.; A computational approach to study water hammer and pump pulsation phenomena, Computational Methods in Applied Science and Engineering, USA, Chapter 22, Nova Science Publications, ISBN:978-1-60876-052-7, USA, 349–363, 2010,https://www.novapublishers.com/catalog/
12. Hariri Asli, K.; Nagiyev, F.B.; Haghi, A.K.;Some aspects of physical and numerical modeling of water hammer in pipelines. Computational Methods in Applied Science and Engineering, USA, Chapter 23, Nova Science Publications, ISBN:978-1-60876-052-7, USA, 365–387, 2010, https://www.novapublishers.com/catalog/
13. Hariri Asli, K.; Nagiyev, F.B.; Haghi, A.K.; A computational approach to study fluid movement,Nanomaterials Yearbook – 2009, From Nanostructures, Nanomaterials and Nanotechnologies to Nanoindustry, Chapter 16, Nova Science Publications, USA, ISBN:978-1-60876-451-8, USA, 181–196,2010, https://www.novapublishers.com/catalog/product_info.php?products_id=11587
14. Hariri Asli, K.; Nagiyev, F.B.; Haghi, A.K.; Physical modeling of fluid movement in pipelines,Nanomaterials Yearbook – 2009, From Nanostructures, Nanomaterials and Nanotechnologies to Nanoindustry, Chapter 17,Nova Science Publications, USA, ISBN:978-1-60876-451-8, USA, 197–214, 2010, https://www.novapublishers.com/catalog/product_info.php?products_id=11587
15. Hariri Asli, K.; Nagiyev, F.B.;Haghi, A.K.; Aliyev, S.A.; Improved modeling for prediction of water transmission failure, Recent Progress in Research in Chemistry and Chemical Engineering, Chapter 2, Nova Science Publications, ISBN: 978-1-61668-501-0, Nova Science Publications, USA, 28–36, 2010, https://www.novapublishers.com/catalog/product_info.php?products_id=13174
16. Hariri Asli, K.; Nagiyev, F.B ;Haghi, A.K.; Aliyev, S.A. Pure Oxygen penetration in wastewater flow, Recent Progress in Research in Chemistry and Chemical Engineering, Chapter 3, Nova Science Publications, ISBN: 978-1-61668-501-0, Nova Science Publications,USA, 17–27, 2010, https://www.novapublishers.com/catalog/product_info.php?products_id=13174
17. Adams, T.M.; Abdel-Khalik, S.I.; Jeter, S.M.; Qureshi, Z.H. AN Experimental investigation of single-phase Forced Convection in Microchannels, Int. J. Heat Mass Transfer, 41, 1998, 851–857.
18. Allievi, L. "General Theory of Pressure Variation in Pipes", Ann. D. Ing., 1982, 166–171.
19. Apoloniusz Kodura, Katarzyna Weinerowska," Some Aspects of Physical and Numerical Modeling of Water Hammer in Pipelines", 2005, 126–132.
20. Anuchina, N.N.; Volkov, V.I.; Gordeychuk, V.A.; Eskov, N.S., Ilyutina, O.S.; Kozyrev, O.M. "Numerical simulations of Rayleigh-Taylor and Richtmyer-Meshkov instability using mah-3 code " J. Comput. Appl. Math. 168, 2004, 11.
21. Bergeron, L. "Water hammer in Hydraulics and Wave Surge in Electricity", John Wiley and Sons, Inc., N.Y.; 1961, 102–109.
22. Bergant Anton Discrete Vapour Cavity Model with Improved Timing of Opening and Collapse of Cavities, 1980, 1–11.

23. Bracco, A.; McWilliams, J.C.; Murante, G.; Provenzale, A.; Weiss, J.B.; "Revisiting freely decaying two-dimensional turbulence at millennial resolution," Phys. Fluids, Issue 11, 2000, v.12, 2931–2941.
24. Brunone, B.; Karney, B.W.; Mecarelli, M.; Ferrante, M. "Velocity Profiles and Unsteady Pipe Friction in Transient Flow" Journal of Water Resources Planning and Management, ASCE, 126(4), Jul., 2000, 236–244.
25. Cabot, W.H.; Cook, A.W.; Miller, L.; Laney, D.E.; Miller, M.C.; Childs, H.R.; "Large eddy simulation of Rayleigh-Taylor instability," Phys. Fluids, September, 2005, vol. 17, 91–106.
26. Cabot, W. University of California, Lawrence Livermore National laboratory, Livermore, CA, Physics of Fluids, 2006, 94–550.
27. Chaudhry, M.H. "Applied Hydraulic Transients", Van Nostrand Reinhold Co., N.Y.; 1979, 1322–1324.
28. Chaudhry, M.H.; Yevjevich, V. "Closed Conduit Flow", Water Resources Publication, USA, 1981, 255–278.
29. Chaudhry, M.H. Applied Hydraulic Transients, Van Nostrand Reinhold, New York, USA, 1987, p.165–167.
30. Clark, T.T.; "A numerical study of the statistics of a two-dimensional Rayleigh-Taylor mixing layer", Phys. Fluids 15, 2003, 2413 p.
31. Cook, A.W.; Cabot, W.; Miller, L.; "The mixing transition in Rayleigh-Taylor instability", J. Fluid Mech. 511,2004, 333 p.
32. Choi, S.B.; Barren, R.R.; Warrington, R.O.; Fluid Flow and Heat Transfer in Microtubes, ASME DSC 40, 1991, 89–93.
33. Dimonte, G.; Youngs, D.; Dimits, A.; Weber, S.; Marinak, M. "A comparative study of the Rayleigh-Taylor instability using high-resolution three-dimensional numerical simulations: the Alpha group collaboration," Phys. Fluids 16,2004, 1668
34. Dimotakis, E.; "The mixing transition in turbulence", J. Fluid Mech. 409,2000, 69
35. Elansari, A.S.; Silva, W.; Chaudhry, M.H. "Numerical and Experimental Investigation of Transient Pipe Flow", Journal of Hydraulic Research, 32,1994. 689
36. Filion, Y.; Karney, B.W.; "A Numerical Exploration of Transient Decay Mechanisms in Water Distribution Systems", Proceedings of the ASCE Environmental Water Resources Institute Conference, American Society of Civil Engineers, Roanoke, Virginia, 2002, 30
37. Fok, A.; "Design Charts for Air Chamber on Pump Pipelines", J. of Hyd. Div., ASCE, Sept., 1978, p.15–74.
38. Fok, A.; Ashamalla, A.; Aldworth, G.; "Considerations in Optimizing Air Chamber for Pumping Plants", Symposium on Fluid Transients and Acoustics in the Power Industry, San Francisco, U.S.A. Dec., 1978, p.112–114.
39. Fok, A.; "Design Charts for Surge Tanks on Pump Discharge Lines", BHRA 3rd Int. Conference on Pressure Surges, Bedford, England, Mar., 1980, p.23–34.
40. Fok, A.; "Water hammer and Its Protection in Pumping Systems", Hydro technical Conference, CSCE, Edmonton,May, 1982, p.45–55.
41. Fok, A.; "A contribution to the Analysis of Energy Losses in Transient Pipe Flow", PhD. Thesis, University of Ottawa, 1987, p.176–182.
42. Fox, J.A.; "Hydraulic Analysis of Unsteady Flow in Pipe Network", Wiley, N.Y.; 1977, p.78–89.
43. Fedorov, A.G.; Viskanta, R.; Three-dimensional Conjugate Heat Transfer into Microchannel Heat Sink for Electronic Packaging, Int. J. Heat Mass Transfer 43, 2000, 399–415.
44. Hamam, M.A.; Mc Corquodale, J.A.; "Transient Conditions in the Transition from Gravity to Surcharged Sewer Flow", Canadian, J. of Civil Eng., Canada, Sep., 1982, p.65–98.
45. Hariri Asli, K.; Nagiyev, F.B.; Water Hammer and fluid condition, Ministry of Energy, Gilan Water and Wastewater Co., Research Week Exhibition, Tehran, Iran, December, 2007, p. 132–148, http://isrc.nww.co.ir.
46. Hariri Asli, K.; Nagiyev, F.B.; Water Hammer analysis and formulation, Ministry of Energy, Gilan Water and Wastewater Co., Research Week Exhibition, Tehran, Iran, December, 2007, p. 111–131, http://isrc.nww.co.ir.
47. Hariri Asli, K.; Nagiyev, F.B.; Water Hammer and hydrodynamics instabilities, Inter-

penetration of two fluids at parallel between plates and turbulent moving in pipe, Ministry of Energy, Guilan Water and Wastewater Co., Research Week Exhibition, Tehran, Iran, December, 2007, p.90–110, http://isrc.nww.co.ir.

48. Hariri Asli, K.; Nagiyev, F.B.; Water Hammer and pump pulsation, Ministry of Energy, Guilan Water and Wastewater Co., Research Week Exhibition, Tehran, Iran, December, 2007, p. 51–72, http://isrc.nww.co.ir.
49. Hariri Asli, K.; Nagiyev, F.B.; Reynolds number and hydrodynamics' instability", Ministry of Energy, Guilan Water and Wastewater Co., Research Week Exhibition, Tehran, Iran, December, 2007, p.31–50, http://isrc.nww.co.ir.
50. Hariri Asli, K.; Nagiyev, F.B.; Water Hammer and valves, Ministry of Energy, Guilan Water and Wastewater Co., Research Week Exhibition, Tehran, Iran, December, 2007, p.20–30, http://isrc.nww.co.ir.
51. Hariri Asli, K.; Nagiyev, F.B.; "Interpenetration of two fluids at parallel between plates and turbulent moving in pipe", Ministry of Energy, Guilan Water and Wastewater Co., Research Week Exhibition, Tehran, Iran, December, 2007, p.73–89, http://isrc.nww.co.ir.
52. Hariri Asli, K.; Nagiyev, F.B.; Decreasing of Unaccounted For Water "UFW" by Geographic Information System"GIS" in Rasht urban water system, civil engineering organization of Guilan, Technical and Art Journal, 2007, p.3–7,http://www.art-of-music.net/.
53. Hariri Asli, K.; Portable Flow meter Tester Machine Apparatus, Certificate on registration of invention, Tehran, Iran. #010757, Series a/82, 24/11/2007, 1–3
54. Hariri Asli, K.; Nagiyev, F.B.; Haghi, A.K.; "Interpenetration of two fluids at parallel between plates and turbulent moving in pipe", 9th Conference on Ministry of Energetic works at research week, Tehran, Iran, 2008, p.73–89, http://isrc.nww.co.ir.
55. Hariri Asli, K.; Nagiyev, F.B.; Haghi, A.K.; "Water hammer and valves", 9th Conference on Ministry of Energetic works at research week, Tehran, Iran, 2008, p.20–30, http://isrc.nww.co.ir.
56. Hariri Asli, K.; Nagiyev, F.B. Haghi, A.K.; "Water hammer and hydrodynamics instability", 9th Conference on Ministry of Energetic works at research week, Tehran, Iran, 2008, p.90–110, http://isrc.nww.co.ir.
57. Hariri Asli, K.; Nagiyev, F.B., Haghi, A.K.; "Water hammer analysis and formulation", 9th Conference on Ministry of Energetic works at research week, Tehran, Iran, 2008, p. 27–42, http://isrc.nww.co.ir.
58. Hariri Asli, K.; Nagiyev, F.B.; Haghi, A.K.; "Water hammer & fluid condition", 9th Conference on Ministry of Energetic works at research week, Tehran, Iran, 2008, p.27–43, http://isrc.nww.co.ir.
59. Hariri Asli, K.; Nagiyev, F.B.; Haghi, A.K.; "Water hammer and pump pulsation", 9th Conference on Ministry of Energetic works at research week, Tehran, Iran, 2008, p.27–44, http://isrc.nww.co.ir.
60. Hariri Asli, K.; Nagiyev, F.B.; Haghi, A.K.; "Reynolds number and hydrodynamics instability", 9th Conference on Ministry of Energetic works at research week, Tehran, Iran, 2008, p. 27–45, http://isrc.nww.co.ir.
61. Hariri Asli, K.; Nagiyev, F.B.; Haghi, A.K.; "Water hammer and fluid Interpenetration", 9th Conference on Ministry of Energetic works at research week, Tehran, Iran, 2008, p.27–47, http://isrc.nww.co.ir.
62. Hariri Asli, K.; GIS and water hammer disaster at earthquake in Rasht water pipeline, civil engineering organization of Guilan, Technical and Art Journal, 2008, 14–17, http://www.art-of-music.net/.
63. Hariri Asli, K.; GIS and water hammer disaster at earthquake in Rasht water pipeline, 3rd International Conference on Integrated Natural Disaster Management, Tehran university, ISSN: 1735–5540, 18–19 Feb., INDM, Tehran, Iran, 2008, №13,53/1–12, http://www.civil ca.com/Paper-INDM03-INDM03_001.html
64. Hariri Asli, K.; Nagiyev, F.B.; Bubbles characteristics and convective effects in the binary mixtures. Transactions issue mathematics and mechanics series of physical-technical and mathematics science, ISSN: 0002–3108, Azerbaijan, Baku, 2009, 68–74, http://www.imm.science.az/journals.html
65. Hariri Asli, K.; Nagiyev, F.B.; Haghi, A.K.; Aliyev, S.A.; Three-Dimensional conjugate

heat transfer in porous media, 1st Festival on Water and Wastewater Research and Technology, Tehran, Iran, 12–17 Dec.2009, 26,28, http://isrc.nww.co.ir.
66. Hariri Asli, K.; Nagiyev, F.B.; Haghi, A.K.; Aliyev, S.A.; Some Aspects of Physical and Numerical Modeling of water hammer in pipelines,1st Festival on Water and Wastewater Research and Technology, Tehran, Iran, 12–17 Dec.2009, 26,29, http://isrc.nww.co.ir
67. Hariri Asli, K.; Nagiyev, F.B.; Haghi, A.K.; Aliyev, S.A.; Modeling for Water Hammer due to valves: From theory to practice, 1st Festival on Water and Wastewater Research and Technology, Tehran, Iran, 12–17 Dec.2009, 26,30, http://isrc.nww.co.ir.
68. Hariri Asli, K.; Nagiyev, F.B.; Haghi, A.K.; Aliyev, S.A.; Water hammer and hydrodynamics instabilities modeling: From Theory to Practice, 1st Festival on Water and Wastewater Research and Technology, Tehran, Iran, 12–17 Dec.2009, 26,31, http://isrc.nww.co.ir
69. Hariri Asli, K.; Nagiyev, F.B.; Haghi, A.K.; Aliyev, S.A.; A computational approach to study fluid movement, 1st Festival on Water and Wastewater Research and Technology, Tehran, Iran, 12–17 Dec.2009, 27–32, http://isrc.nww.co.ir.
70. Hariri Asli, K.; Nagiyev, F.B.; Haghi, A.K.; Aliyev, S.A.; Water Hammer Analysis: Some Computational Aspects and practical hints, 1st Festival on Water and Wastewater Research and Technology, Tehran, Iran, 12–17 Dec.2009, 27–33, http://isrc.nww.co.ir
71. Hariri Asli, K.; Nagiyev, F.B.; Haghi, A.K.; Aliyev, S.A.; Water Hammer and Fluid condition: A computational approach, 1st Festival on Water and Wastewater Research and Technology, Tehran, Iran, 12–17 Dec.,2009, 27–34, http://isrc.nww.co.ir.
72. Hariri Asli, K.; Nagiyev, F.B.; Haghi, A.K.; Aliyev, S.A.; A computational Method to Study Transient Flow in Binary Mixtures, 1st Festival on Water and Wastewater Research and Technology, Tehran, Iran, 12–17 Dec.2009, 27–35, http://isrc.nww.co.ir.
73. Hariri Asli, K.; Nagiyev, F.B.; Haghi, A.K.; Physical modeling of fluid movement in pipelines, 1st Festival on Water and Wastewater Research and Technology, Tehran, Iran, 12–17 Dec.2009, 27–36, http://isrc.nww.co.ir.
74. Hariri Asli, K.; Nagiyev, F.B.; Haghi, A.K.; Aliyev, S.A.; Interpenetration of two fluids at parallel between plates and turbulent moving, 1st Festival on Water and Wastewater Research and Technology, Tehran, Iran, 12–17 Dec. 2009, 27–37, http://isrc.nww.co.ir.
75. Hariri Asli, K.; Nagiyev, F.B.; Haghi, A.K.; Aliyev, S.A.; Modeling of fluid interaction produced by water hammer, 1st Festival on Water and Wastewater Research and Technology, Tehran, Iran, 12–17 Dec.2009, 27–38, http://isrc.nww.co.ir.
76. Hariri Asli, K.; Nagiyev, F.B.; Haghi, A.K.; Aliyev, S.A.; GIS and water hammer disaster at earthquake in Rasht pipeline, 1st Festival on Water and Wastewater Research and Technology, Tehran, Iran, 12–17 Dec.2009, 27–39, http://isrc.nww.co.ir.
77. Hariri Asli, K.; Nagiyev, F.B.; Haghi, A.K.; Aliyev, S.A.; Interpenetration of two fluids at parallel between plates and turbulent moving,1st Festival on Water and Wastewater Research and Technology, Tehran, Iran, 12–17 Dec.2009, 27–40, http://isrc.nww.co.ir.
78. Hariri Asli, K.; Nagiyev, F.B.; Haghi, A.K.; Aliyev, S.A.; Water hammer and hydrodynamics' instability, 1st Festival on Water and Wastewater Research and Technology, Tehran, Iran, 12–17 Dec.2009, 27–41, http://isrc.nww.co.ir.
79. Hariri Asli, K.; Nagiyev, F.B.; Haghi, A.K.; Aliyev, S.A.; Water hammer analysis and formulation, 1st Festival on Water and Wastewater Research and Technology, Tehran, Iran, 12–17 Dec.2009, 27–42, http://isrc.nww.co.ir.
80. Hariri Asli, K.; Nagiyev, F.B.; Haghi, A.K.; Aliyev, S.A.; Water hammer and fluid condition, 1st Festival on Water and Wastewater Research and Technology, Tehran, Iran, 12–17 Dec.2009, 27–43, http://isrc.nww.co.ir.
81. Hariri Asli, K.; Nagiyev, F.B.; Haghi, A.K.; Aliyev, S.A.; Water hammer and pump pulsation,1st Festival on Water and Wastewater Research and Technology, Tehran, Iran, 12–17 Dec.2009, 27–44, http://isrc.nww.co.ir.

82. Hariri Asli, K.; Nagiyev, F.B.; Haghi, A.K.; Aliyev, S.A.; Reynolds number and hydrodynamics instabilities, 1st Festival on Water and Wastewater Research and Technology, Tehran, Iran, 12–17 Dec.2009, 27–45, http://isrc.nww.co.ir.
83. Hariri Asli, K.; Nagiyev, F.B.; Haghi, A.K.; Aliyev, S.A.; water hammer and valves, 1st Festival on Water and Wastewater Research and Technology, Tehran, Iran, 12–17 Dec.2009, 27–46, http://isrc.nww.co.ir.
84. Hariri Asli, K.; Nagiyev, F.B.; Haghi, A.K.; Aliyev, S.A.; "Water hammer and fluid Interpenetration", 1st Festival on Water and Wastewater Research and Technology, Tehran, Iran, 12–17 Dec.2009, 27–47, http://isrc.nww.co.ir.
85. Hariri Asli, K.; Nagiyev, F.B.; Modeling of fluid interaction produced by water hammer, International Journal of Chemoinformatics and Chemical Engineering, IGI, ISSN: 2155–4110, EISSN: 2155–4129, USA, 2010, 29–41, http://www.igi-global.com/journals/details.asp?ID=34654
86. Hariri Asli, K.; Nagiyev, F.B.; Haghi, A.K.; Water hammer and fluid condition; a computational approach, Computational Methods in Applied Science and Engineering, USA, Chapter 5, Nova Science Publications, ISBN:978-1-60876-052-7, USA, 2010, 73–94, https://www.novapublishers.com/catalog/
87. Hariri Asli, K.; Nagiyev, F.B.; Haghi, A.K.; Some aspects of physical and numerical modeling of water hammer in pipelines. Computational Methods in Applied Science and Engineering, USA, Chapter 23, Nova Science Publications, ISBN:978-1-60876-052-7, USA, 2010, 365–387, https://www.novapublishers.com/catalog/
88. Hariri Asli, K.; Nagiyev, F.B.; Haghi, A.K.; Modeling for water hammer due to valves; from theory to practice, Computational Methods in Applied Science and Engineering, USA, Chapter 11, Nova Science Publications ISBN:978-1-60876-052-7, USA, 2010, 229–236, https://www.novapublishers.com/catalog/
89. Hariri Asli, K.; Nagiyev, F.B.; Haghi, A.K.; A computational method to Study transient flow in binary mixtures, Computational Methods in Applied Science and Engineering, USA, Chapter 13, Nova Science Publications ISBN:978-1-60876-052-7, USA, 2010, 229–236, https://www.novapublishers.com/catalog/
90. Hariri Asli, K.; Nagiyev, F.B.; Haghi, A.K.; Water hammer analysis; some computational aspects and practical hints, Computational Methods in Applied Science and Engineering, USA, Chapter 16, Nova Science Publications ISBN:978-1-60876-052-7, USA, 2010, 263–281, https://www.novapublishers.com/catalog/
91. Hariri Asli, K.; Nagiyev, F.B., Haghi, A.K.; Water hammer and hydrodynamics instabilities modeling, Computational Methods in Applied Science and Engineering, USA, Chapter 17, From Theory to Practice, Nova Science Publications ISBN 978-1-60876-052-7, USA, 2010, 283–301, https://www.novapublishers.com/catalog/
92. Hariri Asli, K.; Nagiyev, F.B.; Haghi, A.K.; A computational approach to study water hammer and pump pulsation phenomena, Computational Methods in Applied Science and Engineering, USA, Chapter 22, Nova Science Publications, ISBN 978-1-60876-052-7, USA, 2010, 349–363, https://www.novapublishers.com/catalog/
93. Hariri Asli, K.; Nagiyev, F.B.; Haghi, A.K.; A computational approach to study fluid movement, Nanomaterials Yearbook – 2009, From Nanostructures, Nanomaterials and Nanotechnologies to Nanoindustry, Chapter 16, Nova Science Publications, USA, ISBN: 978-1-60876-451-8, USA, 2010, 181–196, https://www.novapublishers.com/catalog/product_info.php?products_id=11587
94. Hariri Asli, K.; Nagiyev, F.B.; Haghi, A.K.; Physical modeling of fluid movement in pipelines, Nanomaterials Yearbook – 2009, From Nanostructures, Nanomaterials and Nanotechnologies to Nanoindustry, Chapter 17, Nova Science Publications, USA, ISBN: 978-1-60876-451-8, USA, 2010, 197–214, https://www.novapublishers.com/catalog/product_info.php?products_id=11587
95. Hariri Asli, K.; Nagiyev, F.B.; Haghi, A.K.; "Some Aspects of Physical and Numerical Modeling of water hammer in pipelines", Nonlinear Dynamics An International Journal of Nonlinear Dynamics and Chaos in Engineering Systems, ISSN: 1573–269X

(electronic version) Journal no. 11071 Springer, Netherlands, 2009, ISSN: 0924–090X (print version), Springer, Heidelberg, Germany, Number 4 / June, 2010, Volume 60, 677–701, http://www.springerlink.com/openurl.aspgenre=article&id=doi:10.1007/s11071–009–9624–7.
96. Hariri Asli, K.; Nagiyev, F.B.; Haghi, A.K.; Interpenetration of two fluids at parallel between plates and turbulent moving in pipe; a case study, Computational Methods in Applied Science and Engineering, USA, Chapter 7, Nova Science Publications, ISBN:978-1-60876-052-7, USA, 2010, 107–133, https://www.novapublishers.com/catalog/
97. Hariri Asli, K.; Nagiyev, F.B.; Beglou, M.J.; Haghi, A.K.; Kinetic analysis of convective drying, International Journal of the Balkan Tribological Association, ISSN: 1310–4772, Sofia, Bulgaria, 2009, Vol. 15, No 4, 546–556, jbalkta@gmail.com
98. Hariri Asli, K.; Nagiyev, F.B.; Haghi, A.K.; Three-dimensional Conjugate Heat Transfer in Porous Media, International Journal of the Balkan Tribological Association, ISSN: 1310–4772, Sofia, Bulgaria, 2009, Vol. 15, No 3, 336–346, jbalkta@gmail.com
99. Hariri Asli, K.; Nagiyev, F.B.; Haghi, A.K.; Aliyev, S.A.; Pure Oxygen penetration in wastewater flow, Recent Progress in Research in Chemistry and Chemical Engineering, Nova Science Publications, ISBN: 978-1-61668–501–0, Nova Science Publications, USA, 2010, 17–27, https://www.novapublishers.com/catalog/product_info.php?products_id=13174110100. Hariri Asli, K.; Nagiyev, F.B.; Haghi, A.K.; Aliyev, S.A.; Improved modeling for prediction of water transmission failure, Recent Progress in Research in Chemistry and Chemical Engineering, Nova Science Publications, ISBN: 978-1-61668–501–0, Nova Science Publications, USA, 2010, 28–36, https://www.novapublishers.com/catalog/product_info.php?products_id=13174
101. Harms, T.M.; Kazmierczak, M.J.; Cerner, F.M.; Holke, A.; Henderson, H.T.; Pilchowski, H.T.; Baker, K.; Experimental Investigation of Heat Transfer and Pressure Drop through Deep Micro channels in a (100) Silicon Substrate, in: Proceedings of the ASME. Heat Transfer Division, HTD 351, 1997, 347–357.
102. Holland, F.A.; Bragg, R.; Fluid Flow for Chemical Engineers, Edward Arnold Publishers, London, 1995, 1–3.
103. George, E.; Glimm, J.; "Self-similarity of Rayleigh-Taylor mixing rates," Phys. Fluids 17, 054101, 2005, 1–3.
104. Goncharov, V.N.; "Analytical model of nonlinear, single-mode, classical Rayleigh-Taylor instability at arbitrary Atwood numbers", Phys. Rev. Letters 88, 134502, 2002, 10–15.
105. Ishii, M.; Thermo-Fluid Dynamic Theory of Two-Phase Flow, Collection de, D.R. Liles and, W.H. Reed, "A Sern-Implict Method for Two-Phase Fluid la Direction des Etudes et. Recherché d'Electricite de France, Vol. 22 Dynamics," Journal of Computational Physics 26, Paris, 1975, 390–407.
106. Jaime Suárez, A.;"Generalized water hammer algorithm for piping systems with unsteady friction" 2005, 72–77.
107. Jaeger, C.; "Fluid Transients in Hydro-Electric Engineering Practice", Blackie and Son Ltd., 1977, 87–88.
108. Joukowski, N.; Paper to Polytechnic Soc. Moscow, spring of 1898, English translation by Miss, O. Simin. Proc. AWWA, 1904, 57–58.
109. Kodura, A.; Weinerowska, K.; the influence of the local pipeline leak on water hammer properties, Materials of the II Polish Congress of Environmental Engineering, Lublin, 2005, 125–133.
110. Kane, J.; Arnett, D.; Remington, B.A.; Glendinning, S.G.; Baz'an, G.; "Two-dimensional versus three-dimensional supernova hydrodynamic instability growth", Astrophys. J.; 2000, 528–989.
111. Karassik, I.J.; "Pump Handbook – Third Edition", McGraw-Hill, 2001, 19–22.
112. Koelle, E.; Luvizotto Jr. E.; Andrade, J.P.G.; "Personality Investigation of Hydraulic Networks using MOC – Method of Characteristics" Proceedings of the 7th International Conference on Pressure Surges and Fluid Transients, Harrogate Durham, United Kingdom, 1996, 1–8.

113. Kerr, S.L.; "Minimizing service interruptions due to transmission line failures: Discussion, " Journal of the American Water Works Association, 41, 634, July 1949, 266–268.
114. Kerr, S.L.; "Water hammer control, " Journal of the American Water Works Association, 43, December 1951, 985–999.
115. Kraichnan, R.H.; Montgomery, D.; "Two-dimensional turbulence," Rep. Prog. Phys. 43, 547, 1967, 1417–1423.
116. Leon Arturo, S.; "An efficient second-order accurate shock-capturing scheme for modeling one and two-phase water hammer flows" PhD Thesis, March 29, 2007, 4–44.
117. Lee TS, Pejovic S (1996) Air influence on similarity of hydraulic transients and vibrations. ASME J. Fluid Eng. 118(4), 706–709.
118. Li, J.; McCorquodale, A.; "Modeling Mixed Flow in Storm Sewers", Journal of Hydraulic Engineering, ASCE, Vol. 125, No. 11, 1999, 1170–1180.
119. Mala, G.; Li, D.; Dale, J.D.; Heat Transfer and Fluid Flow in Microchannels, J. Heat Transfer, 40, 1997, 3079–3088.
120. Miller, L.; Cabot, W.H.; Cook, A.W.; "Which way is up? A fluid dynamics riddle," Phys. Fluids 17, 091110, 2005, 1–26.
121. Minnaert, M.; on musical air bubbles and the sounds of running water. Phil. Mag., 1933, v. 16, №7, 235–248.
122. Moeng, C.H.; McWilliams, J.C. Rotunno, R.; Sullivan, P.; Weil, J.; "Investigating 2D modeling of atmospheric convection in the PBL," J. Atm. Sci. 61, 2004, 889–903.
123. Moody, L.F.; "Friction Factors for Pipe Flow", Trans. ASME, 1944. Vol. 66, 671–684.
124. Nagiyev, F.B.; Dynamics, heat and mass transfer of vapor-gas bubbles in a two-component liquid. Turkey-Azerbaijan petrol semin., Ankara, Turkey, 1993, 32–40.
125. Nagiyev, F.B.; The method of creation effective coolness liquids, Third Baku international Congress. Baku, Azerbaijan Republic, 1995, 19–22.
126. Neshan, H.; Water Hammer, pumps Iran Co. Tehran, Iran, 1985, 1–60.
127. Nigmatulin, R.I.; Khabeev, N.S ; Nagiyev, F.B.; Dynamics, heat and mass transfer of vapor-gas bubbles in a liquid. Int. J. Heat Mass Transfer,.vol.24, N6, Printed in Great Britain, 1981, 1033–1044.
128. Oron. D.; Arazi, L.; Kartoon, D.; Rikanati, A.; Alon, U.; Shvarts, D.; "Dimensionality dependence of the Rayleigh-Taylor and Richtmyer-Meshkov instability late-time scaling laws", Phys. Plasmas 8, 2001, 2883
129. Parmakian, J.; "Water hammer Design Criteria", J. of Power Div., ASCE, Sept., 1957, 456–460.
130. Parmakian, J.; "Water hammer Analysis", Dover Publications, Inc., New York, New York, 1963, 51–58.
131. Perry. R.H.; Green, D.W.; Maloney, J.O.; Perry's Chemical Engineers Handbook, 7th Edition, McGraw-Hill, New York, 1997, 1–61.
132. Peng, X.F.; Peterson, G.P.; Convective Heat Transfer and Flow Friction for Water Flow in Microchannel Structure, Int. J. Heat Mass Transfer 36, 1996, 2599–2608.
133. Pickford, J.; "Analysis of Surge", Macmillian, London, 1969, 153–156.
134. Pipeline Design for Water and Wastewater, American Society of Civil Engineers, New York, 1975, 54
135. Qu, W.; Mala, G.M.; Li, D.; Heat Transfer for Water Flow in Trapezoidal Silicon Microchannels, 1993, 399–404.
136. Quick, R.S.; "Comparison and Limitations of Various Water hammer Theories", J. of Hyd. Div., ASME, May, 1933, 43–45.
137. Rayleigh, On the pressure developed in a liquid during the collapse of a spherical cavity. Philos. Mag. Ser. 6, v. 34, N 200, 1917, 94–98.
138. Ramaprabhu , Andrews, M.J.; "Experimental investigation of Rayleigh-Taylor mixing at small Atwood numbers," J. Fluid Mech. 502, 2004, 233
139. Rich, G.R.; "Hydraulic Transients", Dover, USA, 1963, 148–154.
140. Savic, D.A.; Walters, G.A.; ' Genetic Algorithms Techniques for Calibrating Network Models", Report No. 95/12, Centre for Systems and Control Engineering, 1995, 137–146.
141. Savic, D.A.; Walters, G.A.; Genetic Algorithms Techniques for Calibrating Network Models, University of Exeter, Exeter, United Kingdom, 1995, 41–77.

142. Shvarts, D.; Oron, D.; Kartoon, D.; Rikanati, A.; Sadot, O.; "Scaling laws of nonlinear Rayleigh-Taylor and Richtmyer-Meshkov instabilities in two and three dimensions," C. R. Acad. Sci. Paris IV, 719, 2000, 312
143. Sharp, B.; "Water hammer Problems and Solutions", Edward Arnold Ltd., London, 1981, 43–55.
144. Skousen ,"Valve Handbook", McGraw Hill, New York, HAMMER Theory and Practice, 1998, 687–721.
145. Song, C.C. et al.,"Transient Mixed-Flow Models for Storm Sewers", J. of Hyd. Div., Nov., 1983, Vol. 109, 458–530.
146. Stephenson, D.; "Pipe Flow Analysis", Elsevier, Vol. 19, S.A.; 1984, 670–788.
147. Streeter, V.L.; Lai, C.; "Water hammer Analysis Including Fluid Friction." Journal of Hydraulics Division, ASCE, 88, 1962, 79
148. Streeter, V.L.; Wylie, E.B.; "Fluid Mechanics", McGraw-Hill Ltd., USA,1979, 492–505.
149. Streeter, V.L.; Wylie, E.B.; "Fluid Mechanics", McGraw-Hill Ltd., USA, 1981, 398–420.
150. Tijsseling, "Alan E Vardy Time scales and FSI in unsteady liquid-filled pipe flow", 1993, 5–12.
151. Tuckerman, D.B.; R.F.W Pease, high performance heat sinking for VLSI, IEEE Electron device letter, DEL-2, 1981, 126–129.
152. Tennekes, H.; Lumley, J.L.; A First Course in Turbulence, the MIT Press, 1972, 410–466.
153. Thorley, A.R.D.; "Fluid Transients in Pipeline Systems", D. and, L. George, Herts, England, 1991, 231–242.
154. Tullis, J.P.; "Control of Flow in Closed Conduits", Fort Collins, Colorado, 1971, 315–340.
155. Tuckerman, D.B.; Heat transfer microstructures for integrated circuits, Ph.D. thesis, Stanford University, 1984, 10–120.
156. Vallentine, H.R.; "Rigid Water Column Theory for Uniform Gate Closure", J. of Hyd. Div. ASCE, July, 1965, 55–243.
157. Waddell, J.T.; Niederhaus, C.E.; Jacobs, J.W.; "Experimental study of Rayleigh-Taylor instability: Low Atwood number liquid systems with single-mode initial perturbations," Phys. Fluids 13, 2001, 1263–1273.
158. Watters, G.Z.; "Modern Analysis and Control of Unsteady Flow in Pipelines", Ann Arbor Sci., 2nd Ed., 1984, 1098–1104.
159. Walski, T.M.; Lutes, T.L.; "Hydraulic Transients Cause Low-Pressure Problems", Journal of the American Water Works Association, 75(2), 1994, 58 160. Weber, S.V.; Dimonte, G.; Marinak, M.M.; "Arbitrary Lagrange-Eulerian code simulations of turbulent Rayleigh-Taylor instability in two and three dimensions," Laser and Particle Beams 21, 2003, 455 .
161. Wood, D.J.; Dorsch, R.G.; Lightener, C.; "Wave-Plan Analysis of Unsteady Flow in Closed Conduits", Journal of Hydraulics Division, ASCE, 92, 1966, 83–110.
162. Wood, D.J.; S.E. Jones., "Water hammer charts for various types of valves", Journal of the Hydraulics Division, Proceedings of the American Society of Civil Engineers, January, 1973, 167–178.
163. Wood, F.M.; "History of Water hammer", Civil Engineering Research Report, #65, Queens University, Canada, 1970, 66–70.
164. Wu, Z.Y.; Simpson, A.R.; Competent genetic-evolutionary optimization of water distribution systems. J. Comput. Civ. Eng. 15(2), 2001, 89–101.
165. Wylie, E.B.; Talpin, L.B.; Matched impedance to control fluid transients. Trans. ASME 105, 2, 1983, 219–224.
166. Wylie, E.B.; Streeter, V.L.; Fluid Transients in Systems, Prentice-Hall, Englewood Cliffs, New Jersey, 1993, 4
167. Wylie, E.B.; Streeter, V.L.; Fluid Transients, Feb Press, Ann Arbor, MI, 1982, corrected copy, 1983, 158
168. Wu, Y.; Little, W.A.; Measurement of friction factor for flow of gases in very fine channels used for micro miniature, Joule Thompson refrigerators, Cryogenics 24 (8), 1983, 273–277.
169. Xu, B.; Ooi, K.T.; Mavriplis, C.; Zaghloul, M.E.; Viscous dissipation effects for liquid flow in microchannels, Micorsystems, 2002, 53–57.
170. Young, Y.N.; Tufo, H.; Dubey, A.; Rosner, R.; "On the miscible Rayleigh-Taylor instability: two and three dimensions," J. Fluid Mech. 447, 377, 2001, 2003–2500.

171. Wood, D.J.; Dorsch, R.G.; Lightener, C.; "Wave-Plan Analysis of Unsteady Flow in Closed Conduits", Journal of Hydraulics Division, ASCE, 92, 1966, 83–110.
172. Wylie, E.B.; Streeter, V.L.; Fluid Transients in Systems, Prentice-Hall, Englewood Cliffs, New Jersey, 1993, 4
173. Brunone, B.; Karney, B.W.; Mecarelli, M.; Ferrante, M. "Velocity Profiles and Unsteady Pipe Friction in Transient Flow" Journal of Water Resources Planning and Management, ASCE, 126(4), Jul., 2000, 236–244.
174. Koelle, E.; Luvizotto Jr. E.; Andrade, J.P.G.; "Personality Investigation of Hydraulic Networks using MOC – Method of Characteristics" Proceedings of the 7th International Conference on Pressure Surges and Fluid Transients, Harrogate Durham, United Kingdom, 1996, 1–8.
175. Filion, Y.; Karney, B.W.; "A Numerical Exploration of Transient Decay Mechanisms in Water Distribution Systems", Proceedings of the ASCE Environmental Water Resources Institute Conference, American Society of Civil Engineers, Roanoke, Virginia, 2002, 30.
176. Hamam, M.A.; Mc Corquodale, J.A.; "Transient Conditions in the Transition from Gravity to Surcharged Sewer Flow", Canadian, J. Civil Eng., Canada, Sep., 1982, p.65–98.
177. Savic, D.A.; Walters, G.A.; "Genetic Algorithms Techniques for Calibrating Network Models", Report No. 95/12, Centre for Systems and Control Engineering, 1995, 137–146.
178. Walski, T.M.; Lutes, T.L.; "Hydraulic Transients Cause Low-Pressure Problems", Journal of the American Water Works Association, 75(2), 1994, 58.
179. Lee, T.S.; Pejovic, S. (1996) Air influence on similarity of hydraulic transients and vibrations. ASME J. Fluid Eng. 118(4), 706–709.

REFERENCES FOR CHAPTER 7

1. Hariri Asli K., Sahleh H., Aliyev S.A., Mathematical Concepts and Computational Approaches on Hydrodynamics Instability, Mathematical Concepts for Mechanical Engineering Design, Toronto, Canada, Published by Apple Academic Press, Inc., Exclusive worldwide distribution by CRC Press, A Taylor & Francis Group, Print ISBN: 9781926895628, 2013, www.AppleAcademicPress.com.
2 Hariri Asli K., Nagiyev F.B , Beglou M. J., Haghi A.K., Kinetic analysis of convective drying, International Journal of the Balkan Tribological Association, ISSN: 1310-4772, Sofia, Bulgaria, Vol. 15, no. 4, 546–556, 2009, jbalkta@gmail.com.
3. Hariri Asli K., Nagiyev F.B., Bubbles characteristics and convective effects in the binary mixtures. Transactions issue mathematics and mechanics series of physical-technical and mathematics science, ISSN 0002–3108, Azerbaijan, Baku, 215–220, 2008, www.imm.science.az/journals/AMEA_xeberleri/./215–220.pdf.
4. Hariri Asli K., Nagiyev F.B., Haghi A.K., Three-dimensional Conjugate Heat Transfer in Porous Media, International Journal of the Balkan Tribological Association, ISSN: 1310-4772, Sofia, Bulgaria, Vol. 15, No 3, 336–346, 2009, jbalkta@gmail.com.
5. Hariri Asli K., Nagiyev F.B., Haghi A.K., Water Hammer and Fluid Condition; A computational approach, Computational Methods in Applied Science and Engineering, USA, Chapter 5, Nova Science Publications, ISBN: 978-1-60876-052-7, USA, 73–94, 2010, https://www.novapublishers.com/catalog/.
6. Hariri Asli K., Nagiyev F.B., Haghi A.K., Interpenetration of Two Fluids at Parallel Between Plates and Turbulent Moving in pipe; A case study, Computational Methods in Applied Science and Engineering, USA, Chapter 7, Nova Science Publications, ISBN: 978-1-60876-052-7, USA, 107–133, 2010, https://www.novapublishers.com/catalog/.
7. Hariri Asli K., Nagiyev F.B., Haghi A.K., Modeling for water hammer due to valves; from theory to practice, Computational Methods in Applied Science and Engineering, USA, Chapter 11, Nova Science Publications, ISBN: 978-1-60876-052-7, USA, 229–236, 2010, https://www.novapublishers.com/catalog/.

8. Hariri Asli K., Nagiyev F.B., Haghi A.K., A computational method to Study transient flow in binary mixtures, Computational Methods in Applied Science and Engineering, USA, Chapter 13, Nova Science Publications, ISBN: 978-1-60876-052-7, USA, 229–236, 2010, https://www.novapublishers.com/catalog/.
9. Hariri Asli K., Nagiyev F.B., Haghi A.K., Water Hammer analysis; some computational aspects and practical hints, Computational Methods in Applied Science and Engineering, USA, Chapter 16, Nova Science Publications, ISBN: 978-1-60876-052-7, USA, 263–281, 2010, https://www.novapublishers.com/catalog/.
10. Hariri Asli K., Nagiyev F.B., Haghi A.K., Water Hammer and Hydrodynamics Instabilities Modeling, Computational Methods in Applied Science and Engineering, USA, Chapter 17, From Theory to Practice, Nova Science Publications, ISBN: 978-1-60876-052-7, USA, 283–301, 2010, https://www.novapublishers.com/catalog/.
11. Hariri Asli K., Nagiyev F.B., Haghi A.K., A computational approach to study water hammer and pump pulsation phenomena, Computational Methods in Applied Science and Engineering, USA, Chapter 22, Nova Science Publications, ISBN: 978-1-60876-052-7, USA, 349–363, 2010, https://www.novapublishers.com/catalog/.
12. Hariri Asli K., Nagiyev F.B., Haghi A.K., Some aspects of physical and numerical modeling of water hammer in pipelines. Computational Methods in Applied Science and Engineering, USA, Chapter 23, Nova Science Publications, ISBN: 978-1-60876-052-7, USA, 365–387, 2010, https://www.novapublishers.com/catalog/.
13. Hariri Asli K., Nagiyev F.B., Haghi A.K., A computational approach to study fluid movement, Nanomaterials Yearbook – 2009, From Nanostructures, Nanomaterials and Nanotechnologies to Nanoindustry, Chapter 16, Nova Science Publications, USA, ISBN: 978-1-60876-451-8, USA, 181–196, 2010. https://www.novapublishers.com/catalog/product_info.php?products_id=11587.
14. Hariri Asli K., Nagiyev F.B., Haghi A.K., Physical Modeling of Fluid Movement in Pipelines, Nanomaterials Yearbook – 2009, From Nanostructures, Nanomaterials and Nanotechnologies to Nanoindustry, Chapter 17, Nova Science Publications, USA, ISBN: 978-1-60876-451-8, USA, 197–214, 2010. https://www.novapublishers.com/catalog/product_info.php?products_id=11587.
15. Hariri Asli K., Nagiyev F.B., Haghi A.K., Aliyev S.A., Improved modeling for prediction of water transmission failure, Recent Progress in Research in Chemistry and Chemical Engineering, Chapter 2, Nova Science Publications, ISBN: 978-1-61668-501-0, Nova Science Publications, USA, 28–36, 2010. https://www.novapublishers.com/catalog/product_info.php?products_id=13174.
16. Hariri Asli K., Nagiyev F.B., Haghi A.K., Aliyev S.A. Pure Oxygen penetration in wastewater flow, Recent Progress in Research in Chemistry and Chemical Engineering, Chapter 3, Nova Science Publications, ISBN: 978-1-61668-501-0, Nova Science Publications, USA, 17–27, 2010, https://www.novapublishers.com/catalog/product_info.php?products_id=13174.
17. Adams T.M., Abdel-Khalik S.I., Jeter S.M., Qureshi Z. H. An Experimental investigation of single-phase Forced Convection in Microchannels, Int. J. Heat Mass Transfer, 41, 1998, 851–857.
18. Allievi L. "General Theory of Pressure Variation in Pipes," Ann. D. Ing., 1982, 166–171.
19. Apoloniusz Kodura, Katarzyna Weinerowska," Some Aspects of Physical and Numerical Modeling of Water Hammer in Pipelines," 2005, 126–132.
20. Anuchina N.N., Volkov V.I., Gordeychuk V.A., Eskov N.S., Ilyutina O.S., Kozyrev O.M. "Numerical simulations of Rayleigh-Taylor and Richtmyer-Meshkov instability using mah-3 code," J. Comput. Appl. Math. 168, 2004, 11.
21. Bergeron L. "Water Hammer in Hydraulics and Wave Surge in Electricity," John Wiley & Sons, Inc., N.Y., 1961, 102–109.
22. Bergant Anton Discrete Vapour Cavity Model with Improved Timing of Opening and Collapse of Cavities, 1980, 1–11.
23. Bracco A., McWilliams J.C., Murante G., Provenzale A., Weiss J.B., "Revisiting free-

ly decaying two-dimensional turbulence at millennial resolution," Phys. Fluids, Issue 11, 2000, v.12, 2931–2941.

24. Brunone B., Karney B.W., Mecarelli M., Ferrante M., "Velocity Profiles and Unsteady Pipe Friction in Transient Flow" Journal of Water Resources Planning and Management, ASCE, 126(4), Jul., 2000, 236–244.
25. Cabot W.H., Cook A.W., Miller P.L., Laney D.E., Miller M.C., Childs H.R., "Large eddy simulation of Rayleigh-Taylor instability," Phys. Fluids, September, 2005, vol. 17, 91–106.
26. Cabot W. University of California, Lawrence Livermore National laboratory, Livermore, CA, Physics of Fluids, 2006, 94–550.
27. Chaudhry M.H. "Applied Hydraulic Transients," Van Nostrand Reinhold Co., N.Y., 1979, 1322–1324.
28. Chaudhry M.H., Yevjevich V. "Closed Conduit Flow," Water Resources Publication, USA, 1981, 255–278.
29. Chaudhry M. H. Applied Hydraulic Transients, Van Nostrand Reinhold, New York, USA, 1987, 165–167.
30. Clark T.T., "A numerical study of the statistics of a two-dimensional Rayleigh-Taylor mixing layer," Phys. Fluids 15, 2003, 2413 p.
31. Cook A.W., Cabot W., Miller P.L., "The mixing transition in Rayleigh-Taylor instability," J. Fluid Mech. 511, 2004, 333 p.
32. Choi S.B., Barren R.R., Warrington R.O., Fluid Flow and Heat Transfer in Microtubes, ASME DSC 40, 1991, 89–93.
33. Dimonte G., Youngs D., Dimits A., Weber S., Marinak M. "A comparative study of the Rayleigh-Taylor instability using high-resolution three-dimensional numerical simulations: the Alpha group collaboration," Phys. Fluids 16, 2004, 1668 p.
34. Dimotakis P.E., "The mixing transition in turbulence," J. Fluid Mech. 409, 2000, 69 p.
35. Elansari A. S., Silva W., Chaudhry M. H. "Numerical and Experimental Investigation of Transient Pipe Flow," Journal of Hydraulic Research, 32, 1994. 689 p.
36. Filion Y., Karney B. W., "A Numerical Exploration of Transient Decay Mechanisms in Water Distribution Systems," Proceedings of the ASCE Environmental Water Resources Institute Conference, American Society of Civil Engineers, Roanoke, Virginia, 2002, 30 p.
37. Fok A., "Design Charts for Air Chamber on Pump Pipelines,' J. Hyd. Div., ASCE, Sept., 1978, 15–74.
38. Fok A., Ashamalla A., Aldworth G., "Considerations in Optimizing Air Chamber for Pumping Plants," Symposium on Fluid Transients and Acoustics in the Power Industry, San Francisco, U.S.A. Dec., 1978, 112–114.
39. Fok A., "Design Charts for Surge Tanks on Pump Discharge Lines," BHRA 3rd Int. Conference on Pressure Surges, Bedford, England, Mar., 1980, 23–34.
40. Fok A., "Water Hammer and Its Protection in Pumping Systems," Hydro technical Conference, CSCE, Edmonton, May, 1982, 45–55.
41. Fok A., "A contribution to the Analysis of Energy Losses in Transient Pipe Flow," PhD. Thesis, University of Ottawa, 1987, 176–182.
42. Fox J.A., "Hydraulic Analysis of Unsteady Flow in Pipe Network," Wiley, N.Y., 1977, 78–89.
43. Fedorov A.G., Viskanta R., Three-dimensional Conjugate Heat Transfer into Microchannel Heat Sink for Electronic Packaging, Int. J. Heat Mass Transfer 43, 2000, 399–415.
44. Hamam M.A., Mc Corquodale, J.A., "Transient Conditions in the Transition from Gravity to Surcharged Sewer Flow," Canadian J. Civil Eng., Canada, Sep., 1982, 65–98.
45. Hariri Asli K., Nagiyev F.B., Water Hammer and Fluid Condition, 2007, 132–148.
46. Hariri Asli K., Nagiyev F.B., Water Hammer Analysis and Formulation, 2007, 111–131.
47. Hariri Asli K., Nagiyev F.B., Water Hammer and hydrodynamics instabilities, Interpenetration of Two Fluids at Parallel Between Plates and Turbulent Moving in pipe, 2007, 90–110.
48. Hariri Asli K., Nagiyev F.B., Water Hammer and Pump Pulsation, 2007, 51–72.
49. Hariri Asli K., Nagiyev F.B., Reynolds number and hydrodynamics' instability,"

Ministry of Energy, Research Week Exhibition, Tehran, Iran, December, 2007, 31–50.
50. Hariri Asli K., Nagiyev F.B., Water Hammer and Valves, 2007, 20–30.
51. Hariri Asli K., Nagiyev F.B., "Interpenetration of Two Fluids at Parallel Between Plates and Turbulent Moving in pipe," 2007, 73–89.
52. Hariri Asli K., Water Hammer and Surge Wave Modeling, Water Hammer Research; Advances in Nonlinear Dynamics Modeling, PhD. Thesis of Kaveh Hariri Asli, Toronto, Canada, Published by Apple Academic Press, Inc., Exclusive worldwide distribution by CRC Press, A Taylor & Francis Group, Print ISBN: 9781926895314, eBook: 978-1-46-656887-7, 2013, www.AppleAcademicPress.com.
53. Hariri Asli K., Portable Flow meter Tester Machine Apparatus, Certificate on registration of invention.
54. Hariri Asli K., Nagiyev F.B., Haghi A.K., "Interpenetration of Two Fluids at Parallel Between Plates and Turbulent Moving in pipe," 2008, 73–89.
55. Hariri Asli K., Nagiyev F.B., Haghi A.K., "Water Hammer and Valves," 2008, 20–30.
56. Hariri Asli K., Nagiyev F.B., Haghi A.K., "Water Hammer and Hydrodynamics Instability," 2008, 90–110.
57. Hariri Asli K., Nagiyev F.B., Haghi A.K., "Water Hammer analysis and Formulation," 2008, 27–42.
58. Hariri Asli K., Nagiyev F.B., Haghi A.K., "Water Hammer and Fluid Condition," 2008, 27–43.
59. Hariri Asli K., Nagiyev F.B., Haghi A.K., "Water Hammer and Pump Pulsation," 2008, 27–44.
60. Hariri Asli K., Nagiyev F.B., Haghi A.K., "Reynolds number and Hydrodynamics Instability," 2008, 27–45.
61. Hariri Asli K., Nagiyev F.B., Haghi A.K., "Water Hammer and fluid Interpenetration," 2008, 27–47.
62. Hariri Asli K., Computer Models for Fluid Interpenetration, Water Hammer Research; Advances in Nonlinear Dynamics Modeling, PhD. Thesis of Kaveh Hariri Asli, Toronto, Canada, Published by Apple Academic Press, Inc., Exclusive worldwide distribution by CRC Press, A Taylor & Francis Group, Print ISBN: 9781926895314, eBook: 978-1-46-656887-7, 2013, www.AppleAcademicPress.com.
63. Hariri Asli K., Heat Flow and Porous Materials, Water Hammer Research; Advances in Nonlinear Dynamics Modeling, PhD. Thesis of Kaveh Hariri Asli, Toronto, Canada, Published by Apple Academic Press, Inc., Exclusive worldwide distribution by CRC Press, A Taylor & Francis Group, Print ISBN: 9781926895314, eBook: 978-1-46-656887-7, 2013, www.AppleAcademicPress.com.
64. Hariri Asli K., Nagiyev F.B., Bubbles characteristics and convective effects in the binary mixtures. Transactions issue mathematics and mechanics series of physical-technical and mathematics science, ISSN: 0002–3108, Azerbaijan, Baku, 2009, 68–74, http://www.imm.science.az/journals.html.
65. Hariri Asli K., Nagiyev F.B., Haghi A.K., Aliyev S.A., Three-Dimensional Conjugate Heat Transfer in Porous Media, 12–17 Dec.2009, 26, 28.
66. Hariri Asli K., Nagiyev F.B., Haghi A.K., Aliyev S.A., Some Aspects of Physical and Numerical Modeling of Water Hammer in Pipelines, 2009, 26, 29.
67. Hariri Asli K., Nagiyev F.B., Haghi A.K., Aliyev S.A., Modeling for Water Hammer due to valves: From theory to practice, 2009, 26, 30.
68. Hariri Asli K., Nagiyev F.B., Haghi A.K., Aliyev S.A., Water Hammer and Hydrodynamics Instabilities Modeling: From Theory to Practice, 2009, 26, 31.
69. Hariri Asli K., Nagiyev F.B., Haghi A.K., Aliyev S.A., A computational approach to study fluid movement, 2009, 27–32.
70. Hariri Asli K., Nagiyev F.B., Haghi A.K., Aliyev S.A., Water Hammer Analysis: Some Computational Aspects and practical hints, 2009, 27–33.
71. Hariri Asli K., Nagiyev F.B., Haghi A.K., Aliyev S.A., Water Hammer and Fluid condition: A computational approach, 2009, 27–34.
72. Hariri Asli K., Nagiyev F.B., Haghi A.K., Aliyev S.A., A Computational Method to Study Transient Flow in Binary Mixtures, 12–17 Dec.2009, 27–35.

73. Hariri Asli K., Nagiyev F.B., Haghi A.K., Physical Modeling of Fluid Movement in Pipelines, 2009, 27–36.
74. Hariri Asli K., Nagiyev F.B., Haghi A.K., Aliyev S.A., Interpenetration of Two Fluids at Parallel Between Plates and Turbulent Moving, 12–17 Dec. 2009, 27–37.
75. Hariri Asli K., Nagiyev F.B., Haghi A.K., Aliyev S.A., Modeling of Fluid Interaction Produced by Water Hammer, 2009, 27–38.
76. Hariri Asli K., Nagiyev F.B., Haghi A.K., Aliyev S.A., Improved modeling for prediction of water transmission failure, Recent Progress in Research in Chemistry and Chemical Engineering, Nova Science Publications, ISBN: 978-1-61668-501-0, Nova Science Publications, USA, 2010, 28–36, https://www.novapublishers.com/catalog/product_info.php?products_id=13174.
77. Hariri Asli K., Nagiyev F.B., Haghi A.K., Aliyev S.A., Interpenetration of Two Fluids at Parallel Between Plates and Turbulent Moving, 2009, 27–40.
78. Hariri Asli K., Nagiyev F.B., Haghi A.K., Aliyev S.A., Water Hammer and hydrodynamics' instability, 2009, 27–41.
79. Hariri Asli K., Nagiyev F.B., Haghi A.K., Aliyev S.A., Water Hammer analysis and Formulation, 2009, 27–42.
80. Hariri Asli K., Nagiyev F.B., Haghi A.K., Aliyev S.A., Water Hammer and Fluid Condition, 2009, 27–43.
81. Hariri Asli K., Nagiyev F.B., Haghi A.K., Aliyev S.A., Water Hammer and Fump Pulsation, 12–17 Dec.2009, 27–44.
82. Hariri Asli K., Nagiyev F.B., Haghi A.K., Aliyev S.A., Reynolds Number and Hydrodynamics Instabilities, 2009, 27–45.
83. Hariri Asli K., Nagiyev F.B., Haghi A.K., Aliyev S.A., Water Hammer and Valves, 12–17 Dec.2009, 27–46.
84. Hariri Asli K., Nagiyev F.B., Haghi A.K., Aliyev S.A., "Water Hammer and fluid Interpenetration," 2009, 27–47.
85. Hariri Asli K., Nagiyev F.B., Modeling of Fluid Interaction Produced by Water Hammer, International Journal of Chemoinformatics and Chemical Engineering, IGI, ISSN: 2155-4110, EISSN: 2155–4129, USA, 2010, 29–41, http://www.igi-global.com/journals/details.asp?ID=34654.
86. Hariri Asli K., Nagiyev F.B. Haghi A.K., Water Hammer and Fluid Condition; A computational approach, Computational Methods in Applied Science and Engineering, USA, Chapter 5, Nova Science Publications, ISBN: 978-1-60876-052-7, USA, 2010, 73–94, https://www.novapublishers.com/catalog/.
87. Hariri Asli K., Nagiyev F.B., Haghi A.K., Some aspects of physical and numerical modeling of water hammer in pipelines. Computational Methods in Applied Science and Engineering, USA, Chapter 23, Nova Science Publications, ISBN: 978-1-60876-052-7, USA, 2010, 365–387. https://www.novapublishers.com/catalog/.
88. Hariri Asli K., Nagiyev F.B., Haghi A.K., Modeling for water hammer due to valves; from theory to practice, Computational Methods in Applied Science and Engineering, USA, Chapter 11, Nova Science Publications, ISBN: 978-1-60876-052-7, USA, 2010, 229–236, https://www.novapublishers.com/catalog/.
89. Hariri Asli K., Nagiyev F.B., Haghi A.K., A computational method to Study transient flow in binary mixtures, Computational Methods in Applied Science and Engineering, USA, Chapter 13, Nova Science Publications, ISBN: 978-1-60876-052-7, USA, 2010, 229–236, https://www.novapublishers.com/catalog/.
90. Hariri Asli K., Nagiyev F.B., Haghi A.K., Water Hammer analysis; some computational aspects and practical hints, Computational Methods in Applied Science and Engineering, USA, Chapter 16, Nova Science Publications, ISBN: 978-1-60876-052-7, USA, 2010, 263–281. https //www.novapublishers.com/catalog/.
91. Hariri Asli K., Nagiyev F.B., Haghi A.K., Water Hammer and Hydrodynamics Instabilities Modeling, Computational Methods in Applied Science and Engineering, USA, Chapter 17, From Theory to Practice, Nova Science Publications, ISBN: 978-1-60876-052-7, USA, 2010 283–301, https://www.novapublishers.com/catalog/.
92. Hariri Asli K., Nagiyev F.B., Haghi A.K., A computational approach to study water hammer and pump pulsation phenomena, Computational Methods in Applied Science

and Engineering, USA, Chapter 22, Nova Science Publications, ISBN: 978-1-60876-052-7, USA, 2010, 349–363, https://www.novapublishers.com/catalog/.
93. Hariri Asli K., Nagiyev F.B., Haghi A.K., A computational approach to study fluid movement, Nanomaterials Yearbook – 2009, From Nanostructures, Nanomaterials and Nanotechnologies to Nanoindustry, Chapter 16, Nova Science Publications, USA, ISBN: 978-1-60876-451-8, USA, 2010, 181–196, https://www.novapublishers.com/catalog/product_info.php?products_id=11587.
94. Hariri Asli K., Nagiyev F.B., Haghi A.K., Physical Modeling of Fluid Movement in Pipelines, Nanomaterials Yearbook – 2009, From Nanostructures, Nanomaterials and Nanotechnologies to Nanoindustry, Chapter 17, Nova Science Publications, USA, ISBN: 978-1-60876-451-8, USA, 2010, 197–214. https://www.novapublishers.com/catalog/product_info.php?products_id=11587.
95. Hariri Asli K., Thermal Environment, Water Hammer Research; Advances in Nonlinear Dynamics Modeling, Published by Apple Academic Press, Inc., Exclusive worldwide distribution by CRC Press, A Taylor & Francis Group, Print ISBN: 9781926895314, eBook: 978-1-46-656887-7, 2013, www.AppleAcademicPress.com.
96. Hariri Asli K., Nagiyev F.B., Haghi A.K., Interpenetration of Two Fluids at Parallel Between Plates and Turbulent Moving in pipe; A case study, Computational Methods in Applied Science and Engineering, USA, Chapter 7, Nova Science Publications, ISBN: 978-1-60876-052-7, USA, 2010, 107–133, https://www.novapublishers.com/catalog/.
97. Hariri Asli K., Nagiyev F.B., Beglou M. J., Haghi A.K., Kinetic analysis of convective drying, International Journal of the Balkan Tribological Association, ISSN: 1310-4772, Sofia, Bulgaria, 2009, Vol. 15, No 4, 546–556, jbalkta@gmail.com.
98. Hariri Asli K., Nagiyev F.B., Haghi A.K., Three-dimensional Conjugate Heat Transfer in Porous Media, International Journal of the Balkan Tribological Association, ISSN: 1310-4772, Sofia, Bulgaria, 2009, Vol. 15, No 3, 336–346, jbalkta@gmail.com.
99. Hariri Asli K., Nagiyev F.B., Haghi A.K., Aliyev S.A., Pure Oxygen penetration in wastewater flow, Recent Progress in Research in Chemistry and Chemical Engineering, Nova Science Publications, ISBN: 978-1-61668-501-0, Nova Science Publications, USA, 2010, 17–27. https://www.novapublishers.com/catalog/product_info.php?products_id=13174110
100. Hariri Asli K., Nagiyev F.B., Haghi A.K., Aliyev S.A., Improved modeling for prediction of water transmission failure, Recent Progress in Research in Chemistry and Chemical Engineering, Nova Science Publications, ISBN: 978-1-61668-501-0, Nova Science Publications, USA, 2010, 28–36, https://www.novapublishers.com/catalog/product_info.php?products_id=13174.
101. Harms T.M., Kazmierczak M.J., Cerner F.M., Holke A., Henderson H.T., Pilchowski H.T., Baker K., Experimental Investigation of Heat Transfer and Pressure Drop through Deep Micro channels in a (100) Silicon Substrate, in: Proceedings of the ASME. Heat Transfer Division, HTD 351, 1997, 347–357.
102. Holland F.A., Bragg R., Fluid Flow for Chemical Engineers, Edward Arnold Publishers, London, 1995, 1–3.
103. George E., Glimm J., "Self-similarity of Rayleigh-Taylor mixing rates," Phys. Fluids 17, 054101, 2005, 1–3.
104. Goncharov V.N., "Analytical model of nonlinear, single-mode, classical Rayleigh-Taylor instability at arbitrary Atwood numbers," Phys. Rev. Letters 88, 134502, 2002, 10–15.
105. Ishii M., Thermo-Fluid Dynamic Theory of Two-Phase Flow, Collection de D. R. Liles and W. H. Reed, "A Sern-Implict Method for Two-Phase Fluid la Direction des Etudes et. Recherché d'Electricite de France, Vol. 22 Dynamics," Journal of Computational Physics 26, Paris, 1975, 390–407.
106. Jaime Suárez A., "Generalized water hammer algorithm for piping systems with unsteady friction" 2005, 72–77.
107. Jaeger C., "Fluid Transients in Hydro-Electric Engineering Practice," Blackie & Son Ltd., 1977, 87–88.

108. Joukowski N., Paper to Polytechnic Soc. Moscow, spring of 1898, English translation by Miss O. Simin. Proc. AWWA, 1904, 57–58.
109. Kodura A., Weinerowska K., the influence of the local pipeline leak on water hammer properties, Materials of the II Polish Congress of Environmental Engineering, Lublin, 2005, 125–133.
110. Kane J., Arnett D., Remington B.A., Glendinning S.G., Baz′an G., "Two-dimensional versus three-dimensional supernova hydrodynamic instability growth," Astrophys. J., 2000, 528–989.
111. Karassik I.J., "Pump Handbook– Third Edition," McGraw-Hill, 2001, 19–22.
112. Koelle E., Luvizotto Jr. E., Andrade J.P.G., "Personality Investigation of Hydraulic Networks using MOC – Method of Characteristics" Proceedings of the 7th International Conference on Pressure Surges and Fluid Transients, Harrogate Durham, United Kingdom, 1996, 1–8.
113. Kerr S.L., "Minimizing service interruptions due to transmission line failures: Discussion," Journal of the American Water Works Association, 41, 634, July 1949, 266–268.
114. Kerr S.L., "Water Hammer control," Journal of the American Water Works Association, 43, December 1951, 985–999.
115. Kraichnan R.H., Montgomery D., "Two-dimensional turbulence," Rep. Prog. Phys. 43, 547, 1967, 1417–1423.
116. Leon Arturo S., "An efficient second-order accurate shock-capturing scheme for modeling one and two-phase water hammer flows" PhD Thesis, March 29, 2007, 4–44.
117. Lee T.S., Pejovic S. Air influence on similarity of hydraulic transients and vibrations. ASME J. Fluid Eng. 1996, 118(4), 706–709.
118. Li J., McCorquodale A., "Modeling Mixed Flow in Storm Sewers," Journal of Hydraulic Engineering, ASCE, Vol. 125, no. 11, 1999, 1170–1180.
119. Mala G., Li D., Dale J.D., Heat Transfer and Fluid Flow in Microchannels, J. Heat Transfer, 40, 1997, 3079–3088.
120. Miller P.L., Cabot W.H., Cook A.W., "Which way is up? A fluid dynamics riddle," Phys. Fluids 17, 091110, 2005, 1–26.
121. Minnaert M., on musical air bubbles and the sounds of running water. Phil. Mag., 1933, v. 16, №7, 235–248.
122. Moeng C.H., McWilliams J.C., Rotunno R., Sullivan P.P., Weil J., "Investigating 2D modeling of atmospheric convection in the PBL," J. Atm. Sci. 61, 2004, 889–903.
123. Moody L. F., "Friction Factors for Pipe Flow," Trans. ASME, 1944. Vol. 66, 671–684.
124. Nagiyev F.B., Dynamics, heat and mass transfer of vapor-gas bubbles in a two-component liquid. Turkey-Azerbaijan petrol semin., Ankara, Turkey, 1993, 32–40.
125. Nagiyev F.B., The method of creation effective coolness liquids, Third Baku international Congress. Baku, Azerbaijan Republic, 1995, 19–22.
126. Neshan H., Water Hammer, pumps Iran Co. Tehran, Iran, 1985, 1–60.
127. Nigmatulin R.I., Khabeev N.S., Nagiyev F.B., Dynamics, heat and mass transfer of vapor-gas bubbles in a liquid. Int. J. Heat Mass Transfer,.vol.24, N6, Printed in Great Britain, 1981, 1033–1044.
128. Oron. D., Arazi L., Kartoon D., Rikanati A., Alon U., Shvarts D., 'Dimensionality dependence of the Rayleigh-Taylor and Richtmyer-Meshkov instability late-time scaling laws," Phys. Plasmas 8, 2001, 2883 p.
129. Parmakian J., "Water Hammer Design Criteria," J. Power Div., ASCE, Sept., 1957, 456–460.
130. Parmakian J., "Water Hammer Analysis," Dover Publications, Inc., New York, New York, 1963, 51–58.
131. Perry. R.H., Green D.W., Maloney J.O., Perry's Chemical Engineers Handbook, 7th Edition, McGraw-Hill, New York, 1997, 1–61.
132. Peng X.F., Peterson G.P., Convective Heat Transfer and Flow Friction for Water Flow in Microchannel Structure, Int. J. Heat Mass Transfer 36, 1996, 2599–2608.
133. Pickford J., "Analysis of Surge," Macmillian, London, 1959, 153–155.
134. Pipeline Design for Water and Wastewater, American Society of Civil Engineers, New York, 1975, 54 p.

135. Qu W., Mala G.M., Li D., Heat Transfer for Water Flow in Trapezoidal Silicon Microchannels, 1993, 399–404.
136. Quick R.S., "Comparison and Limitations of Various Water Hammer Theories," J. Hyd. Div., ASME, May, 1933, 43–45.
137. Rayleigh, On the pressure developed in a liquid during the collapse of a spherical cavity. Philos. Mag. Ser. 6, v. 34, no. 200, 1917, 94–98.
138. Ramaprabhu P., Andrews M.J., "Experimental investigation of Rayleigh-Taylor mixing at small Atwood numbers," J. Fluid Mech. 502, 2004, 233 p.
139. Rich G.R., "Hydraulic Transients," Dover, USA, 1963, 148–154.
140. Savic D.A., Walters G.A., "Genetic Algorithms Techniques for Calibrating Network Models," Report no. 95/12, Centre for Systems and Control Engineering, 1995, 137–146.
141. Savic D.A., Walters G.A., Genetic Algorithms Techniques for Calibrating Network Models, University of Exeter, Exeter, United Kingdom, 1995, 41–77.
142. Shvarts D., Oron D., Kartoon D., Rikanati A., Sadot O., "Scaling laws of nonlinear Rayleigh-Taylor and Richtmyer-Meshkov instabilities in two and three dimensions," C. R. Acad. Sci. Paris IV, 719, 2000, 312 p.
143. Sharp B., "Water Hammer Problems and Solutions," Edward Arnold Ltd., London, 1981, 43–55.
144. Skousen P., "Valve Handbook," McGraw Hill, New York, HAMMER Theory and Practice, 1998, 687–721.
145. Song C.C. et al, "Transient Mixed-Flow Models for Storm Sewers," J. Hyd. Div., Nov., 1983, Vol. 109, 458–530.
146. Stephenson D., "Pipe Flow Analysis," Elsevier, Vol. 19, S.A., 1984, 670–788.
147. Streeter V.L., Lai C., "Water Hammer Analysis Including Fluid Friction." Journal of Hydraulics Division, ASCE, 88, 1962, 79 p.
148. Streeter V.L., Wylie E.B., "Fluid Mechanics," McGraw-Hill Ltd., USA, 1979, 492–505.
149. Streeter V.L., Wylie E.B., "Fluid Mechanics," McGraw-Hill Ltd., USA, 1981,398–420.
150. Tijsseling, "Alan E Vardy Time scales and FSI in unsteady liquid-filled pipe flow," 1993, 5–12.
151. Tuckerman D.B., R.F.W Pease, high performance heat sinking for VLSI, IEEE Electron device letter, DEL-2, 1981, 126–129.
152. Tennekes H., Lumley J.L., A First Course in Turbulence, the MIT Press, 1972, 410–466.
153. Thorley A.R.D., "Fluid Transients in Pipeline Systems," D. & L. George, Herts, England, 1991, 231–242.
154. Tullis J.P., "Control of Flow in Closed Conduits," Fort Collins, Colorado, 1971, 315–340.
155. Tuckerman D.B., Heat transfer microstructures for integrated circuits, Ph.D. thesis, Stanford University, 1984, 10–120.
156. Vallentine H.R., "Rigid Water Column Theory for Uniform Gate Closure," J. Hyd. Div. ASCE, July, 1965, 55–243.
157. Waddell J.T., Niederhaus C.E., Jacobs J.W., "Experimental study of Rayleigh-Taylor instability: Low Atwood number liquid systems with single-mode initial perturbations," Phys. Fluids 13, 2001, 1263–1273.
158. Watters G.Z., "Modern Analysis and Control of Unsteady Flow in Pipelines," Ann Arbor Sci., 2nd Ed., 1984, 1098–1104.
159. Walski T.M., Lutes T.L., "Hydraulic Transients Cause Low-Pressure Problems," Journal of the American Water Works Association, 75(2), 1994, 58, 160. Weber S.V., Dimonte G., Marinak M.M., "Arbitrary Lagrange-Eulerian code simulations of turbulent Rayleigh-Taylor instability in two and three dimensions," Laser and Particle Beams 21, 2003, 455 p.
161. Wood D.J., Dorsch R.G., Lightener C., "Wave-Plan Analysis of Unsteady Flow in Closed Conduits," Journal of Hydraulics Division, ASCE, 92, 1966, 83–110.
162. Wood D.J., Jones, S.E., "Water Hammer charts for various types of valves," Journal of the Hydraulics Division, Proceedings of the American Society of Civil Engineers, January, 1973, 167–178.
163. Wood F.M., "History of Water hammer," Civil Engineering Research Report, #65, Queens University, Canada, 1970, 66–70.
164. Wu Z.Y., Simpson A.R., Competent genetic-evolutionary optimization of water

distribution systems. J. Comput. Civ. Eng. 15(2), 2001, 89–101.

165. Wylie E.B., Talpin L.B., Matched impedance to control fluid transients. Trans. ASME 105, 2, 1983, 219–224.
166. Wylie E. B., Streeter V. L., Fluid Transients in Systems, Prentice-Hall, Englewood Cliffs, New Jersey, 1993, 4 p.
167. Wylie E. B., Streeter V. L., Fluid Transients, Feb Press, Ann Arbor, MI, 1982, corrected copy, 1983, 158 p.
168. Wu P.Y., Little W.A., measurement of friction factor for flow of gases in very fine channels used for micro miniature, Joule Thompson refrigerators, Cryogenics 24 (8), 1983, 273–277.
169. Xu B., Ooi K.T., Mavriplis C., Zaghloul M. E., Viscous dissipation effects for liquid flow in microchannels, Micorsystems, 2002, 53–57.
170. Young Y.N., Tufo H., Dubey A., Rosner R., "On the miscible Rayleigh-Taylor instability: two and three dimensions," J. Fluid Mech. 447, 377, 2001, 2003–2500.
171. Wood D.J., Dorsch R.G., Lightener C., "Wave-Plan Analysis of Unsteady Flow in Closed Conduits," Journal of Hydraulics Division, ASCE, 92, 1966, 83–110.
172. Wylie E. B., Streeter V. L., Fluid Transients in Systems, Prentice-Hall, Englewood Cliffs, New Jersey, 1993, 4 p.
173. Brunone B., Karney B.W., Mecarelli M., Ferrante M., "Velocity Profiles and Unsteady Pipe Friction in Transient Flow" Journal of Water Resources Planning and Management, ASCE, 126(4), Jul., 2000, 236–244.
174. Koelle E., Luvizotto Jr. E., Andrade J.P.G., "Personality Investigation of Hydraulic Networks using MOC – Method of Characteristics" Proceedings of the 7th International Conference on Pressure Surges and Fluid Transients, Harrogate Durham, United Kingdom, 1996, 1–8.
175. Filion Y., Karney B. W., "A Numerical Exploration of Transient Decay Mechanisms in Water Distribution Systems," Proceedings of the ASCE Environmental Water Resources Institute Conference, American Society of Civil Engineers, Roanoke, Virginia, 2002, 30.
176. Hamam M.A., Mc Corquodale, J.A., "Transient Conditions in the Transition from Gravity to Surcharged Sewer Flow," Canadian J. Civil Eng., Canada, Sep., 1982, 65–98.
177. Savic D.A., Walters G.A., "Genetic Algorithms Techniques for Calibrating Network Models," Report no. 95/12, Centre for Systems and Control Engineering, 1995, 137–146.
178. Walski T.M., Lutes T.L., "Hydraulic Transients Cause Low-Pressure Problems," Journal of the American Water Works Association, 75(2), 1994, 58.
179. Lee T.S., Pejovic S. Air influence on similarity of hydraulic transients and vibrations. ASME J. Fluid Eng. 1996, 118(4), 706–709.
180. Till Mathis Wagner, A very short introduction to the Finite Element Method, Technical University of Munich, JASS 2004, St Petersburg, May 4, 2004.

REFERENCES FOR CHAPTER 8

1. Leon, S. A., Improved Modeling of Unsteady Free Surface. Pressurized and Mixed Flows in Storm-Sewer Systems, Submitted in Partial Fulfillment of the Requirements for the degree of Doctor of Philosophy in Civil Engineering in the Graduate College of the University of Illinois at Urbana-Champaign, 2007 57–58.
2. Hariri Asli K., Nagiyev F.B , Haghi A.K., Physical and Numerical Modeling of Fluid Flow in Pipelines. A computational approach, International J. the Balkan Tribological Association, Thomson Reuters Master Journal List, ISSN: 1310-4772, Sofia, Bulgaria, Vol. 16, No 19, 20–34, (2009).
3. Hariri Asli K., Nagiyev F.B., Haghi A.K., Computational Methods in Applied Science and Engineering, Interpenetration of Two Fluids at Parallel between Plates and Turbulent Moving in Pipe, Nova Science, New York, USA, 2009, 115–128, chapter 7.
4. Wylie E.B., Streeter V.L., Fluid Transients in Systems Prentice Hall, 1993, corrected copy, 1982, 166–171.
5. Kodura A., Weinerowska K., Some Aspects of Physical and Numerical Modeling

of Water Hammer in Pipelines Ottenstein, Austria, 2005, 125–133.

6. Hariri Asli K., Sahleh H., Aliyev S.A., Mathematical Concepts and Computational Approaches on Hydrodynamics Instability, Mathematical Concepts for Mechanical Engineering Design, Toronto, Canada, Published by Apple Academic Press, Inc., Exclusive worldwide distribution by CRC Press, A Taylor & Francis Group, Print ISBN: 9781926895628, 2013, www.AppleAcademicPress.com.
7. Hariri Asli K., Nagiyev F.B., Haghi A.K., Computational Methods in Applied Science and Engineering, Water Hammer analysis; some computational aspects and practical hints Nova Science Publications, New York, USA, 2009, chapter 16.
8. Apoloniusz, Kodura, Katarzyna, Weinerowska: Some Aspects of Physical and Numerical Modeling of Water Hammer in Pipelines, International symposium on water management and hydraulic engineering, 4th–7th September, paper no.11.05, 126–132, Ottenstein, Austria, 2005.
9. Fedorov A.G., Viskanta R., Three-dimensional Conjugate Heat Transfer into Microchannel Heat Sink for Electronic Packaging, Int. J. Heat Mass Transfer 43, 2000, 399–415.
10. Tuckerman D.B., Heat transfer microstructures for integrated circuits, Ph.D. thesis, Stanford University, 1984, 10–120.
11. Harms T.M., Kazmierczak M.J., Cerner F.M., Holke A., Henderson H.T., Pilchowski H.T., Baker K., Experimental Investigation of Heat Transfer and Pressure Drop through Deep Micro channels in a (100) Silicon Substrate, in: Proceedings of the ASME. Heat Transfer Division, HTD 351, 1997, 347–357.
12. Holland F.A., Bragg R., Fluid Flow for Chemical Engineers, Edward Arnold Publishers, London, 1995, 1–3.
13. Lee T.S., Pejovic S. Air influence on similarity of hydraulic transients and vibrations. ASME J. Fluid Eng. 1996, 118(4), 706–709.
14. Li J., McCorquodale A., "Modeling Mixed Flow in Storm Sewers," Journal of Hydraulic Engineering, ASCE, Vol. 125, no. 11, 1999, 1170–1180.
15. Minnaert M., on musical air bubbles and the sounds of running water. Phil. Mag., 1933, v. 16, №7, 235–248.
16. Moeng C.H., McWilliams J.C., Rotunno R., Sullivan P.P., Weil J., "Investigating 2D modeling of atmospheric convection in the PBL," J. Atm. Sci. 61, 2004, 889–903.
17. Tuckerman D.B., R.F.W Pease, high performance heat sinking for VLSI, IEEE Electron device letter, DEL-2, 1981, 126–129.
18. Nagiyev F.B., Khabeev N.S, Bubble dynamics of binary solutions. High Temperature, 1988, v. 27, № 3, 528–533.
19. Shvarts D., Oron D., Kartoon D., Rikanati A., Sadot O., "Scaling laws of nonlinear Rayleigh-Taylor and Richtmyer-Meshkov instabilities in two and three dimensions," C. R. Acad. Sci. Paris IV, 719, 2000, 312 p.
20. Cabot W.H., Cook A.W., Miller P.L., Laney D.E., Miller M.C., Childs H.R., "Large eddy simulation of Rayleigh-Taylor instability," Phys. Fluids, September, 2005, vol. 17, 91–106.
21. Cabot W., University of California, Lawrence Livermore National laboratory, Livermore, CA, Physics of Fluids, 2006, 94–550.
22. Goncharov V.N., "Analytical model of nonlinear, single-mode, classical Rayleigh-Taylor instability at arbitrary Atwood numbers," Phys. Rev. Letters 88, 134502, 2002, 10–15.
23. Ramaprabhu P., Andrews M.J., "Experimental investigation of Rayleigh-Taylor mixing at small Atwood numbers," J. Fluid Mech. 502, 2004, 233 p.
24. Clark T.T., "A numerical study of the statistics of a two-dimensional Rayleigh-Taylor mixing layer," Phys. Fluids 15, 2003, 2413.
25. Cook A.W., Cabot W., Miller P.L., "The mixing transition in Rayleigh-Taylor instability," J. Fluid Mech. 511, 2004, 333.
26. Waddell J.T., Niederhaus C.E., Jacobs J.W., "Experimental study of Rayleigh-Taylor instability: Low Atwood number liquid systems with single-mode initial perturbations," Phys. Fluids 13, 2001, 1263–1273.
27. Weber S.V., Dimonte G., Marinak M.M., "Arbitrary Lagrange-Eulerian code simulations of turbulent Rayleigh-Taylor instability in two and three dimensions," Laser and Particle Beams 21, 2003, 455 p.

28. Dimonte G., Youngs D., Dimits A., Weber S., Marinak M. "A comparative study of the Rayleigh-Taylor instability using high-resolution three-dimensional numerical simulations: the Alpha group collaboration," Phys. Fluids 16, 2004, 1668.
29. Young Y.N., Tufo H., Dubey A., Rosner R., "On the miscible Rayleigh-Taylor instability: two and three dimensions," J. Fluid Mech. 447, 377, 2001, 2003–2500.
30. George E., Glimm J., "Self-similarity of Rayleigh-Taylor mixing rates," Phys. Fluids 17, 054101, 2005, 1–3.
31. Oron. D., Arazi L., Kartoon D., Rikanati A., Alon U., Shvarts D., "Dimensicnality dependence of the Rayleigh-Taylor and Richtmyer-Meshkov instability late-time scaling laws," Phys. Plasmas 8, 2001, 2883 p.
32. Nigmatulin R.I., Nagiyev F.B., Khabeev N.S., Effective heat transfer coefficients of the bubbles in the liquid radial pulse. Mater. Second-Union. Conf. Heat Mass Transfer, "Heat massoob-men in the biphasic with Minsk", 1980, v.5, 111–115.
33. Nagiyev F.B., Khabeev N.S, Bubble dynamics of binary solutions. High Temperature, 1988, v. 27, № 3, 528–533.
34. Nagiyev F.B., Damping of the cscillations of bubbles boiling binary soluticns. Mater. VIII Resp. Conf. mathematics and mechanics. Baku, October 26–29, 1988, 177–178.
35. Nagyiev F.B., Kadyrov B.A., Small oscillations of the bubbles in a binary mixture in the acoustic field. Math. An Az.SSR Ser. Physico-tech. and mate. Science, 1986, № 1, 23–26.
36. Nagiyev F.B., Dynamics, heat and mass transfer of vapor-gas bubbles in a two-component liquid. Turkey-Azerbaijan petrol semin., Ankara, Turkey, 1993, 32–40.
37. Nagiyev F.B., The method of creation effective coolness liquids, Third Baku international Congress. Baku, Azerbaijan Republic, 1995, 19–22.
38. Nagiyev F.B., The linear theory of disturbances in binary liquids bubble solution. Dep. In VINITI, 1986, № 405, in 86, 76–79.
39. Nagiyev F.B., Structure of stationary shock waves in boiling binary solutions. Math. USSR, Fluid Dynamics, 1989, № 1, 81–87.
40. Rayleigh, On the pressure developed in a liquid during the collapse of a spherical cavity. Philos. Mag. Ser. 6, v. 34, no. 200, 1917, 94–98.
41. Perry. R.H., Green D.W., Maloney J.O., Perry's Chemical Engineers Handbook, 7th Edition, McGraw-Hill, New York, 1997, 1–61.
42. Nigmatulin R.I., Dynamics of Multiphase Media, Moscow, 'Nauka," 1987, v. 1, 2, 12–14.
43. Kane J., Arnett D., Remington B.A., Glendinning S.G., Baz'an G., "Two-dimensional versus three-dimensional supernova hydrodynamic instability growth," Astrophys. J., 2000, 528–989.
44. Quick R.S., "Comparison and Limitations of Various Water Hammer Theories," J. Hyd. Div., ASME, May, 1933, 43–45.
45. Jaeger C., "Fluid Transients in Hydro-Electric Engineering Practice," Blackie & Son Ltd., 1977, 87–88.
46. Jaime Suárez A., 'Generalized water hammer algorithm for piping systems with unsteady friction" 2005, 72–77.
47. Fok A., Ashamalla A., Aldworth G., "Considerations in Optimizing Air Chamber for Pumping Plants," Symposium on Fluid Transients and Acoustics in the Power Industry, San Francisco, U.S.A. Dec., 1978, 112–114.
48. Fok A., "Design Charts for Surge Tanks on Pump Discharge Lines," BHRA 3rd Int. Conference on Pressure Surges, Bedford, England, Mar., 1980, 23–34.
49. Fok A., "Water Hammer and Its Protection in Pumping Systems," Hydro technical Conference, CSCE, Edmonton, May, 1982, 45–55.
50. Fok A., "A contribution to the Analysis of Energy Losses in Transient Pipe Flow," PhD. Thesis, University of Ottawa, 1987, 176–182.
51. Hariri Asli K., Nagiyev F.B., "Interpenetration of Two Fluids at Parallel Between Plates and Turbulent Moving in pipe," 2007, 73–89.
52. Hariri Asli K., Water Hammer and Surge Wave Modeling, Water Hammer Research; Advances in Nonlinear Dynamics Modeling, PhD. Thesis of Kaveh Hariri Asli, Toronto, Canada, Published by Apple Academic Press, Inc., Exclusive worldwide distribution by CRC Press, A Taylor & Francis

Group, Print ISBN: 9781926895314, eBook: 978-1-46-656887-7, 2013, www.AppleAcademicPress.com.
53. Hariri Asli K., Portable Flow meter Tester Machine Apparatus, Certificate on registration of invention, 2008, 98–115.
54. Hariri Asli K., Nagiyev F.B., Haghi A.K., "Interpenetration of Two Fluids at Parallel Between Plates and Turbulent Moving in pipe," 2008, 73–89.
55. Hariri Asli K., Nagiyev F.B., Haghi A.K., "Water Hammer and Valves," 2008, 20–30.
56. Hariri Asli K., Nagiyev F.B., Haghi A.K., "Water Hammer and Hydrodynamics Instability," 2008, 90–110.
57. Hariri Asli K., Nagiyev F.B., Haghi A.K., "Water Hammer analysis and Formulation," 2008, 27–42.
58. Hariri Asli K., Nagiyev F.B., Haghi A.K., "Water Hammer and Fluid Condition," 2008, 27–43.
59. Hariri Asli K., Nagiyev F.B., Haghi A.K., "Water Hammer and Pump Pulsation," 2008, 27–44.
60. Hariri Asli K., Nagiyev F.B., Haghi A.K., "Reynolds number and Hydrodynamics Instability," 2008, 27–45.
61. Hariri Asli K., Nagiyev F.B., Haghi A.K., "Water Hammer and fluid Interpenetration," 2008, 27–47.
62. Hariri Asli K., Computer Models for Fluid Interpenetration, Water Hammer Research; Advances in Nonlinear Dynamics Modeling, PhD. Thesis of Kaveh Hariri Asli, Toronto, Canada, Published by Apple Academic Press, Inc., Exclusive worldwide distribution by CRC Press, A Taylor & Francis Group, Print ISBN: 9781926895314, eBook: 978-1-46-656887-7, 2013, www.AppleAcademicPress.com.
63. Hariri Asli K., Heat Flow and Porous Materials, Water Hammer Research; Advances in Nonlinear Dynamics Modeling, PhD. Thesis of Kaveh Hariri Asli, Toronto, Canada, Published by Apple Academic Press, Inc., Exclusive worldwide distribution by CRC Press, A Taylor & Francis Group, Print ISBN: 9781926895314, eBook: 978-1-46-656887-7, 2013, www.AppleAcademicPress.com.
64. Hariri Asli K., Nagiyev F.B., Bubbles characteristics and convective effects in the binary mixtures. Transactions issue mathematics and mechanics series of physical-technical and mathematics science, ISSN: 0002–3108, Azerbaijan, Baku, 2009, 68–74, http://www.imm.science.az/journals.html.
65. Hariri Asli K., Nagiyev F.B., Haghi A.K., Aliyev S.A., Three-Dimensional Conjugate Heat Transfer in Porous Media, 12–17 Dec.2009, 26, 28.
66. Hariri Asli K., Nagiyev F.B., Haghi A.K., Aliyev S.A., Some Aspects of Physical and Numerical Modeling of Water Hammer in Pipelines, 2009, 26, 29.
67. Hariri Asli K., Nagiyev F.B., Haghi A.K., Aliyev S.A., Modeling for Water Hammer due to valves: From theory to practice, 2009, 26, 30.
68. Hariri Asli K., Nagiyev F.B., Haghi A.K., Aliyev S.A., Water Hammer and Hydrodynamics Instabilities Modeling: From Theory to Practice, 2009, 26, 31.
69. Hariri Asli K., Nagiyev F.B., Haghi A.K., Aliyev S.A., A computational approach to study fluid movement, 2009, 27–32.
70. Hariri Asli K., Nagiyev F.B., Haghi A.K., Aliyev S.A., Water Hammer Analysis: Some Computational Aspects and practical hints, 2009, 27–33.
71. Hariri Asli K., Nagiyev F.B., Haghi A.K., Aliyev S.A., Water Hammer and Fluid condition: A computational approach, 2009, 27–34.
72. Hariri Asli K., Nagiyev F.B., Haghi A.K., Aliyev S.A., A Computational Method to Study Transient Flow in Binary Mixtures, 12–17 Dec.2009, 27–35.
73. Hariri Asli K., Nagiyev F.B., Haghi A.K., Physical Modeling of Fluid Movement in Pipelines, 2009, 27–36.
74. Hariri Asli K., Nagiyev F.B., Haghi A.K., Aliyev S.A., Interpenetration of Two Fluids at Parallel Between Plates and Turbulent Moving, 12–17 Dec. 2009, 27–37.
75. Hariri Asli K., Nagiyev F.B., Haghi A.K., Aliyev S.A., Modeling of Fluid Interaction Produced by Water Hammer, 2009, 27–38.
76. Hariri Asli K., Nagiyev F.B., Water Hammer and Valves, 2007, 20–30.
77. Kodura A., Weinerowska K., the influence of the local pipeline leak on water hammer properties, Materials of the II Polish Con-

gress of Environmental Engineering, Lublin, 2005, 125–133.
78. Apoloniusz, Kodura, Katarzyna, Weinerowska: Some Aspects of Physical and Numerical Modeling of Water Hammer in Pipelines, International symposium on water management and hydraulic engineering, 4th–7th September, paper no.11.05, 126–132, Ottenstein, Austria, 2005.
79. Joukowski N., Paper to Polytechnic Soc. Moscow, spring of 1898, English translation by Miss O. Simin. Proc. AWWA, 1904, 57–58.
80. Hariri Asli K., Nagiyev F.B., Haghi A.K., A computational method to Study transient flow in binary mixtures, Computational Methods in Applied Science and Engineering, USA, Chapter 13, Nova Science Publications, ISBN: 978-1-60876-052-7, USA, 2010, 229–236, https://www.novapublishers.com/catalog/.
81. Parmakian J., "Water Hammer Design Criteria," J. Power Div., ASCE, Sept., 1957, 456–460.
82. Wood D.J., Dorsch R.G., Lightener C., "Wave-Plan Analysis of Unsteady Flow in Closed Conduits," Journal of Hydraulics Division, ASCE, 92, 1966, 83–110.
83. Wylie E. B., Streeter V. L., Fluid Transients in Systems, Prentice-Hall, Englewood Cliffs, New Jersey, 1993, 4 p.
84. Brunone B., Karney B.W., Mecarelli M., Ferrante M., "Velocity Profiles and Unsteady Pipe Friction in Transient Flow" Journal of Water Resources Planning and Management, ASCE, 126(4), Jul., 2000, 236–244.
85. Koelle E., Luvizotto Jr. E., Andrade J.P.G., "Personality Investigation of Hydraulic Networks using MOC – Method of Characteristics" Proceedings of the 7th International Conference on Pressure Surges and Fluid Transients, Harrogate Durham, United Kingdom, 1996, 1–8.
86. Filion Y., Karney B. W., "A Numerical Exploration of Transient Decay Mechanisms in Water Distribution Systems," Proceedings of the ASCE Environmental Water Resources Institute Conference, American Society of Civil Engineers, Roanoke, Virginia, 2002, 30.
87. Hamam M.A., Mc Corquodale, J.A., "Transient Conditions in the Transition from Gravity to Surcharged Sewer Flow," Canadian J. Civil Eng., Canada, Sep., 1982, 65–98.
88. Savic D.A., Walters G.A., "Genetic Algorithms Techniques for Calibrating Network Models," Report no. 95/12, Centre for Systems and Control Engineering, 1995, 137–146.
89. Walski T.M., Lutes T.L., "Hydraulic Transients Cause Low-Pressure Problems," Journal of the American Water Works Association, 75(2), 1994, 58.
90. Lee T.S., Pejovic S. Air influence on similarity of hydraulic transients and vibrations. ASME J. Fluid Eng. 1996, 118(4), 706–709.
91. Chaudhry M.H., "Applied Hydraulic Transients," Van Nostrand Reinhold Co., N Y., 1979, 1322–1324.
92. Parmakian J., "Water Hammer Analysis," Dover Publications, Inc., New York, New York, 1963, 51–58.
93. Streeter V.L., Wylie E.B., "Fluid Mechanics," McGraw-Hill Ltd., USA, 1979, 492–505.
94. Leon Arturo S., "An efficient second-order accurate shock-capturing scheme for modeling one and two-phase water hammer flows" PhD Thesis, March 29, 2007, 4–44.
95. Tuckerman D.B., R.F.W Pease, high performance heat sinking for VLSI, IEEE Electron device letter, DEL-2, 1981, 126–129.
96. Wu P.Y., Little W.A., measurement of friction factor for flow of gases in very fine channels used for micro miniature, Joule Thompson refrigerators, Cryogenics 24 (8), 1983, 273–277.
97. Harms T.M., Kazmierczak M.J., Cerner F.M., Holke A., Henderson H.T., Pilchowski H.T., Baker K., Experimental Investigation of Heat Transfer and Pressure Drop through Deep Micro channels in a (100) Silicon Substrate, in: Proceedings of the ASME. Heat Transfer Division, HTD 351, 1997, 347–357.
98. Fedorov A.G., Viskanta R., Three-dimensional Conjugate Heat Transfer into Microchannel Heat Sink for Electronic Packaging, Int. J. Heat Mass Transfer 43, 2000, 399–415.

99. Qu W., Mala G.M., Li D., Heat Transfer for Water Flow in Trapezoidal Silicon Microchannels, 1993, 399–404.
100. Choi S.B., Barren R.R., Warrington R.O., Fluid Flow and Heat Transfer in Microtubes, ASME DSC 40, 1991, 89–93.
101. Adams T.M., Abdel-Khalik S.I., Jeter S.M., Qureshi Z. H. An Experimental investigation of single-phase Forced Convection in Microchannels, Int. J. Heat Mass Transfer, 41, 1998, 851–857.
102. Peng X.F., Peterson G.P., Convective Heat Transfer and Flow Friction for Water Flow in Microchannel Structure, Int. J. Heat Mass Transfer 36, 1996, 2599–2608.
103. Mala G., Li D., Dale J.D., Heat Transfer and Fluid Flow in Microchannels, J. Heat Transfer, 40, 1997, 3079–3088.
104. Xu B., Ooi K.T., Mavriplis C., Zaghloul M. E., Viscous dissipation effects for liquid flow in microchannels, Micorsystems, 2002, 53–57.
105. Tuckerman D.B., R.F.W Pease, high performance heat sinking for VLSI, IEEE Electron device letter, DEL-2, 1981, 126–129.
106. Incropera, F. P., Dewitt, D.P., Fundamentals of Heat and Mass Transfer, second ed., Wiley, New York, 1985.
107. Ghali, K., Jones, B., Tracy, E., Modeling Heat and Mass Transfer in Fabrics, Int. J. Heat Mass Transfer 38(1), 13–21 (1995).
108. Flory, P.J., Statistical Mechanics of Chain Molecules, Interscience Pub. NY, 1969.
109. Hadley, G. R., Numerical Modeling of the Drying of Porous Materials, in: Proceedings of The Fourth International Drying Symposium, 1984, vol. 1, 151–158.
110. Hong, K., Hollies, N. R. S., Spivak, S. M., Dynamic Moisture Vapour Transfer Through Textiles, Part I: Clothing Hygrometry and the Influence of Fiber Type, Textile Res. J. 1988, 58(12), 697–706.
111. Chen, C. S. and Johnson, W. H., Kinetics of Moisture Movement in Hygroscopic Materials, In: Theoretical Considerations of Drying Phenomenon, Trans. ASAE., 12, 109–113 (1969).
112. Barnes, J., Holcombe, B., Moisture Sorption and Transport in Clothing During Wear, Textile Res. J. 66 (12), 777–786 (1996).
113. Chen, and Pei, D., A Mathematical Model of Drying Process, Int. J. Heat Mass Transfer 31 (12), 2517–2562 (1988).
114. Davis, A., James, D., Slow Flow Through a Model Fibrous Porous Medium, Int. J. Multiphase Flow 22, 969–989 (1996).
115. Jirsak, O., Gok, T., Ozipek, B., Pau, N., Comparing Dynamic and Static Methods for Measuring Thermal Conductive Properties of Textiles, Textile Res. J. 68(1), 47–56 (1998).
116. Kaviany, M., "Principle of Heat Transfer in Porous Media," Springer, New York, 1991.
117. Jackson, J., James, D., The Permeability of Fibrous Porous Media, Can. J. Chem. Eng. 64, 364–374 (1986).
118. Dietl, C. and George, O. P., Bansal, N. K., Modeling of Diffusion in Capillary Porous Materials During the Drying Process, Drying Technol. 13 (1&2), 267–293 (1995).
119. Ea, J. Y., Water Vapour Transfer in Breathable Fabrics for Clothing, PhD thesis, University of Leeds, 1988.
120. Haghi, A.K., Moisture permeation of clothing, JTAC 76, 1035–1055 (2004).
121. Haghi, A.K., Thermal analysis of drying process, JTAC 74, 827–842 (2003).
122. Haghi, A.K., Some Aspects of Microwave Drying, The Annals of Stefan cel Mare University, Year VII, no. 14, 22–25 (2000).
123. Haghi, A.K., A Thermal Imaging Technique for Measuring Transient Temperature Field- An Experimental Approach, The Annals of Stefan cel Mare University, Year VI, no. 12, 73–76 (2000).
124. Haghi, A.K., Experimental Investigations on Drying of Porous Media using Infrared Radiation, Acta Polytechnica, 41(1), 55–57 (2001).
125. Haghi, A.K., A Mathematical Model of the Drying Process, Acta Polytechnica 2001, 41(3), 20–23.
126. Haghi, A.K., Simultaneous Moisture and Heat Transfer in Porous System, Journal of Computational and Applied Mechanics 2(2), 195–204 (2001).
127. Haghi, A.K., A Detailed Study on Moisture Sorption of Hygroscopic Fiber, Journal of Theoretical and Applied Mechanics 2002, 32(2), 47–62.
128. Flory, P.J., Statistical Mechanics of Chain Molecules, Interscience Pub. NY, 1969.

REFERENCES FOR CHAPTER 9

1. Incropera, F.P., Dewitt D.P., Fundamentals of Heat and Mass Transfer, second ed., Wiley, New York, 1985.
2. Ghali K., Jones B., Tracy E., Modeling Heat and Mass Transfer in Fabrics, Int. J. Heat Mass Transfer 38(1), 13–21 (1995).
3. Flory P.J., Statistical Mechanics of Chain Molecules, Interscience Pub. NY, 1969.
4. Hadley, G.R., Numerical Modeling of the Drying of Porous Materials, in: Proceedings of The Fourth International Drying Symposium, 1984, vol. 1, 151–158.
5. Hong K., Hollies, N. R.S., Spivak, S.M., Dynamic Moisture Vapour Transfer Through Textiles, Part I: Clothing Hygrometry and the Influence of Fiber Type, Textile Res. J. 1988, 58(12), 697–706.
6. Chen, C. S. and Johnson, W.H., Kinetics of Moisture Movement in Hygroscopic Materials, In: Theoretical Considerations of Drying Phenomenon, Trans. ASAE., 12, 109–113 (1969).
7. Barnes J., Holcombe B., Moisture Sorption and Transport in Clothing During Wear, Textile Res. J. 66 (12), 777–786 (1996).
8. Chen, and Pei D., A Mathematical Model of Drying Process, Int. J. Heat Mass Transfer 31 (12), 2517–2562 (1988).
9. Davis A., James D., Slow Flow Through a Model Fibrous Porous Medium, Int. J. Multiphase Flow 22, 969–989 (1996).
10. Jirsak O., Gok T., Ozipek B., Pau N., Comparing Dynamic and Static Methods for Measuring Thermal Conductive Properties of Textiles, Textile Res. J. 68(1), 47–56 (1998).
11. Kaviany M., "Principle of Heat Transfer in Porous Media," Springer, New York, 1991.
12. Jackson J., James D., The Permeability of Fibrous Porous Media, Can. J. Chem. Eng. 64, 364–374 (1986).
13. Dietl, C. and George, O.P., Bansal, N.K., Modeling of Diffusion in Capillary Porous Materials During the Drying Process, Drying Technol. 13 (1&2), 267–293 (1995).
14. Ea, J.Y., Water Vapour Transfer in Breathable Fabrics for Clothing, PhD thesis, University of Leeds, 1988.
15. Haghi A.K., Moisture permeation of clothing, JTAC 76, 1035–1055(2004).
16. Haghi A.K., Thermal analysis of drying process, JTAC 74, 827–842(2003).
17. Haghi A.K., Some Aspects of Microwave Drying, The Annals of Stefan cel Mare University, Year VII, no. 14, 22–25 (2000).
18. Haghi A.K., A Thermal Imaging Technique for Measuring Transient Temperature Field-An Experimental Approach, The Annals of Stefan cel Mare University, Year VI, no. 12, 73–76 (2000).
19. Haghi A.K., Experimental Investigations on Drying of Porous Media using Infrared Radiation, Acta Polytechnica, 41(1), 55–57 (2001).
20. Haghi A.K., A Mathematical Model of the Drying Process, Acta Polytechnica 2001, 41(3), 20–23.
21. Haghi A.K., Simultaneous Moisture and Heat Transfer in Porous System, Journal of Computational and Applied Mechanics 2(2), 195–204 (2001).
22. Haghi A.K., A Detailed Study on Moisture Sorption of Hygroscopic Fiber, Journal of Theoretical and Applied Mechanics 2002, 32(2), 47–62.
23. Dietl, C., George, O.P., Bansal, N.K., Modeling of Diffusion in Capillary Porous Materials During the Drying Process, Drying Technol. 13 (1&2), 267–293 (1995).
24. Ea, J.Y., Water Vapour Transfer in Breathable Fabrics for Clothing, PhD thesis, University of Leeds, 1988.
25. Flory P.J., Statistical Mechanics of Chain Molecules, Interscience Pub. NY, 1969.
26. Nigmatulin R.I., Khabeev N.S., Nagiyev F.B., Dynamics, heat and mass transfer of vapor-gas bubbles in a liquid. Int. J. Heat Mass Transfer, vol.24, N6, Printed in Great Britain, 1981, 1033–1044.
27. Vargaftik N.B., Handbook of thermo-physical properties of gases and liquids. Oxford: Pergamon Press, 1972, 98.
28. Laman B.F., Hydro pumps and installation, 1988, 278.
29. Nagiyev F.B., Kadyrov B.A., Heat transfer and the dynamics of vapor bubbles in a liquid binary solution. DAn Azerbaijani SSR, 1986, № 4, 10–13.

30. Alyshev V.M., Hydraulic calculations of open channels on your PC.– Part 1 Tutorial.– Moscow: MSUE, 2003, 185.
31. Streeter V.L., Wylie E.B., "Fluid Mechanics," McGraw-Hill Ltd., USA, 1979, 492–505.
32. Sharp B., "Water Hammer Problems and Solutions," Edward Arnold Ltd., London, 1981, 43–55.
33. Skousen P., "Valve Handbook," McGraw Hill, New York, HAMMER Theory and Practice, 1998, 687–721.
34. Shaking N.I., Water Hammer to break the continuity of the flow in pressure conduits pumping stations: Dis. on Kharkov, 1988, 225.
35. Tijsseling, "Alan E Vardy Time scales and FSI in unsteady liquid-filled pipe flow," 1993, 5–12.
36. Wu P.Y., Little W.A., measurement of friction factor for flow of gases in very fine channels used for micro miniature, Joule Thompson refrigerators, Cryogenics 24 (8), 1983, 273–277.
37. Song C.C. et al, "Transient Mixed-Flow Models for Storm Sewers," J. Hyd. Div., Nov., 1983, Vol. 109, 458–530.
38. Stephenson D., "Pipe Flow Analysis," Elsevier, Vol. 19, S.A., 1984, 670–788.
39. Chaudhry M.H., "Applied Hydraulic Transients," Van Nostrand Reinhold Co., N.Y., 1979, 1322–1324.
40. Chaudhry M.H., Yevjevich V. "Closed Conduit Flow," Water Resources Publication, USA, 1981, 255–278.
41. Chaudhry M. H., Applied Hydraulic Transients, Van Nostrand Reinhold, New York, USA, 1987, 165–167.
42. Kerr S.L., "Minimizing service interruptions due to transmission line failures: Discussion," Journal of the American Water Works Association, 41, 634, July 1949, 266–268.
43. Kerr S.L., "Water Hammer control," Journal of the American Water Works Association, 43, December 1951, 985–999.
44. Apoloniusz Kodura, Katarzyna Weinerowska," Some Aspects of Physical and Numerical Modeling of Water Hammer in Pipelines," 2005, 126–132.
45. Anuchina N.N., Volkov V.I., Gordeychuk V.A., Es'kov N.S., Ilyutina O.S., Kozyrev O.M. "Numerical simulations of Rayleigh-Taylor and Richtmyer-Meshkov instability using mah-3 code," J. Comput. Appl. Math. 168, 2004, 11.
46. Fox J.A., "Hydraulic Analysis of Unsteady Flow in Pipe Network," Wiley, N.Y., 1977, 78–89.
47. Karassik I.J., "Pump Handbook– Third Edition," McGraw-Hill, 2001, 19–22.
48. Fok A., "Design Charts for Air Chamber on Pump Pipelines," J. Hyd. Div., ASCE, Sept., 1978, 15–74.
49. Fok A., Ashamalla A., Aldworth G., "Considerations in Optimizing Air Chamber for Pumping Plants," Symposium on Fluid Transients and Acoustics in the Power Industry, San Francisco, U.S.A. Dec., 1978, 112–114.
50. Fok A., "Design Charts for Surge Tanks on Pump Discharge Lines," BHRA 3rd Int. Conference on Pressure Surges, Bedford, England, Mar., 1980, 23–34.
51. Fok A., "Water Hammer and Its Protection in Pumping Systems," Hydro technical Conference, CSCE, Edmonton, May, 1982, 45–55.
52. Fok A., "A contribution to the Analysis of Energy Losses in Transient Pipe Flow," PhD. Thesis, University of Ottawa, 1987, 176–182.
53. Hariri Asli K., Portable Flow meter Tester Machine Apparatus, Certificate on registration of invention, 2008, 27–47.
54. Hariri Asli K., Nagiyev F.B., Haghi A.K., "Interpenetration of Two Fluids at Parallel Between Plates and Turbulent Moving in pipe," 2008, 73–89.
55. Hariri Asli K., Nagiyev F.B., Haghi A.K., "Water Hammer and Valves," 2008, 20–30.
56. Hariri Asli K., Nagiyev F.B., Haghi A.K., "Water Hammer and Hydrodynamics Instability," 2008, 90–110.
57. Hariri Asli K., Nagiyev F.B., Haghi A.K., "Water Hammer analysis and Formulation," 2008, 27–42.
58. Hariri Asli K., Nagiyev F.B., Haghi A.K., "Water Hammer and Fluid Condition," 2008, 27–43.
59. Hariri Asli K., Nagiyev F.B., Haghi A.K., "Water Hammer and Pump Pulsation," 2008, 27–44.

60. Hariri Asli K., Nagiyev F.B., Haghi A.K., "Reynolds number and Hydrodynamics Instability," 2008, 27–45.
61. Hariri Asli K., Nagiyev F.B., Haghi A.K., "Water Hammer and fluid Interpenetration," 2008, 27–47.
62. Hariri Asli K., Computer Models for Fluid Interpenetration, Water Hammer Research; Advances in Nonlinear Dynamics Modeling, PhD. Thesis of Kaveh Hariri Asli, Toronto, Canada, Published by Apple Academic Press, Inc., Exclusive worldwide distribution by CRC Press, A Taylor & Francis Group, Print ISBN: 9781926895314, eBook: 978-1-46-656887-7, 2013, www.AppleAcademicPress.com.
63. Hariri Asli K., Heat Flow and Porous Materials, Water Hammer Research; Advances in Nonlinear Dynamics Modeling, PhD. Thesis of Kaveh Hariri Asli, Toronto, Canada, Published by Apple Academic Press, Inc., Exclusive worldwide distribution by CRC Press, A Taylor & Francis Group, Print ISBN: 9781926895314, eBook: 978-1-46-656887-7, 2013, www.AppleAcademicPress.com.
64. Hariri Asli K., Nagiyev F.B., Bubbles characteristics and convective effects in the binary mixtures. Transactions issue mathematics and mechanics series of physical-technical and mathematics science, ISSN: 0002–3108, Azerbaijan, Baku, 2009, 68–74, http://www.imm.science.az/journals.html.
65. Hariri Asli K., Nagiyev F.B., Haghi A.K., Aliyev S.A., Three-Dimensional Conjugate Heat Transfer in Porous Media, 12–17 Dec.2009, 26, 28.
66. Hariri Asli K., Nagiyev F.B., Haghi A.K., Aliyev S.A., Some Aspects of Physical and Numerical Modeling of Water Hammer in Pipelines, 2009, 26, 29.
67. Hariri Asli K., Nagiyev F.B., Haghi A.K., Aliyev S.A., Modeling for Water Hammer due to valves: From theory to practice, 2009, 26, 30.
68. Hariri Asli K., Nagiyev F.B., Haghi A.K., Aliyev S.A., Water Hammer and Hydrodynamics Instabilities Modeling: From Theory to Practice, 2009, 26, 31.
69. Hariri Asli K., Nagiyev F.B., Haghi A.K., Aliyev S.A., A computational approach to study fluid movement, 2009, 27–32.
70. Hariri Asli K., Nagiyev F.B., Haghi A.K., Aliyev S.A., Water Hammer Analysis: Some Computational Aspects and practical hints, 2009, 27–33.
71. Hariri Asli K., Nagiyev F.B. Haghi A.K., Aliyev S.A., Water Hammer and Fluid condition: A computational approach, 2009, 27–34.
72. Hariri Asli K., Nagiyev F.B., Haghi A.K., Aliyev S.A., A Computational Method to Study Transient Flow in Binary Mixtures, 12–17 Dec.2009, 27–35.
73. Hariri Asli K., Nagiyev F.B., Haghi A.K., Physical Modeling of Fluid Movement in Pipelines, 2009, 27–36.
74. Hariri Asli K., Nagiyev F.B., Haghi A.K., Aliyev S.A., Interpenetration of Two Fluids at Parallel Between Plates and Turbulent Moving, 12–17 Dec. 2009, 27–37.
75. Hariri Asli K., Nagiyev F.B., Haghi A.K., Aliyev S.A., Modeling of Fluid Interaction Produced by Water Hammer, 2009, 27–38
76. Hariri Asli K., Thermal Environment, Water Hammer Research; Advances in Nonlinear Dynamics Modeling, Published by Apple Academic Press, Inc., Exclusive worldwide distribution by CRC Press, A Taylor & Francis Group, Print ISBN: 9781926895314, eBook: 978-1-46-656887-7, 2013, www.AppleAcademicPress.com.
77. Hariri Asli K., Nagiyev F.B., Haghi A.K., Aliyev S.A., Interpenetration of Two Fluids at Parallel Between Plates and Turbulent Moving, 2009, 27–40.
78. Hariri Asli K., Nagiyev F.B., Haghi A.K., Aliyev S.A., Water Hammer and hydrodynamics' instability, 2009, 27–41.
79. Hariri Asli K., Nagiyev F.B., Haghi A.K., Aliyev S.A., Water Hammer analysis and Formulation, 2009, 27–42.
80. Hariri Asli K., Nagiyev F.B., Haghi A.K., Aliyev S.A., Water Hammer and Fluid Condition, 2009, 27–43.
81. Hariri Asli K., Nagiyev F.B., Haghi A.K., Aliyev S.A., Water Hammer and Pump Pulsation, 12–17 Dec.2009, 27–44.
82. Hariri Asli K., Nagiyev F.B., Haghi A.K., Aliyev S.A., Reynolds Number and Hydrodynamics Instabilities, 2009, 27–45.
83. Hariri Asli K., Nagiyev F.B., Haghi A.K., Aliyev S.A., Water Hammer and Valves, 12–17 Dec.2009, 27–46.

84. Hariri Asli K., Nagiyev F.B., Haghi A.K., Aliyev S.A., "Water Hammer and fluid Interpenetration," 2009, 27–47.
85. Hariri Asli K., Nagiyev F.B., Modeling of Fluid Interaction Produced by Water Hammer, International Journal of Chemoinformatics and Chemical Engineering, IGI, ISSN: 2155-4110, EISSN: 2155–4129, USA, 2010, 29–41, http://www.igi-global.com/journals/details.asp?ID=34654.
86. Hariri Asli K., Nagiyev F.B., Haghi A.K., Water Hammer and Fluid Condition; A computational approach, Computational Methods in Applied Science and Engineering, USA, Chapter 5, Nova Science Publications, ISBN: 978-1-60876-052-7, USA, 2010, 73–94, https://www.novapublishers.com/catalog/.
87. Hariri Asli K., Nagiyev F.B., Haghi A.K., Some aspects of physical and numerical modeling of water hammer in pipelines. Computational Methods in Applied Science and Engineering, USA, Chapter 23, Nova Science Publications, ISBN: 978-1-60876-052-7, USA, 2010, 365–387. https://www.novapublishers.com/catalog/.
88. Hariri Asli K., Nagiyev F.B., Haghi A.K., Modeling for water hammer due to valves; from theory to practice, Computational Methods in Applied Science and Engineering, USA, Chapter 11, Nova Science Publications, ISBN: 978-1-60876-052-7, USA, 2010, 229–236, https://www.novapublishers.com/catalog/.
89. Hariri Asli K., Nagiyev F.B., Haghi A.K., A computational method to Study transient flow in binary mixtures, Computational Methods in Applied Science and Engineering, USA, Chapter 13, Nova Science Publications, ISBN: 978-1-60876-052-7, USA, 2010, 229–236, https://www.novapublishers.com/catalog/.
90. Hariri Asli K., Nagiyev F.B., Haghi A.K., Water Hammer analysis; some computational aspects and practical hints, Computational Methods in Applied Science and Engineering, USA, Chapter 16, Nova Science Publications, ISBN: 978-1-60876-052-7, USA, 2010, 263–281. https://www.novapublishers.com/catalog/.
91. Hariri Asli K., Nagiyev F.B., Haghi A.K., Water Hammer and Hydrodynamics Instabilities Modeling, Computational Methods in Applied Science and Engineering, USA, Chapter 17, From Theory to Practice, Nova Science Publications, ISBN: 978-1-60876-052-7, USA, 2010, 283–301, https://www.novapublishers.com/catalog/.
92. Hariri Asli K., Nagiyev F.B., Haghi A.K., A computational approach to study water hammer and pump pulsation phenomena, Computational Methods in Applied Science and Engineering, USA, Chapter 22, Nova Science Publications, ISBN: 978-1-60876-052-7, USA, 2010, 349–363, https://www.novapublishers.com/catalog/.
93. Hariri Asli K., Nagiyev F.B., Modeling of Fluid Interaction Produced by Water Hammer, International Journal of Chemoinformatics and Chemical Engineering, IGI, ISSN: 2155-4110, EISSN: 2155–4129, USA, 2010, 29–41, http://www.igi-global.com/journals/details.asp?ID=34654.
94. Hariri Asli K., Nagiyev F.B., Haghi A.K., Water Hammer and Fluid Condition; A computational approach, Computational Methods in Applied Science and Engineering, USA, Chapter 5, Nova Science Publications, ISBN: 978-1-60876-052-7, USA, 2010, 73–94, https://www.novapublishers.com/catalog/.
95. Hariri Asli K., Nagiyev F.B., Haghi A.K., Some aspects of physical and numerical modeling of water hammer in pipelines. Computational Methods in Applied Science and Engineering, USA, Chapter 23, Nova Science Publications, ISBN: 978-1-60876-052-7, USA, 2010, 365–387, https://www.novapublishers.com/catalog/.
96. Hariri Asli K., Nagiyev F.B., Haghi A.K., Modeling for water hammer due to valves; from theory to practice, Computational Methods in Applied Science and Engineering, USA, Chapter 11, Nova Science Publications, ISBN: 978-1-60876-052-7, USA, 2010, 229–236, https://www.novapublishers.com/catalog/.
97. Hariri Asli K., Nagiyev F.B., Haghi A.K., A computational method to Study transient flow in binary mixtures, Computational Methods in Applied Science and Engineering, USA, Chapter 13, Nova Science Publications, ISBN: 978-1-60876-052-7, USA,

2010, 229–236, https://www.novapublishers.com/catalog/.

98. Hariri Asli K., Nagiyev F.B., Haghi A.K., Water Hammer analysis; some computational aspects and practical hints, Computational Methods in Applied Science and Engineering, USA, Chapter 16, Nova Science Publications, ISBN: 978-1-60876-052-7, USA, 2010, 263–281, https://www.novapublishers.com/catalog/.
99. Hariri Asli K., Nagiyev F.B., Haghi A.K., Water Hammer and Hydrodynamics Instabilities Modeling, Computational Methods in Applied Science and Engineering, USA, Chapter 17, From Theory to Practice, Nova Science Publications, ISBN: 978-1-60876-052-7, USA, 2010, 283–301, https://www.novapublishers.com/catalog/.
100. Hariri Asli K., Nagiyev F.B., Haghi A.K., A computational approach to study water hammer and pump pulsation phenomena, Computational Methods in Applied Science and Engineering, USA, Chapter 22, Nova Science Publications, ISBN: 978-1-60876-052-7, USA, 2010, 349–363, https://www.novapublishers.com/catalog/.
101. Hariri Asli K., Nagiyev F.B., Haghi A.K., A computational approach to study fluid movement, Nanomaterials Yearbook – 2009, From Nanostructures, Nanomaterials and Nanotechnologies to Nanoindustry, Chapter 16, Nova Science Publications, USA, ISBN: 978-1-60876-451-8, USA, 2010, 181–196, https://www.novapublishers.com/catalog/product_info.php?products_id=11587.
102. Hariri Asli K., Nagiyev F.B., Haghi A.K., Physical Modeling of Fluid Movement in Pipelines, Nanomaterials Yearbook – 2009, From Nanostructures, Nanomaterials and Nanotechnologies to Nanoindustry, Chapter 17, Nova Science Publications, USA, ISBN: 978-1-60876-451-8, USA, 2010, 197–214, https://www.novapublishers.com/catalog/product_info.php?products_id=11587.
103. Hariri Asli K., Sahleh H., Aliyev S.A., Mathematical Concepts for Mechanical Engineering Design, Toronto, Canada, Published by Apple Academic Press, Inc., Exclusive worldwide distribution by CRC Press, A Taylor & Francis Group, Print ISBN: 9781926895628, 2013, www.AppleAcademicPress.com.
104. Hariri Asli K., Nagiyev F.B., Haghi A.K., Interpenetration of Two Fluids at Parallel Between Plates and Turbulent Moving in pipe; A case study, Computational Methods in Applied Science and Engineering, USA, Chapter 7, Nova Science Publications, ISBN: 978-1-60876-052-7, USA, 2010, 107–133, https://www.novapublishers.com/catalog/.
105. Hariri Asli K., Nagiyev F.B., Beglou M. J., Haghi A.K., Kinetic analysis of convective drying, International Journal of the Balkan Tribological Association, ISSN: 1310-4772, Sofia, Bulgaria, 2009, Vol. 15, No 4, 546–556, jbalkta@gmail.com.
106. Hariri Asli K., Nagiyev F.B., Haghi A.K., Three-dimensional Conjugate Heat Transfer in Porous Media, International Journal of the Balkan Tribological Association, ISSN: 1310-4772, Sofia, Bulgaria, 2009, Vol. 15, No 3, 336–346, jbalkta@gmail.com.
107. Hariri Asli K., Nagiyev F.B., Haghi A.K., Aliyev S.A., Pure Oxygen penetration in wastewater flow, Recent Progress in Research in Chemistry and Chemical Engineering, Nova Science Publications, ISBN: 978-1-61668-501-0, Nova Science Publications, USA, 2010, 17–27, https://www.novapublishers.com/catalog/product_info.php?products_id=13174110100.
108. Hariri Asli K., Nagiyev F.B., Haghi A.K., Aliyev S.A., Improved modeling for prediction of water transmission failure, Recent Progress in Research in Chemistry and Chemical Engineering, Nova Science Publications, ISBN: 978-1-61668-501-0, Nova Science Publications, USA, 2010, 28–36, https://www.novapublishers.com/catalog/product_info.php?products_id=13174.

REFERENCES FOR CHAPTER 10

1. Qu W., Mala G.M., Li D., Heat Transfer for Water Flow in Trapezoidal Silicon Microchannels, 1993, 399–404.
2. Hariri Asli K., Nagiyev F.B., Haghi A.K., Aliyev S.A., A computational approach to study fluid movement, 2009, 27–32.

3. Peng X.F., Peterson G.P., Convective Heat Transfer and Flow Friction for Water Flow in Microchannel Structure, Int. J. Heat Mass Transfer 36, 1996, 2599–2608.
4. Bergant Anton, Discrete Vapour Cavity Model with Improved Timing of Opening and Collapse of Cavities, 1980, 1–11.
5. Ishii M., Thermo-Fluid Dynamic Theory of Two-Phase Flow, Collection de D. R. Liles and W. H. Reed, "A Sern-Implict Method for Two-Phase Fluid la Direction des Etudes et. Recherché d'Electricite de France, Vol. 22 Dynamics," Journal of Computational Physics 26, Paris, 1975, 390–407.
6. Hariri Asli K., Nagiyev F.B., Haghi A.K., Aliyev S.A., A computational approach to study fluid movement, 2009, 27–32.
7. Pickford J., "Analysis of Surge," Macmillian, London, 1969, 153–156.
8. Pipeline Design for Water and Wastewater, American Society of Civil Engineers, New York, 1975, 54 p.
9. Xu B., Ooi K.T., Mavriplis C., Zaghloul M. E., Viscous dissipation effects for liquid flow in microchannels, Micorsystems, 2002, 53–57.
10. Fedorov A.G., Viskanta R., Three-dimensional Conjugate Heat Transfer into Microchannel Heat Sink for Electronic Packaging, Int. J. Heat Mass Transfer 43, 2000, 399–415.
11. Tuckerman D.B., Heat transfer microstructures for integrated circuits, Ph.D. thesis, Stanford University, 1984, 10–120.
12. Harms T.M., Kazmierczak M.J., Cerner F.M., Holke A., Henderson H.T., Pilchowski H.T., Baker K., Experimental Investigation of Heat Transfer and Pressure Drop through Deep Micro channels in a (100) Silicon Substrate, in: Proceedings of the ASME. Heat Transfer Division, HTD 351, 1997, 347–357.
13. Holland F.A., Bragg R., Fluid Flow for Chemical Engineers, Edward Arnold Publishers, London, 1995, 1–3.
14. Lee T.S., Pejovic S. Air influence on similarity of hydraulic transients and vibrations. ASME J. Fluid Eng. 1996, 118(4), 706–709.
15. Li J., McCorquodale A., "Modeling Mixed Flow in Storm Sewers," Journal of Hydraulic Engineering, ASCE, Vol. 125, no. 11, 1999, 1170–1180.
16. Minnaert M., on musical air bubbles and the sounds of running water. Phil. Mag., 1933, v. 16, №7, 235–248.
17. Moeng C.H., McWilliams J.C., Rotunno R., Sullivan P.P., Weil J., "Investigating 2D modeling of atmospheric convection in the PBL," J. Atm. Sci. 61, 2004, 889–903.
18. Tuckerman D.B., R.F.W Pease, high performance heat sinking for VLSI, IEEE Electron device letter, DEL-2, 1981, 126–129.
19. Nagiyev F.B., Khabeev N.S, Bubble dynamics of binary solutions. High Temperature, 1988, v. 27, № 3, 528–533.
20. Shvarts D., Oron D., Kartoon D., Rikanati A., Sadot O., "Scaling laws of nonlinear Rayleigh-Taylor and Richtmyer-Meshkov instabilities in two and three dimensions," C. R. Acad. Sci. Paris IV, 719, 2000, 312 p.
21. Cabot W.H., Cook A.W., Miller P.L., Laney D.E., Miller M.C., Childs H.R., "Large eddy simulation of Rayleigh-Taylor instability," Phys. Fluids, September, 2005, vol. 17, 91–106.
22. Cabot W., University of California, Lawrence Livermore National laboratory, Livermore, CA, Physics of Fluids, 2006, 94–550.
23. Goncharov V.N., "Analytical model of nonlinear, single-mode, classical Rayleigh-Taylor instability at arbitrary Atwood numbers," Phys. Rev. Letters 88, 134502, 2002, 10–15.
24. Ramaprabhu P., Andrews M.J., "Experimental investigation of Rayleigh-Taylor mixing at small Atwood numbers," J. Fluid Mech. 502, 2004, 233 p.
25. Clark T.T., "A numerical study of the statistics of a two-dimensional Rayleigh-Taylor mixing layer," Phys. Fluids 15, 2003, 2413.
26. Cook A.W., Cabot W., Miller P.L., "The mixing transition in Rayleigh-Taylor instability," J. Fluid Mech. 511, 2004, 333.
27. Waddell J.T., Niederhaus C.E., Jacobs J.W., "Experimental study of Rayleigh-Taylor instability: Low Atwood number liquid systems with single-mode initial perturbations," Phys. Fluids 13, 2001, 1263–1273.
28. Weber S.V., Dimonte G., Marinak M.M., "Arbitrary Lagrange-Eulerian code simulations of turbulent Rayleigh-Taylor instability in two and three dimensions," Laser and Particle Beams 21, 2003, 455 p.

29. Dimonte G., Youngs D., Dimits A., Weber S., Marinak M. "A comparative study of the Rayleigh-Taylor instability using high-resolution three-dimensional numerical simulations: the Alpha group collaboration," Phys. Fluids 16, 2004, 1668.
30. Young Y.N., Tufo H., Dubey A., Rosner R., "On the miscible Rayleigh-Taylor instability: two and three dimensions," J. Fluid Mech. 447, 377, 2001, 2003–2500.
31. George E., Glimm J., "Self-similarity of Rayleigh-Taylor mixing rates," Phys. Fluids 17, 054101, 2005, 1–3.
32. Oron. D., Arazi L., Kartoon D., Rikanati A., Alon U., Shvarts D., "Dimensionality dependence of the Rayleigh-Taylor and Richtmyer-Meshkov instability late-time scaling laws," Phys. Plasmas 8, 2001, 2883 p.
33. Nigmatulin R.I., Nagiyev F.B., Khabeev N.S., Effective heat transfer coefficients of the bubbles in the liquid radial pulse. Mater. Second-Union. Conf. Heat Mass Transfer, "Heat massoob-men in the biphasic with Minsk", 1980, v.5, 111–115.
34. Nagiyev F.B., Khabeev N.S, Bubble dynamics of binary solutions. High Temperature, 1988, v. 27, № 3, 528–533.
35. Nagiyev F.B., Damping of the oscillations of bubbles boiling binary solutions. Mater. VIII Resp. Conf. mathematics and mechanics. Baku, October 26–29, 1988, 177–178.
36. Nagyiev F.B., Kadyrov B.A., Small oscillations of the bubbles in a binary mixture in the acoustic field. Math. An Az.SSR Ser. Physico-tech. and mate. Science, 1986, № 1, 23–26.
37. Nagiyev F.B., Dynamics, heat and mass transfer of vapor-gas bubbles in a two-component liquid. Turkey-Azerbaijan petrol semin., Ankara, Turkey, 1993, 32–40.
38. Nagiyev F.B., The method of creation effective coolness liquids, Third Baku international Congress. Baku, Azerbaijan Republic, 1995, 19–22.
39. Nagiyev F.B., The linear theory of disturbances in binary liquids bubble solution. Dep. In VINITI, 1986, № 405, in 86, 76–79.
40. Nagiyev F.B., Structure of stationary shock waves in boiling binary solutions. Math. USSR, Fluid Dynamics, 1989, № 1, 81–87.
41. Rayleigh, On the pressure developed in a liquid during the collapse of a spherical cavity. Philos. Mag. Ser. 6, v. 34, no. 200, 1917, 94–98.
42. Perry. R.H., Green D.W., Maloney J.O., Perry's Chemical Engineers Handbook, 7th Edition, McGraw-Hill, New York, 1997, 1–61.
43. Nigmatulin R.I., Dynamics of Multiphase Media, Moscow, "Nauka," 1987, v. 1, 2, 12–14.
44. Kodura A., Weinerowska K., the influence of the local pipeline leak on water hammer properties, Materials of the II Polish Congress of Environmental Engineering, Lublin, 2005, 125–133.
45. Kane J., Arnett D., Remington B.A., Glendinning S.G., Baz′an G., "Two-dimensional versus three-dimensional supernova hydrodynamic instability growth," Astrophys. J., 2000, 528–989.
46. Quick R.S., "Comparison and Limitations of Various Water Hammer Theories," J. Hyd. Div., ASME May, 1933, 43–45.
47. Jaeger C., "Fluid Transients in Hydro-Electric Engineering Practice," Blackie & Son Ltd., 1977, 87–88
48. Jaime Suárez A., "Generalized water hammer algorithm for piping systems with unsteady friction" 2005, 72–77.
49. Fok A., Ashamalla A., Aldworth G., "Considerations in Optimizing Air Chamber for Pumping Plants," Symposium on Fluid Transients and Acoustics in the Power Industry, San Francisco, U.S.A. Dec., 1978, 112–114.
50. Fok A., "Design Charts for Surge Tanks on Pump Discharge Lines," BHRA 3rd Int. Conference on Pressure Surges, Bedford, England, Mar., 1980, 23–34.
51. Fok A., "Water Hammer and Its Protection in Pumping Systems," Hydro technical Conference, CSCE, Edmonton, May, 1982, 45–55.
52. Fok A., "A contribution to the Analysis of Energy Losses in Transient Pipe Flow," PhD. Thesis, University of Ottawa, 1987, 176–182.
53. Hariri Asli K., Portable Flow meter Tester Machine Apparatus, Certificate on registration of invention, 2008, 27–43.
54. Hariri Asli K., Nagiyev F.B., Haghi A.K., "Interpenetration of Two Fluids at Parallel

Between Plates and Turbulent Moving in pipe," 2008, 73–89.
55. Hariri Asli K., Nagiyev F.B., Haghi A.K., "Water Hammer and Valves," 2008, 20–30.
56. Hariri Asli K., Nagiyev F.B., Haghi A.K., "Water Hammer and Hydrodynamics Instability," 2008, 90–110.
57. Hariri Asli K., Nagiyev F.B., Haghi A.K., "Water Hammer analysis and Formulation," 2008, 27–42.
58. Hariri Asli K., Nagiyev F.B., Haghi A.K., "Water Hammer and Fluid Condition," 2008, 27–43. 59. Hariri Asli K., Nagiyev F.B., Haghi A.K., Aliyev S.A., Pure Oxygen penetration in wastewater flow, Recent Progress in Research in Chemistry and Chemical Engineering, Nova Science Publications, ISBN: 978-1-61668-501-0, Nova Science Publications, USA, 2010, 17–27. https://www.novapublishers.com/catalog/product_info.php?products_id=13174110100.
60. Hariri Asli K., Nagiyev F.B., Haghi A.K., "Reynolds number and Hydrodynamics Instability," 2008, 27–45.
61. Hariri Asli K., Nagiyev F.B., Haghi A.K., "Water Hammer and fluid Interpenetration," 2008, 27–47.
62. Hariri Asli K., Computer Models for Fluid Interpenetration, Water Hammer Research; Advances in Nonlinear Dynamics Modeling, PhD. Thesis of Kaveh Hariri Asli, Toronto, Canada, Published by Apple Academic Press, Inc., Exclusive worldwide distribution by CRC Press, A Taylor & Francis Group, Print ISBN: 9781926895314, eBook: 978-1-46-656887-7, 2013, www.AppleAcademicPress.com.
63. Hariri Asli K., Heat Flow and Porous Materials, Water Hammer Research; Advances in Nonlinear Dynamics Modeling, PhD. Thesis of Kaveh Hariri Asli, Toronto, Canada, Published by Apple Academic Press, Inc., Exclusive worldwide distribution by CRC Press, A Taylor & Francis Group, Print ISBN: 9781926895314, eBook: 978-1-46-656887-7, 2013, www.AppleAcademicPress.com.
64. Hariri Asli K., Nagiyev F.B., Bubbles characteristics and convective effects in the binary mixtures. Transactions issue mathematics and mechanics series of physical-technical and mathematics science, ISSN: 0002–3108, Azerbaijan, Baku, 2009, 68–74, http://www.imm.science.az/journals.html.
65. Hariri Asli K., Nagiyev F.B., Haghi A.K., Aliyev S.A., Three-Dimensional Conjugate Heat Transfer in Porous Media, 12–17 Dec.2009, 26, 28.
66. Hariri Asli K., Nagiyev F.B., Haghi A.K., Aliyev S.A., Some Aspects of Physical and Numerical Modeling of Water Hammer in Pipelines, 2009, 26, 29.
67. Hariri Asli K., Nagiyev F.B., Haghi A.K., Aliyev S.A., Modeling for Water Hammer due to valves: From theory to practice, 2009, 26, 30.
68. Hariri Asli K., Nagiyev F.B., Haghi A.K., Aliyev S.A., Water Hammer and Hydrodynamics Instabilities Modeling: From Theory to Practice, 2009, 26, 31.
69. Hariri Asli K., Nagiyev F.B., Haghi A.K., Aliyev S.A., A computational approach to study fluid movement, 2009, 27–32.
70. Hariri Asli K., Handbook of Research for Mechanical Engineering, Vol. 1 Fluid Mechanics and Heat Transfer Published by Apple Academic Press, Inc., Exclusive worldwide distribution by CRC Press, A Taylor & Francis Group, 2013, (USA, Canada), www.AppleAcademicPress.com.
71. Hariri Asli K., Handbook of Research for Mechanical Engineering, Vol. 2- flow control in complex systems Published by Apple Academic Press, Inc., Exclusive worldwide distribution by CRC Press, A Taylor & Francis Group, 2013, (USA, Canada), www.AppleAcademicPress.com.
72. Hariri Asli K., Nagiyev F.B., Bubbles characteristics and convective effects in the binary mixtures. Transactions issue mathematics and mechanics series of physical-technical and mathematics science, ISSN: 0002–3108, Azerbaijan, Baku, 2009, 68–74, http://www.imm.science.az/journals.html.
73. Hariri Asli K., Nagiyev F.B., Haghi A.K., Physical Modeling of Fluid Movement in Pipelines, 2009, 27–36.
74. Hariri Asli K., Nagiyev F.B., Haghi A.K., Aliyev S.A., Interpenetration of Two Fluids at Parallel Between Plates and Turbulent Moving, 12–17 Dec. 2009, 27–37.

75. Hariri Asli K., Nagiyev F.B., Haghi A.K., Aliyev S.A., Modeling of Fluid Interaction Produced by Water Hammer, 2009, 27–38.
76. Hariri Asli K., Nagiyev F.B., Haghi A.K., Three-dimensional Conjugate Heat Transfer in Porous Media, International Journal of the Balkan Tribological Association, ISSN: 1310-4772, Sofia, Bulgaria, 2009, Vol. 15, No 3, 336–346, jbalkta@gmail.com.
77. Hariri Asli K., Nagiyev F.B., Haghi A.K., Aliyev S.A., Interpenetration of Two Fluids at Parallel Between Plates and Turbulent Moving, 2009, 27–40.
78. Hariri Asli K., Nagiyev F.B., Haghi A.K., Aliyev S.A., Water Hammer and hydrodynamics' instability, 2009, 27–41.
79. Hariri Asli K., Nagiyev F.B., Haghi A.K., Aliyev S.A., Water Hammer analysis and Formulation, 2009, 27–42.
80. Hariri Asli K., Nagiyev F.B., Haghi A.K., Aliyev S.A., Water Hammer and Fluid Condition, 2009, 27–43.
81. Hariri Asli K., Nagiyev F.B., Haghi A.K., Aliyev S.A., Water Hammer and Pump Pulsation, 12–17 Dec.2009, 27–44.
82. Hariri Asli K., Nagiyev F.B., Haghi A.K., Aliyev S.A., Reynolds Number and Hydrodynamics Instabilities, 2009, 27–45.
83. Hariri Asli K., Nagiyev F.B., Haghi A.K., Aliyev S.A., Water Hammer and Valves, 12–17 Dec.2009, 27–46.
84. Hariri Asli K., Nagiyev F.B., Haghi A.K., Aliyev S.A., "Water Hammer and fluid Interpenetration," 2009, 27–47.
85. Hariri Asli K., Nagiyev F.B., Modeling of Fluid Interaction Produced by Water Hammer, International Journal of Chemoinformatics and Chemical Engineering, IGI, ISSN: 2155-4110, EISSN: 2155–4129, USA, 2010, 29–41, http://www.igi-global.com/journals/details.asp?ID=34654.
86. Hariri Asli K., Nagiyev F.B., Haghi A.K., Water Hammer and Fluid Condition; A computational approach, Computational Methods in Applied Science and Engineering, USA, Chapter 5, Nova Science Publications, ISBN: 978-1-60876-052-7, USA, 2010, 73–94, https://www.novapublishers.com/catalog/.
87. Hariri Asli K., Nagiyev F.B., Haghi A.K., Some aspects of physical and numerical modeling of water hammer in pipelines. Computational Methods in Applied Science and Engineering, USA, Chapter 23, Nova Science Publications, ISBN: 978-1-60876-052-7, USA, 2010, 365–387, https://www.novapublishers.com/catalog/.
88. Hariri Asli K., Nagiyev F.B., Haghi A.K., Modeling for water hammer due to valves; from theory to practice, Computational Methods in Applied Science and Engineering, USA, Chapter 11, Nova Science Publications, ISBN: 978-1-60876-052-7, USA, 2010, 229–236, https://www.novapublishers.com/catalog/.
89. Hariri Asli K., Nagiyev F.B., Haghi A.K., A computational method to Study transient flow in binary mixtures, Computational Methods in Applied Science and Engineering, USA, Chapter 13, Nova Science Publications, ISBN: 978-1-60876-052-7, USA, 2010, 229–236, https://www.novapublishers.com/catalog/.
90. Hariri Asli K., Nagiyev F.B., Haghi A.K., Water Hammer analysis; some computational aspects and practical hints, Computational Methods in Applied Science and Engineering, USA, Chapter 16, Nova Science Publications, ISBN: 978-1-60876-052-7, USA, 2010, 263–281. https://www.novapublishers.com/catalog/.

INDEX

M

T

U

V